Friedrich Haslinger

The $\bar{\partial}$-Neumann Problem and Schrödinger Operators

De Gruyter Expositions in Mathematics

Volume 59

Friedrich Haslinger

The $\bar{\partial}$-Neumann Problem and Schrödinger Operators

2nd edition

DE GRUYTER

Mathematics Subject Classification 2020
Primary: 32W05, 32A25, 32A36; Secondary: 35P05, 35J10

Author
Prof. Dr. Friedrich Haslinger
Universität Wien
Fakultät für Mathematik
Oskar-Morgenstern-Platz 1
1090 Vienna
Austria
friedrich.haslinger@univie.ac.at

ISBN 978-3-11-118290-2
e-ISBN (PDF) 978-3-11-118292-6
e-ISBN (EPUB) 978-3-11-118467-8
ISSN 0938-6572

Library of Congress Control Number: 2023940864

Bibliographic information published by the Deutsche Nationalbibliothek
The Deutsche Nationalbibliothek lists this publication in the Deutsche Nationalbibliografie; detailed bibliographic data are available on the Internet at http://dnb.dnb.de.

Typesetting: VTeX UAB, Lithuania
Printing and binding: CPI books GmbH, Leck

www.degruyter.com

Preface to the first edition

The subject of this book is complex analysis in several variables and its connections to partial differential equations and functional analysis. The first sections of each chapter contain prerequisites from functional analysis, Sobolev spaces, partial differential equations, and spectral analysis, which are used in the following sections devoted to the main topic of the book. In this way the book becomes self-contained, with only one exception, where we do not provide all details in the proof of the general spectral theorem for unbounded self-adjoint operators.

We concentrate on the Cauchy–Riemann equation ($\overline{\partial}$-equation) and investigate the properties of the canonical solution operator to $\overline{\partial}$, the solution with minimal L^2-norm and its relationship to the $\overline{\partial}$-Neumann operator. The first chapter contains a discussion of Bergman spaces in one and several complex variables, including basic facts on Hilbert spaces. In the second chapter the solution operator to $\overline{\partial}$ restricted to holomorphic L^2-functions in one complex variable is investigated, pointing out that the Bergman kernel of the associated Hilbert space of holomorphic functions plays an important role. We investigate operator properties like compactness and Schatten class membership, also for the solution operator on weighted spaces of entire functions (Fock spaces). In the third chapter, we generalize the results to several complex variables and explain some new phenomena that do not appear in one variable.

In the following, we consider the general $\overline{\partial}$-complex and derive properties of the complex Laplacian on L^2-spaces of bounded pseudoconvex domains and on weighted L^2-spaces. For this purpose, we first concentrate on basic results about distributions, Sobolev spaces, and unbounded operators on Hilbert spaces. The key result in Kohn's far-reaching method is the Kohn–Morrey formula, which is presented in different versions. Using this formula, the basic property of the $\overline{\partial}$-Neumann operator – the bounded inverse of the complex Laplacian – is proved. In recent years, it has turned out to be useful to investigate an even more general situation, namely the twisted $\overline{\partial}$-complex, where $\overline{\partial}$ is composed with a positive twist factor. In this way, we obtain a rather general basic estimate, from which we get Hörmander's L^2-estimates for the solution of the Cauchy–Riemann equation together with results on related weighted spaces of entire functions, such as that these spaces are infinite dimensional if the eigenvalues of the Levi matrix of the weight function show a certain behavior at infinity. In addition, it is pointed out that some L^2-estimates for $\overline{\partial}$ can be interpreted in the sense of a general Brascamp–Lieb inequality.

The next chapter contains a detailed account of the application of the $\overline{\partial}$-methods to Schrödinger operators, Pauli and Dirac operators, and Witten Laplacians. In this context, spectral analysis plays an important role. Therefore an extensive chapter on spectral analysis was inserted to provide a better understanding for the operator theoretic aspects in the $\overline{\partial}$-Neumann problem, which, in particular, is used to exactly describe the spectrum of complex Laplacian on the Fock space. Returning to the $\overline{\partial}$-Neumann problem, we characterize the compactness of the $\overline{\partial}$-Neumann operator using a description

https://doi.org/10.1515/9783111182926-201

of precompact subsets in L^2-spaces. The compactness of the $\overline{\partial}$-Neumann operator is also related to properties of commutators of the Bergman projection and multiplication operators.

In the last part, we use the $\overline{\partial}$-methods and some spectral theory to settle the question whether certain Schrödinger operators with a magnetic field have a compact resolvent. It is also shown that a large class of Dirac operators fail to have a compact resolvent. Finally, we exhibit some situations where the $\overline{\partial}$-Neumann operator is not compact.

Numerous references for the topics of the text and for additional results are given in the notes at the end of each chapter.

Most of the material of the book stems from various lectures of the author given at the University of Vienna, the Erwin Schrödinger International Institute for Mathematical Physics (ESI) in Vienna and at CIRM, Luminy, during programs on the $\overline{\partial}$-Neumann operator in recent years. The author is indebted to both institutions, ESI and CIRM, for their help and hospitality. I would also like to thank my students Franz Berger, Damir Ferizović, and Tobias Preinerstorfer for their constructive criticisms of the manuscript and also for their help in eliminating a number of typos and minor errors.

Vienna, March 2014 Friedrich Haslinger

Preface to the second edition

We tried to correct some misprints of the first edition. Chapter 1 contains some new sections on the Segal–Bargmann space, the uncertainty principle, and nonradial weights. In Chapter 11, we included a complete treatment of magnetic Schrödinger operators with compact resolvent due to Iwatsuka [59]. In addition, we give a characterization of compact subsets in the Segal–Bargmann space, which is used in the new chapters. With the help of Muckenhoupt characteristics, we present a concise compactness criterion for the weighted $\bar{\partial}$-operator on $\mathbb{C}$. This criterion shows the remarkable interplay between methods of real analysis (Schrödinger operators) and complex analysis ($\bar{\partial}$ equation). We also included a new chapter on the ∂-complex with the Segal–Bargmann space as an underlying Hilbert space. In many aspects, we can use methods of the $\bar{\partial}$-operator for its analysis. The creation and annihilation operators of quantum mechanics play an important role. The last part is devoted to the generalized ∂-complex, where only partial results can be delivered. The main problem consists in analyzing the basic estimates, which can be seen as generalizations of the uncertainty principle, hence raising many interesting questions for further research.

I was partially supported by the FWF-grant P 36884 of the Austrian Science Fund.

Vienna, June 2023

Friedrich Haslinger

https://doi.org/10.1515/9783111182926-202

Contents

1 Bergman spaces

To investigate the solution to the inhomogeneous $\overline{\partial}$-equation $\overline{\partial}u = g$, we will first consider the case where the right-hand side g is a holomorphic function. Therefore we need an appropriate Hilbert space of holomorphic functions, the Bergman space. We will use standard basic facts about Hilbert spaces, such as the Riesz representation theorem for continuous linear functional and facts about orthogonal projections and complete orthonormal bases.

Let $\Omega \subseteq \mathbb{C}^n$ be a domain. The Bergman space is defined as

$$A^2(\Omega) = \left\{ f : \Omega \longrightarrow \mathbb{C} \text{ holomorphic} : \|f\|^2 = \int_\Omega |f(z)|^2 \, d\lambda(z) < \infty \right\},$$

where λ is the Lebesgue measure on $\mathbb{C}^n$. The inner product is given by

$$(f, g) = \int_\Omega f(z) \overline{g(z)} \, d\lambda(z)$$

for $f, g \in A^2(\Omega)$.

1.1 Elementary properties

For simplicity, we first restrict ourselves to domains $\Omega \subseteq \mathbb{C}$. We consider special continuous linear functionals on $A^2(\Omega)$, the point evaluations. Let $f \in A^2(\Omega)$ and fix $z \in \Omega$. By Cauchy's integral theorem we have

$$f(z) = \frac{1}{2\pi i} \int_{\gamma_s} \frac{f(\zeta)}{\zeta - z} \, d\zeta,$$

where $\gamma_s(t) = z + se^{it}$, $t \in [0, 2\pi]$, $0 < s \leq r$, and $D(z, r) = \{w : |w - z| < r\} \subset \Omega$. Using polar coordinates and integrating the above equality with respect to s between 0 and r, we get

$$f(z) = \frac{1}{\pi r^2} \int_{D(z,r)} f(w) \, d\lambda(w). \tag{1.1}$$

Then, by the Cauchy–Schwarz inequality,

$$|f(z)| \leq \frac{1}{\pi r^2} \int_{D(z,r)} 1 \, . \, |f(w)| \, d\lambda(w)$$

https://doi.org/10.1515/9783111182926-001

$$\leq \frac{1}{\pi r^2}\Big(\int_{D(z,r)} 1^2\, d\lambda(w)\Big)^{1/2}\Big(\int_{D(z,r)} |f(w)|^2\, d\lambda(w)\Big)^{1/2}$$
$$\leq \frac{1}{\pi^{1/2} r}\Big(\int_{\Omega} |f(w)|^2\, d\lambda(w)\Big)^{1/2} \leq \frac{1}{\pi^{1/2} r}\, \|f\|.$$

If K is a compact subset of Ω, then there is $r(K) > 0$ such that for all $z \in K$, we have $D(z, r(K)) \subset \Omega$, and we get

$$\sup_{z\in K} |f(z)| \leq \frac{1}{\pi^{1/2} r(K)}\, \|f\|.$$

If $K \subset \Omega \subset \mathbb{C}^n$, then we can find a polycylinder

$$P(z, r(K)) = \{w \in \mathbb{C}^n : |w_j - z_j| < r(K),\ j = 1, \ldots, n\}$$

such that for all $z \in K$, we have $P(z, r(K)) \subset \Omega$. Hence by iterating the above Cauchy integrals we get the following:

Proposition 1.1. *Let $K \subset \Omega$ be a compact set. Then there exists a constant $C(K)$, only depending on K, such that*

$$\sup_{z\in K} |f(z)| \leq C(K)\, \|f\| \tag{1.2}$$

for all $f \in A^2(\Omega)$.

Proposition 1.2. *$A^2(\Omega)$ is a Hilbert space.*

Proof. If $(f_k)_k$ is a Cauchy sequence in $A^2(\Omega)$, by (1.2) it is also a Cauchy sequence with respect to uniform convergence on compact subsets of Ω. Hence the sequence $(f_k)_k$ has a holomorphic limit f with respect to uniform convergence on compact subsets of Ω. On the other hand, the original L^2-Cauchy sequence has a subsequence that converges pointwise almost everywhere to the L^2-limit of the original L^2-Cauchy sequence (see, for instance, [83]), and so the L^2-limit coincides with the holomorphic function f. Therefore $A^2(\Omega)$ is a closed subspace of $L^2(\Omega)$ and itself a Hilbert space. □

In the sequel, we present basic facts about Hilbert spaces and their consequences for the Bergman spaces.

Proposition 1.3. *Let E be a nonempty, convex, and closed subset of the Hilbert space H, i. e., $tx+(1-t)y \in E$ for all for $x, y \in E$ and $t \in [0,1]$. Then E contains a uniquely determined element of minimal norm.*

Proof. The parallelogram rule says that

$$\|x+y\|^2 + \|x-y\|^2 = 2\,\|x\|^2 + 2\,\|y\|^2, \quad x, y \in H.$$

Let $\delta = \inf\{\|x\| : x \in E\}$. For $x, y \in E$, we have $\frac{1}{2}(x+y) \in E$, and hence

$$1/4\,\|x-y\|^2 = 1/2\,\|x\|^2 + 1/2\,\|y\|^2 - \left\|1/2(x+y)\right\|^2,$$

which implies that

$$\|x-y\|^2 \le 2\,\|x\|^2 + 2\,\|y\|^2 - 4\delta^2.$$

So, if $\|x\| = \|y\| = \delta$, then $x = y$ (uniqueness).

By the definition of δ there exists a sequence $(y_k)_k$ in E such that $\|y_k\| \to \delta$ as $k \to \infty$. The estimate

$$\|y_k - y_m\|^2 \le 2\,\|y_k\|^2 + 2\,\|y_m\|^2 - 4\delta^2$$

implies that $(y_k)_k$ is a Cauchy sequence in H. Since H is complete, there exists $x_0 \in H$ such that $\|y_k - x_0\| \to 0$, and, as E is closed, we have $x_0 \in E$; the mapping $x \mapsto \|x\|$ is continuous, and therefore $\|x_0\| = \lim_{k\to\infty} \|y_k\| = \delta$. □

Theorem 1.4. *Let M be a closed subspace of a Hilbert space H. Then there exist uniquely determined mappings*

$$P : H \longrightarrow M, \quad Q : H \longrightarrow M^\perp$$

such that

(1) $x = Px + Qx \ \forall x \in H$.

(2) *for $x \in M$, we have $Px = x$, and hence $P^2 = P$ and $Qx = 0$; for $x \in M^\perp$, we have $Px = 0$, $Qx = x$, and $Q^2 = Q$.*

(3) *The distance of $x \in H$ to M is given by*

$$\inf\{\|x-y\| : y \in M\} = \|x - Px\|.$$

(4) *For each $x \in H$, we have*

$$\|x\|^2 = \|Px\|^2 + \|Qx\|^2.$$

(5) *P and Q are continuous, linear, and self-adjoint operators, and they are the orthogonal projections of H onto M and $M^\perp$.*

Proof. For each $x \in H$, the set $x+M = \{x+y : y \in M\}$ is convex. Hence, by Proposition 1.3, there exists a uniquely determined element of minimal norm in $x+M$, which is denoted by Qx. We set $Px = x - Qx$ and see that $Px \in M$, since $Qx \in x + M$.

Now we claim that $Qx \in M^\perp$. We have to show that $(Qx, y) = 0$ for all $y \in M$. We can suppose that $\|y\| = 1$, and then we have

$$(Qx, Qx) = \|Qx\|^2 \leq \|Qx - \alpha y\|^2 = (Qx - \alpha y, Qx - \alpha y) \quad \forall \alpha \in \mathbb{C}$$

by the minimality of Qx. Therefore we get

$$0 \leq -\alpha(y, Qx) - \overline{\alpha}(Qx, y) + |\alpha|^2.$$

Setting $\alpha = (Qx, y)$, we obtain $0 \leq -|(Qx, y)|^2$ and $(Qx, y) = 0$; hence $Q : H \longrightarrow M^\perp$.

If $x = x_0 + x_1$ with $x_0 \in M$ and $x_1 \in M^\perp$, then $x_0 - Px = Qx - x_1$, and since $M \cap M^\perp = \{0\}$, we obtain $x_0 = Px$ and $x_1 = Qx$, therefore P and Q are uniquely determined. In a similar way, we get that

$$P(\alpha x + \beta y) - \alpha Px - \beta Py = \alpha Qx + \beta Qy - Q(\alpha x + \beta y).$$

The left side belongs to M, the right side belongs to $M^\perp$, and hence both sides are 0, which proves that P and Q are linear.

Property 3 follows by the definition of Q, and property 4 by the fact that

$$(Px, Qx) = 0 \quad \forall x \in H.$$

In addition, we have

$$\|Q(x - y)\| = \inf\{\|x - y + m\| : m \in M\} \leq \|x - y\|,$$

and hence Q and $P = I - Q$ are continuous.

For $x, y \in H$, we have

$$(Px, y) = (Px, Py + Qy) = (Px, Py) \quad \text{and} \quad (x, Py) = (Px + Qx, Py) = (Px, Py),$$

hence $(Px, y) = (x, Py)$, and so P is self-adjoint. □

Corollary 1.5. *If $M \neq H$ is a closed proper subspace of a Hilbert space H, then there exists an element $y \neq 0$ such that $y \perp M$.*

Proof. Let $x \in H$ be such that $x \notin M$. Set $y = Qx$. Then $x \neq Px$ implies $y \neq 0$. □

The next result is the Riesz representation theorem.

Theorem 1.6. *Let L be a continuous linear functional on a Hilbert space H. Then there exists a uniquely determined element $y \in H$ such that $Lx = (x, y)$ for all $x \in H$.*

Proof. If $L(x) = 0$ for all $x \in H$, then we set $y = 0$. Otherwise, we define $M = \{x : Lx = 0\}$. Then, by the continuity of L, the subspace M of H is closed. By Corollary 1.5 we have $M^\perp \neq 0$. Let $z \in M^\perp$, $z \neq 0$. Then $Lz \neq 0$. Now set $y = \alpha z$, where $\overline{\alpha} = \frac{Lz}{\|z\|^2}$. Then $y \in M^\perp$ and

$$Ly = L(\alpha z) = \frac{\overline{Lz}}{\|z\|^2} Lz = \frac{|Lz|^2}{\|z\|^2} = (y, y) = |\alpha|^2 (z, z).$$

For $x \in H$, we define

$$x' = x - \frac{Lx}{(y,y)}y \quad \text{and} \quad x'' = \frac{Lx}{(y,y)}y.$$

Then we obtain $Lx' = 0$ and $x' \in M$; hence $(x', y) = 0$, and

$$(x,y) = (x'',y) = \left(\frac{Lx}{(y,y)}y, y\right) = Lx.$$

If $(x,y) = (x,y')$ for all $x \in H$, then we get $(x, y - y') = 0$ for all $x \in H$; in particular, $(y - y', y - y') = 0$. Therefore $y = y'$, which shows that y is uniquely determined. □

Corollary 1.7. *Let H be a Hilbert space, and let $L \in H'$ be a continuous linear functional. Then the dual norm*

$$\|L\| = \sup\{|Lx| : \|x\| \leq 1\}$$

can be expressed in the form

$$\|L\| = \sup\{|(x,y)| : \|x\| \leq 1\} = \|y\|,$$

where $y \in H$ corresponds to L.

For fixed $z \in \Omega$, (1.2) also implies that the point evaluation $f \mapsto f(z)$ is a continuous linear functional on $A^2(\Omega)$, and hence, by the Riesz representation, Theorem 1.6, there is a uniquely determined function $k_z \in A^2(\Omega)$ such that

$$f(z) = (f, k_z) = \int_\Omega f(w)\,\overline{k_z(w)}\,d\lambda(w). \tag{1.3}$$

We set $K(z,w) = \overline{k_z(w)}$. Then $w \mapsto \overline{K(z,w)} = k_z(w)$ is an element of $A^2(\Omega)$, and hence the function $w \mapsto K(z,w)$ is antiholomorphic on Ω, and we have

$$f(z) = \int_\Omega K(z,w) f(w)\,d\lambda(w), \quad f \in A^2(\Omega).$$

The function of two complex variables $(z,w) \mapsto K(z,w)$ is called the Bergman kernel of Ω, and the above identity represents the reproducing property of the Bergman kernel.

Now we use the reproducing property for the holomorphic function $z \mapsto k_u(z)$, where $u \in \Omega$ is fixed:

$$k_u(z) = \int_\Omega K(z,w) k_u(w)\,d\lambda(w) = \int_\Omega \overline{k_z(w)}\,\overline{K(u,w)}\,d\lambda(w)$$

$$= \left(\int_\Omega K(u,w) k_z(w)\, d\lambda(w) \right)^{-} = \overline{k_z(u)},$$

and hence we have $k_u(z) = \overline{k_z(u)}$, or $K(z,u) = \overline{K(u,z)}$.

It follows that the Bergman kernel is holomorphic in the first variable and anti-holomorphic in the second variable.

Proposition 1.8. *The Bergman kernel is uniquely determined by the properties that it is an element of $A^2(\Omega)$ in z and that it is conjugate symmetric and reproduces $A^2(\Omega)$.*

Proof. To see this, let $K'(z,w)$ be another kernel with these properties. Then we have

$$\begin{aligned} K(z,w) &= \int_\Omega K'(z,u) K(u,w)\, d\lambda(u) \\ &= \left(\int_\Omega K(w,u) K'(u,z)\, d\lambda(u) \right)^{-} \\ &= \overline{K'(w,z)} \\ &= K'(z,w). \end{aligned}$$

□

Now let $\phi \in L^2(\Omega)$. Since $A^2(\Omega)$ is a closed subspace of $L^2(\Omega)$, there exists a uniquely determined orthogonal projection $P : L^2(\Omega) \longrightarrow A^2(\Omega)$; see Theorem 1.4. For the function $P\phi \in A^2(\Omega)$, using the reproducing property, we obtain

$$P\phi(z) = \int_\Omega K(z,w) P\phi(w)\, d\lambda(w) = (P\phi, k_z) = (\phi, Pk_z) = (\phi, k_z), \tag{1.4}$$

where we still have used that P is a self-adjoint operator and that $Pk_z = k_z$. Hence

$$P\phi(z) = \int_\Omega K(z,w) \phi(w)\, d\lambda(w). \tag{1.5}$$

The operator P is called the Bergman projection.

To compute the Bergman kernel for some particular cases, we need the notion of a complete orthonormal basis.

A subset $\{u_\alpha : \alpha \in A\}$ of a Hilbert space is called orthonormal if $(u_\alpha, u_\beta) = \delta_{\alpha\beta}$ for each $\alpha, \beta \in A$. If $(x_k)_k$ is a linearly independent sequence in H, then there is a standard procedure, called the Gram–Schmidt process, for converting $(x_k)_k$ into an orthonormal sequence $(u_k)_k$ such that the linear span of $(u_k)_{k=1}^N$ equals the linear span of $(x_k)_{k=1}^N$ for all $N \in \mathbb{N}$. We start with $u_1 = x_1/\|x_1\|$. Having defined $u_1, \dots, u_{N-1}$, we set

$$v_N = x_N - \sum_{j=1}^{N-1} (x_N, u_j) u_j.$$

The element v_N is nonzero because x_N is not in the linear span of $x_1, \dots, x_{N-1}$ and hence not in the span of $u_1, \dots, u_{N-1}$. So we can set $u_N = v_N/\|v_N\|$. It is now clear that $(u_k)_{k=1}^N$ has the desired properties.

Next we prove Bessel's inequality.

Proposition 1.9. *If $\{u_\alpha : \alpha \in A\}$ is an orthonormal set in a Hilbert space H, then for all $u \in H$,*

$$\sum_{\alpha\in A} |(u, u_\alpha)|^2 \le \|u\|^2. \tag{1.6}$$

In particular, the set $\{\alpha : (u, u_\alpha) \neq 0\}$ is countable.

Proof. It suffices to show that $\sum_{\alpha\in F} |(u, u_\alpha)|^2 \le \|u\|^2$ for any finite set $F \subset A$. We use the property $\|u_\alpha + u_\beta\|^2 = \|u_\alpha\|^2 + \|u_\beta\|^2$ for $\alpha, \beta \in A$, $\alpha \neq \beta$:

$$\begin{aligned} 0 &\le \Big\| u - \sum_{\alpha\in F} (u, u_\alpha)u_\alpha \Big\|^2 \\ &= \|u\|^2 - 2\Re\Big(u, \sum_{\alpha\in F}(u, u_\alpha)u_\alpha \Big) + \Big\| \sum_{\alpha\in F} (u, u_\alpha)u_\alpha \Big\|^2 \\ &= \|u\|^2 - 2\sum_{\alpha\in F} |(u, u_\alpha)|^2 + \sum_{\alpha\in F} |(u, u_\alpha)|^2 \\ &= \|u\|^2 - \sum_{\alpha\in F} |(u, u_\alpha)|^2. \end{aligned}$$

□

Proposition 1.10. *If $\{u_\alpha : \alpha \in A\}$ is an orthonormal set in a Hilbert space H, then the following conditions are equivalent:*

(1) *Completeness: if $(u, u_\alpha) = 0$ for all $\alpha \in A$, then $u = 0$.*
(2) *Parseval's identity: $\|u\|^2 = \sum_{\alpha\in A} |(u, u_\alpha)|^2$ for all $u \in H$.*
(3) *$u = \sum_{\alpha\in A} (u, u_\alpha)u_\alpha$ for each $u \in H$, where the sum has only countably many nonzero terms and converges in norm to u no matter how these terms are ordered.*

Proof. (1) implies (3): If $u \in H$, let $\alpha_1, \alpha_2, \dots$ be any enumeration of those α for which $(u, u_\alpha) \neq 0$. By Bessel's inequality the series $\sum_{j=1}^\infty |(u, u_{\alpha_j})|^2$ converges, so

$$\Big\| \sum_{j=m}^M (u, u_\alpha)u_\alpha \Big\|^2 = \sum_{j=m}^M |(u, u_\alpha)|^2 \to 0 \quad \text{as } m, M \to \infty.$$

Hence the series $\sum_{j=1}^\infty (u, u_{\alpha_j})u_{\alpha_j}$ converges in H. If $v = u - \sum_{j=1}^\infty (u, u_{\alpha_j})u_{\alpha_j}$, then $(v, u_\alpha) = 0$ for all $\alpha \in A$, so by (1), $v = 0$.

(3) implies (2): As in the proof of Bessel's inequality, we have

$$\|u\|^2 - \sum_{j=1}^{m} |(u, u_{\alpha_j})|^2 = \left\| u - \sum_{j=1}^{m} (u, u_{\alpha_j}) u_{\alpha_j} \right\|^2 \to 0 \quad \text{as } m \to \infty.$$

Finally, that (2) implies (1) is obvious. □

An orthonormal set having the properties of Proposition 1.10 is called an orthonormal basis of H.

Remark 1.11. An application of Zorn's lemma shows that the collection of orthonormal sets in a Hilbert space, ordered by inclusion, has a maximal element. Maximality is equivalent to (1) of Proposition 1.10, and hence for each orthonormal set, there exists an orthonormal basis that contains the given orthonormal set.

Every separable Hilbert space H has a countable orthonormal basis.

Proposition 1.12. *Let $K \subset \Omega$ be a compact subset, and let $\{\phi_j\}$ be an orthonormal basis of $A^2(\Omega)$. Then the series*

$$\sum_{j=1}^{\infty} \phi_j(z)\, \overline{\phi_j(w)}$$

sums uniformly on $K \times K$ to the Bergman kernel $K(z, w)$.

Proof. For the proof of this statement, we use the duality for the sequence space l^2 to get

$$\begin{aligned} \sup_{z \in K} \left(\sum_{j=1}^{\infty} |\phi_j(z)|^2 \right)^{1/2} &= \sup \left\{ \left| \sum_{j=1}^{\infty} a_j \phi_j(z) \right| : \sum_{j=1}^{\infty} |a_j|^2 = 1,\ z \in K \right\} \\ &= \sup\{ |f(z)| : \|f\| = 1,\ z \in K \} \\ &\le C(K), \end{aligned} \tag{1.7}$$

where we have used (1.2) in the last inequality. Now the Cauchy–Schwarz inequality gives

$$\sum_{j=1}^{\infty} |\overline{\phi_j(z)}\, \phi_j(w)| \le \left(\sum_{j=1}^{\infty} |\phi_j(z)|^2 \right)^{1/2} \left(\sum_{j=1}^{\infty} |\phi_j(w)|^2 \right)^{1/2}$$

with uniform convergence in $z, w \in K$. In addition, it follows that $(\phi_j(z))_j \in l^2$ and the function

$$w \mapsto \sum_{j=1}^{\infty} \phi_j(z)\, \overline{\phi_j(w)}$$

belongs to $\overline{A^2(\Omega)}$. Denote the sum of the series by $K'(z, w)$. Notice that $K'(z, w)$ is conjugate symmetric and that for $f \in A^2(\Omega)$, we get

$$\int_{\Omega} K'(z,w)f(w)\,d\lambda(w) = \sum_{j=1}^{\infty} \int_{\Omega} f(w)\overline{\phi_j(w)}\,d\lambda(w)\,\phi_j(z) = f(z)$$

with convergence in the Hilbert space $A^2(\Omega)$. However, (1.7) implies uniform convergence on compact subsets of Ω, and hence

$$f(z) = \int_{\Omega} K'(z,w)f(w)\,d\lambda(w)$$

for all $f \in A^2(\Omega)$, so $K'(z,w)$ is a reproducing kernel. By the uniqueness of the Bergman kernel we obtain $K'(z,w) = K(z,w)$. □

We note that (1.7) implies

$$K(z,z) = \sup\{|f(z)|^2 : f \in A^2(\Omega)\,,\, \|f\| = 1\}. \tag{1.8}$$

1.2 Examples

(a)

The functions $\phi_n(z) = \sqrt{\frac{n+1}{\pi}}\, z^n$, $n = 0,1,2,\ldots$, constitute an orthonormal basis in $A^2(\mathbb{D})$, $\mathbb{D} = \{z \in \mathbb{C} : |z| < 1\}$.

This follows from

$$\int_{\mathbb{D}} z^n\,\overline{z^m}\,d\lambda(z) = \int_0^{2\pi}\int_0^1 r^n e^{in\theta}\, r^m e^{-im\theta}\, r\,dr\,d\theta = \frac{2\pi}{n+m+2}\,\delta_{n,m}.$$

For each $f \in A^2(\mathbb{D})$ with Taylor series expansion $f(z) = \sum_{n=0}^{\infty} a_n z^n$, we get

$$\begin{aligned}(f, z^n) &= \int_{\mathbb{D}} f(z)\overline{z^n}\,d\lambda(z) = \int_0^1\int_0^{2\pi} f(re^{i\theta})r^n e^{-in\theta} r\,dr\,d\theta \\ &= \int_0^1\int_0^{2\pi} \frac{f(re^{i\theta})}{r^{n+1}e^{i(n+1)\theta}} re^{i\theta}\,d\theta\, r^{2n+1}\,dr = 2\pi a_n \int_0^1 r^{2n+1}\,dr = \pi\,\frac{a_n}{n+1},\end{aligned}$$

where we used the fact that

$$a_n = \frac{1}{2\pi i}\int_{\gamma_r} \frac{f(z)}{z^{n+1}}\,dz$$

for $\gamma_r(\theta) = re^{i\theta}$. Hence, by the uniqueness of the Taylor series expansion, we obtain that $(f, \phi_n) = 0$ for each $n = 0,1,2,\ldots$ implies $f \equiv 0$. This means that $(\phi_n)_{n=0}^{\infty}$ constitutes an orthonormal basis for $A^2(\mathbb{D})$, and we get

$$\|f\|^2 = \sum_{n=0}^{\infty} |(f, \phi_n)|^2,$$

which is equivalent to

$$\|f\|^2 = \pi \sum_{n=0}^{\infty} \frac{|a_n|^2}{n+1}, \quad f(z) = \sum_{n=0}^{\infty} a_n z^n.$$

Hence each $f \in A^2(\mathbb{D})$ can be written in the form $f = \sum_{n=0}^{\infty} c_n \phi_n$, where the sum converges in $A^2(\mathbb{D})$ and also uniformly on compact subsets of $\mathbb{D}$. For the coefficients c_n, we have $c_n = (f, \phi_n)$.

Now we compute an explicit formula for the Bergman kernel $K(z,w)$ of $\mathbb{D}$. The function $z \mapsto K(z,w)$, with $w \in \mathbb{D}$ fixed, belongs to $A^2(\mathbb{D})$. Hence we get from the above formula that

$$K(z,w) = \sum_{n=0}^{\infty} c_n \phi_n(z),$$

where $c_n = (K(.,w), \phi_n)$; in other words,

$$\overline{c_n} = (\phi_n, K(.,w)) = \int_{\mathbb{D}} \phi_n(z) K(w,z)\, d\lambda(z) = \phi_n(w)$$

by the reproducing property of the Bergman kernel. This implies that the Bergman kernel is of the form

$$K(z,w) = \sum_{n=0}^{\infty} \phi_n(z)\, \overline{\phi_n(w)}, \tag{1.9}$$

where the sum converges uniformly in z on all compact subsets of $\mathbb{D}$. (This is true for any complete orthonormal system, as is shown above.) A simple computation now gives

$$K(z,w) = \sum_{n=0}^{\infty} \phi_n(z)\, \overline{\phi_n(w)} = \frac{1}{\pi} \sum_{n=0}^{\infty} (n+1)(z\overline{w})^n = \frac{1}{\pi} \frac{1}{(1 - z\overline{w})^2}. \tag{1.10}$$

Hence for each $f \in A^2(\mathbb{D})$, we have

$$f(z) = \frac{1}{\pi} \int_{\mathbb{D}} \frac{1}{(1 - z\overline{w})^2} f(w)\, d\lambda(w).$$

If we fix $z \in \mathbb{D}$ and set $f(w) = 1/(1 - w\overline{z})^2$, then we get the interesting formula

$$\frac{1}{\pi} \int_{\mathbb{D}} \frac{1}{|1 - z\overline{w}|^4}\, d\lambda(w) = \frac{1}{(1 - |z|^2)^2}.$$

(b)

Next, we determine the Bergman kernel of the polycylinder.

Proposition 1.13. *Let $\Omega_j \subset \mathbb{C}^{n_j}, j = 1, 2$, be two bounded domains with Bergman kernels K_{Ω_1} and K_{Ω_2}. Then the Bergman kernel K_Ω of the product domain $\Omega = \Omega_1 \times \Omega_2$ is given by*

$$K_\Omega((z_1, z_2), (w_1, w_2)) = K_{\Omega_1}(z_1, w_1)\, K_{\Omega_2}(z_2, w_2) \tag{1.11}$$

for $(z_1, z_2), (w_1, w_2) \in \Omega_1 \times \Omega_2$.

Proof. To show this, let F denote the function on the right-hand side of (1.11). It is clear that $(z_1, z_2) \mapsto F((z_1, z_2), (w_1, w_2))$ belongs to $A^2(\Omega)$ for each fixed $(w_1, w_2) \in \Omega$ and that F is antiholomorphic in the second variable. The reproducing property

$$f(z_1, z_2) = \int_{\Omega_1 \times \Omega_2} F((z_1, z_2), (w_1, w_2)) f(w_1, w_2)\, d\lambda(w_1, w_2)$$

is a consequence of Fubini's theorem and the corresponding reproducing properties of K_{Ω_1} and K_{Ω_2}. Hence, by the uniqueness property of the Bergman kernel, Proposition 1.8, we obtain $F = K_\Omega$. □

From this we get that the Bergman kernel of the polycylinder $\mathbb{D}^n$ is given by

$$K_{\mathbb{D}^n}(z, w) = \frac{1}{\pi^n} \prod_{j=1}^{n} \frac{1}{(1 - z_j \overline{w}_j)^2}. \tag{1.12}$$

(c)

For the computation of the Bergman kernel $K_{\mathbb{B}^n}$ of the unit ball in $\mathbb{C}^n$, we use the beta and gamma functions

$$\int_0^1 x^k (1 - x)^m\, dx = B(k + 1, m + 1) = \frac{\Gamma(k + 1)\Gamma(m + 1)}{\Gamma(k + m + 2)},$$

where $k, m \in \mathbb{N}$ and that for $0 \le a < 1$,

$$\begin{aligned}\int_0^{\sqrt{1-a^2}} x^{2k+1} \left(1 - \frac{x^2}{1 - a^2}\right)^{m+1} dx &= \frac{1}{2}(1 - a^2)^{k+1} \int_0^1 y^k (1 - y)^{m+1}\, dy \\ &= \frac{1}{2}(1 - a^2)^{k+1} B(k + 1, m + 2) \\ &= \frac{1}{2}(1 - a^2)^{k+1} \frac{\Gamma(k + 1)\Gamma(m + 2)}{\Gamma(k + m + 3)}.\end{aligned}$$

Now we can normalize the orthogonal basis $\{z^\alpha = z_1^{\alpha_1} \dots z_n^{\alpha_n}\}$ in $A^2(\mathbb{B}^n)$ and obtain

$$
\begin{aligned}
\|z^\alpha\|^2 &= \int_{\mathbb{B}^n} |z_1|^{2\alpha_1} \dots |z_n|^{2\alpha_n}\, d\lambda(z) \\
&= \frac{\pi}{\alpha_n+1} \int_{\mathbb{B}^{n-1}} |z_1|^{2\alpha_1} \dots |z_{n-1}|^{2\alpha_{n-1}} (1-|z_1|^2-\dots-|z_{n-1}|^2)^{\alpha_n+1}\, d\lambda \\
&= \frac{\pi}{\alpha_n+1} \int_{\mathbb{B}^{n-1}} |z_1|^{2\alpha_1} \dots |z_{n-2}|^{2\alpha_{n-2}} (1-|z_1|^2-\dots-|z_{n-2}|^2)^{\alpha_n+1} \\
&\quad \cdot |z_{n-1}|^{2\alpha_{n-1}} \left(1-\frac{|z_{n-1}|^2}{1-|z_1|^2-\dots-|z_{n-2}|^2}\right)^{\alpha_n+1} d\lambda \\
&= \frac{\pi}{\alpha_n+1} \frac{\pi\Gamma(\alpha_{n-1}+1)\Gamma(\alpha_n+2)}{\Gamma(\alpha_n+\alpha_{n-1}+3)} \\
&\quad \cdot \int_{\mathbb{B}^{n-2}} |z_1|^{2\alpha_1} \dots |z_{n-2}|^{2\alpha_{n-2}} (1-|z_1|^2-\dots-|z_{n-2}|^2)^{\alpha_n+\alpha_{n-1}+2}\, d\lambda \\
&= \frac{\pi}{\alpha_n+1} \frac{\pi\Gamma(\alpha_{n-1}+1)\Gamma(\alpha_n+2)}{\Gamma(\alpha_n+\alpha_{n-1}+3)} \dots \frac{\pi\Gamma(\alpha_1+1)\Gamma(\alpha_n+\dots+\alpha_2+n)}{\Gamma(\alpha_n+\dots+\alpha_1+n+1)} \\
&= \frac{\pi^n \alpha_1! \dots \alpha_n!}{(\alpha_n+\dots+\alpha_1+n)!}.
\end{aligned}
$$

Hence the Bergman kernel of the unit ball is given by

$$
\begin{aligned}
K_{\mathbb{B}^n}(z,w) &= \sum_\alpha \frac{(\alpha_n+\dots+\alpha_1+n)!}{\pi^n \alpha_1! \dots \alpha_n!} z^\alpha \overline{w}^\alpha \\
&= \frac{1}{\pi^n} \sum_{k=0}^\infty \sum_{|\alpha|=k} \frac{(\alpha_n+\dots+\alpha_1+n)!}{\alpha_1! \dots \alpha_n!} z^\alpha \overline{w}^\alpha \\
&= \frac{1}{\pi^n} \sum_{k=0}^\infty (k+n)(k+n-1)\dots(k+1)(z_1\overline{w}_1+\dots+z_n\overline{w}_n)^k \\
&= \frac{n!}{\pi^n} \frac{1}{(1-(z_1\overline{w}_1+\dots+z_n\overline{w}_n))^{n+1}}.
\end{aligned}
$$

1.3 Biholomorphic maps

Here we describe the behavior of the Bergman kernel under biholomorphic maps.

Proposition 1.14. *Let $F : \Omega_1 \longrightarrow \Omega_2$ be a biholomorphic map between bounded domains in $\mathbb{C}^n$. Let $f_1, \dots, f_n$ be the components of F, and let $F'(z) = (\frac{\partial f_j(z)}{\partial z_k})_{j,k=1}^n$.*

Then

$$
K_{\Omega_1}(z,w) = \det F'(z)\, K_{\Omega_2}(F(z),F(w))\, \overline{\det F'(w)} \tag{1.13}
$$

for all $z, w \in \Omega_1$.

Proof. Notice that, by the Cauchy–Riemann equations for $f_j = u_j + iv_j, j = 1, \dots, n$, the determinant of the Jacobian of the mapping F equals $|\det F'(z)|^2$. The substitution formula for integrals implies that for $g \in L^2(\Omega_2)$, we have

$$\int_{\Omega_2} |g(\zeta)|^2 \, d\lambda(\zeta) = \int_{\Omega_1} |g(F(z))|^2 \, |\det F'(z)|^2 \, d\lambda(z). \tag{1.14}$$

Hence the map $T_F : g \mapsto (g \circ F) \det F'$ establishes an isometric isomorphism from $L^2(\Omega_2)$ to $L^2(\Omega_1)$, with inverse map $T_{F^{-1}}$, which restricts to an isomorphism between $A^2(\Omega_1)$ and $A^2(\Omega_2)$. Now let $f \in A^2(\Omega_1)$ and apply the reproducing property of K_{Ω_2} to the function $T_{F^{-1}} f = (f \circ F^{-1}) \det(F^{-1})'$; setting $F(z) = u$, we get

$$\int_{\Omega_2} K_{\Omega_2}(u, v) T_{F^{-1}} f(v) \, d\lambda(v) = T_{F^{-1}} f(u) = f(z)(\det F'(z))^{-1}. \tag{1.15}$$

Since T_F is an isometry,

$$\int_{\Omega_2} T_{F^{-1}} f(v) [K_{\Omega_2}(v, u)]^- \, d\lambda(v) = \int_{\Omega_1} f(w) [T_F K_{\Omega_2}(., u)(w)]^- \, d\lambda(w). \tag{1.16}$$

From (1.15) and (1.16) we obtain

$$f(z) = \int_{\Omega_1} \det F'(z) \, K_{\Omega_2}(F(z), F(w)) \, \overline{\det F'(w)} f(w) \, d\lambda(w),$$

which means that the right-hand side of (1.13) has the required reproducing property, belongs to $A^2(\Omega_1)$ in the variable z, is antiholomorphic in the variable w, and hence must agree with $K_{\Omega_1}(z, w)$. □

Finally, we derive a useful formula for the corresponding orthogonal projections

$$P_j : L^2(\Omega_j) \longrightarrow A^2(\Omega_j), \quad j = 1, 2.$$

Proposition 1.15. *For all $g \in L^2(\Omega_2)$, we have*

$$P_1(\det F' \, g \circ F) = \det F' \, (P_2(g) \circ F). \tag{1.17}$$

Proof. The left-hand side of (1.17) can be written in the form $P_1(T_F(g))$, and hence, by (1.5), we obtain

$$P_1(T_F(g))(z) = \int_{\Omega_1} K_{\Omega_1}(z, w) T_F(g)(w) \, d\lambda(w), \quad z \in \Omega_1.$$

Now (1.13), together with (1.16), implies that

$$K_{\Omega_1}(w,z) = [T_F(K_{\Omega_2}(.,F(z)))(w)]\,\overline{\det F'(z)},$$

so, since T_F is an isometric isomorphism, we get

$$\begin{aligned}P_1(T_F(g))(z) &= \det F'(z) \int_{\Omega_1} T_F(g)(w)\,[T_F(K_{\Omega_2}(.,F(z)))(w)]^{-}\,d\lambda(w)\\ &= \det F'(z) \int_{\Omega_2} g(v)\,[K_{\Omega_2}(v,F(z))]^{-}\,d\lambda(v)\\ &= \det F'(z)\,(P_2(g))(F(z)),\end{aligned}$$

which proves (1.17). □

Remark 1.16. If $n = 1$ and $\Omega \subsetneq \mathbb{C}$ is a simply connected domain, then there is an interesting connection between the Bergman kernel K_Ω of Ω and the Riemann mapping $F : \Omega \longrightarrow \mathbb{D}$ with the uniqueness properties $F(a) = 0$ and $F'(a) > 0$ for some $a \in \Omega$:

$$F'(z) = \sqrt{\frac{\pi}{K_\Omega(a,a)}}\,K_\Omega(z,a), \quad z \in \Omega. \tag{1.18}$$

By (1.14) the transformation T_F establishes an isometry between $L^2(\Omega)$ and $L^2(\mathbb{D})$, which restricts to an isometry between $A^2(\Omega)$ and $A^2(\mathbb{D})$. Therefore we have

$$(T_F u, T_F u)_\Omega = (u,u)_\mathbb{D}, \quad u \in L^2(\mathbb{D}),$$

where $(\cdot,\cdot)_\Omega$ denotes the inner product in $L^2(\Omega)$, and $(\cdot,\cdot)_\mathbb{D}$ denotes the inner product of $L^2(\mathbb{D})$. Similarly, for $v \in L^2(\Omega)$ and $G = F^{-1}$, we have

$$(T_G v, T_G v)_\mathbb{D} = (v,v)_\Omega.$$

The polarization identity

$$(u_1,u_2) = \frac{1}{4}(\|u_1+u_2\|^2 - \|u_1-u_2\|^2) - \frac{i}{4}(\|u_1+iu_2\|^2 - \|u_1-iu_2\|^2)$$

in an inner product space over $\mathbb{C}$ yields

$$(T_F u_1, T_F u_2)_\Omega = (u_1,u_2)_\mathbb{D}, \quad u_1,u_2 \in L^2(\mathbb{D}), \tag{1.19}$$

and

$$(T_G v_1, T_G v_2)_\mathbb{D} = (v_1,v_2)_\Omega, \quad v_1,v_2 \in L^2(\Omega). \tag{1.20}$$

Since $T_F T_G$ is the identity operator, from (1.19) and (1.20) we obtain

$$(T_F u, v)_\Omega = (T_F u, T_F(T_G v))_\Omega = (u, T_G v)_\mathbb{D}. \tag{1.21}$$

Now let $h \in A^2(\mathbb{D})$ and observe that, by (1.1),

$$(h, 1)_{\mathbb{D}} = \pi h(0).$$

By (1.21) we get that for $f \in A^2(\Omega)$,

$$(f, F')_{\Omega} = (G'(f \circ G), 1)_{\mathbb{D}} = \pi G'(0) f(G(0)) = \frac{\pi}{\overline{F'(a)}} f(a).$$

It follows that the function $k(z,a) = \frac{\overline{F'(a)}}{\pi} F'(z)$ has the reproducing property and belongs to $A^2(\Omega)$, so, by Proposition 1.8, $k(z,a) = K_{\Omega}(z,a)$, and we get

$$F'(z) = \frac{\pi}{\overline{F'(a)}} K_{\Omega}(z,a).$$

Setting $z = a$, we obtain $F'(a)^2 = \pi K_{\Omega}(a,a)$, which proves (1.18).

1.4 The Segal–Bargmann space

In the sequel, we will also consider the Segal–Bargmann space or Fock space $A^2(\mathbb{C}^n, e^{-|z|^2})$ consisting of all entire functions f such that

$$\int_{\mathbb{C}^n} |f(z)|^2 e^{-|z|^2} \, d\lambda(z) < \infty.$$

It is clear that the Fock space is a Hilbert space with the inner product

$$(f, g) = \int_{\mathbb{C}^n} f(z) \overline{g(z)} e^{-|z|^2} \, d\lambda(z).$$

Similarly as in the beginning of this chapter, setting $n = 1$, we obtain for $f \in A^2(\mathbb{C}, e^{-|z|^2})$ that

$$\begin{aligned} |f(z)| &\leq \frac{1}{\pi r^2} \int_{D(z,r)} e^{|w|^2/2} |f(w)| e^{-|w|^2/2} \, d\lambda(w) \\ &\leq \frac{1}{\pi r^2} \Big(\int_{D(z,r)} e^{|w|^2} \, d\lambda(w) \Big)^{1/2} \Big(\int_{D(z,r)} |f(w)|^2 e^{-|w|^2} \, d\lambda(w) \Big)^{1/2} \\ &\leq C \Big(\int_{\mathbb{C}} |f(w)|^2 e^{-|w|^2} \, d\lambda(w) \Big)^{1/2} \\ &\leq C \|f\|, \end{aligned} \tag{1.22}$$

where C is a constant only depending on z. This implies that the Fock space $A^2(\mathbb{C}^n, e^{-|z|^2})$ has the reproducing property. In addition, for each compact subset L of $\mathbb{C}$, there exists a constant $C_L > 0$ such that

$$\sup_{z\in L}|f(z)| \le C_L\|f\| \tag{1.23}$$

for all $f \in A^2(\mathbb{C}, e^{-|z|^2})$.

The monomials $\{z^\alpha\}$ constitute an orthogonal basis, and the norms of the monomials are

$$\begin{aligned}\|z^\alpha\|^2 &= \int_{\mathbb{C}} |z_1|^{2\alpha_1} e^{-|z_1|^2}\, d\lambda(z_1)\cdots \int_{\mathbb{C}} |z_n|^{2\alpha_n} e^{-|z_n|^2}\, d\lambda(z_n)\\ &= (2\pi)^n \int_0^\infty r^{2\alpha_1+1}e^{-r^2}\, dr\cdots \int_0^\infty r^{2\alpha_n+1}e^{-r^2}\, dr\\ &= \pi^n\alpha_1!\dots\alpha_n!.\end{aligned}$$

It follows that each function $f \in A^2(\mathbb{C}^n, e^{-|z|^2})$ can be written in the form

$$f = \sum_\alpha f_\alpha\varphi_\alpha,$$

where

$$\varphi_\alpha(z) = \frac{z^\alpha}{\sqrt{\pi^n\alpha!}} \quad \text{and} \quad \sum_\alpha |f_\alpha|^2 < \infty \tag{1.24}$$

with $\alpha! = \alpha_1!\dots\alpha_n!$.

Hence the Bergman kernel of $A^2(\mathbb{C}^n, e^{-|z|^2})$ is of the form

$$K(z,w) = \sum_\alpha \frac{z^\alpha\overline{w}^\alpha}{\|z^\alpha\|^2} = \frac{1}{\pi^n}\sum_{k=0}^\infty \sum_{|\alpha|=k} \frac{z^\alpha\overline{w}^\alpha}{\alpha_1!\dots\alpha_n!} = \frac{1}{\pi^n}\exp(z_1\overline{w}_1 + \dots + z_n\overline{w}_n). \tag{1.25}$$

In the following, we set $\langle z,w\rangle = z_1\overline{w}_1 + \dots + z_n\overline{w}_n$ and $|z|^2 = |z_1|^2 + \dots + |z_n|^2$.

Let $\phi_u(z) = \frac{1}{\pi^{n/2}}\exp\langle z,u\rangle$ for $u, z \in \mathbb{C}^n$. In simplicity, we set $n = 1$. Then

$$\begin{aligned}(\phi_u, \phi_v) &= \frac{1}{\pi}\left(\sum_{k=0}^\infty \frac{\bar{u}^k}{k!}z^k, \sum_{k=0}^\infty \frac{\bar{v}^k}{k!}z^k\right)\\ &= \frac{1}{\pi}\sum_{k=0}^\infty \frac{1}{k!^2}(v\bar{u})^k(z^k,z^k)\\ &= \exp\langle v,u\rangle.\end{aligned} \tag{1.26}$$

In particular, we have $\|\phi_u\|^2 = \exp(|u|^2)$. The reproducing property implies that

$$f(u) = \frac{1}{\pi^{1/2}}(f, \phi_u) \tag{1.27}$$

for $f \in A^2(\mathbb{C}, e^{-|z|^2})$, and hence

$$|f(z)| = \frac{1}{\pi^{1/2}}|(f, \phi_z)| \le \frac{1}{\pi^{1/2}}\|\phi_z\|\|f\| = \frac{1}{\pi^{1/2}} \exp\left(\frac{|z|^2}{2}\right)\|f\|. \tag{1.28}$$

In a similar way,

$$|f(z)| \le \frac{1}{\pi^{n/2}} e^{|z|^2/2}\|f\| \tag{1.29}$$

for each $f \in A^2(\mathbb{C}^n, e^{-|z|^2})$.

We point out that the space $A^2(\mathbb{C}, e^{-|z|^2})$ serves for a representation of the states in quantum mechanics (see [30]), where

$$a(f) = \frac{df}{dz}$$

is the annihilation operator, and

$$a^*(f) = zf$$

is the creation operator, both of them being densely defined unbounded operators on $A^2(\mathbb{C}, e^{-|z|^2})$. The span of the finite linear combinations of the basis functions φ_a is dense in $A^2(\mathbb{C}, e^{-|z|^2})$. Hence both operators a and a^* are densely defined.

The function

$$F(z) = \sum_{k=2}^{\infty} \frac{\varphi_k(z)}{\sqrt{k(k-1)}} \in A^2(\mathbb{C}, e^{-|z|^2}),$$

but

$$F'(z) = \sum_{k=1}^{\infty} \frac{\varphi_k(z)}{\sqrt{k}} \notin A^2(\mathbb{C}, e^{-|z|^2}),$$

and

$$G(z) = \sum_{k=0}^{\infty} \frac{\varphi_k(z)}{k+1} \in A^2(\mathbb{C}, e^{-|z|^2}),$$

but

$$zG(z) = \sum_{k=1}^{\infty} \frac{\varphi_k(z)}{\sqrt{k}} \notin A^2(\mathbb{C}, e^{-|z|^2}),$$

and hence both operators a and a^* are unbounded operators on $A^2(\mathbb{C}, e^{-|z|^2})$.

However, taking a primitive of a function $f \in A^2(\mathbb{C}, e^{-|z|^2})$ yields a bounded operator

$$T : A^2(\mathbb{C}, e^{-|z|^2}) \longrightarrow A^2(\mathbb{C}, e^{-|z|^2});$$

let

$$f(z) = \sum_{k=0}^{\infty} f_k \frac{z^k}{\sqrt{\pi}\sqrt{k!}}, \quad \sum_{k=0}^{\infty} |f_k|^2 < \infty.$$

Then the function

$$h(z) = \sum_{k=0}^{\infty} f_k \frac{z^{k+1}}{\sqrt{\pi}\sqrt{k+1}\sqrt{(k+1)!}}$$

defines a primitive of f, and we can write

$$T(f) = h = \sum_{k=1}^{\infty} \frac{1}{\sqrt{k!}} (f, \tilde{\varphi}_k)\varphi_k,$$

where $\tilde{\varphi}_k = \varphi_{k-1}$, and the constant term in the Fourier series expansion of the primitive is always 0. This implies immediately that T is even a compact operator.

This is also a special result from the theory of Volterra-type integration operators on the Segal–Bargmann space of the form

$$T_g f(z) = \int_0^z f g' \, d\zeta$$

(see [19]).

We define the domain of the operator a as

$$\operatorname{dom}(a) = \{f \in A^2(\mathbb{C}, e^{-|z|^2}) : f' \in A^2(\mathbb{C}, e^{-|z|^2})\}.$$

Then $\operatorname{dom}(a^*)$ consists of all functions $g \in A^2(\mathbb{C}^n, e^{-|z|^2})$ such that the densely defined linear functional $L(f) = (a(f), g)$ is continuous on $\operatorname{dom}(a)$. This implies that there exists a function $h \in A^2(\mathbb{C}, e^{-|z|^2})$ such that $L(f) = (a(f), g) = (f, h)$.

Next, we show that

$$\operatorname{dom}(a^*) = \{g \in A^2(\mathbb{C}, e^{-|z|^2}) : \; zg \in A^2(\mathbb{C}, e^{-|z|^2})\}.$$

For this purpose, we use Fourier series expansion: let

$$f(z) = \sum_{k=0}^{\infty} f_k \varphi_k(z) \quad \text{and} \quad g(z) = \sum_{k=0}^{\infty} g_k \varphi_k(z),$$

where $(f_k)_k, (g_k)_k \in l^2$. Then we have

$$(a(f), g) = \sum_{k=0}^{\infty} \sqrt{k+1} f_{k+1} \overline{g_k} = (f, a^*(g)).$$

Notice that the commutator satisfies

$$[a, a^*] = I, \tag{1.30}$$

which is important in quantum mechanics; see [30].

For $f \in \operatorname{dom}(a) \cap \operatorname{dom}(a^*)$ with Fourier expansion $f(z) = \sum_{k=0}^{\infty} f_k \varphi_k(z)$, we have

$$\begin{aligned} \|a^*(f)\|^2 &= \sum_{k=0}^{\infty} (k+1)|f_k|^2 \\ &= \sum_{k=0}^{\infty} k|f_k|^2 + \sum_{k=0}^{\infty} |f_k|^2 \\ &= \|a(f)\|^2 + \|f\|^2, \end{aligned} \tag{1.31}$$

which implies that $\operatorname{dom}(a) = \operatorname{dom}(a^*)$.

1.5 The uncertainty principle

If we consider entire functions of several variables, then we get that the operators $a_j(f) = \frac{\partial f}{\partial z_j}$ and $a_j^*(f) = z_j f$ are densely defined unbounded operators on $A^2(\mathbb{C}^n, e^{-|z|^2})$. They are adjoint to each other, $[a_j, a_j^*] = I$, and $\operatorname{dom}(a_j) = \operatorname{dom}(a_j^*)$. Using (1.31), we obtain

$$\|f\|^2 \leq \|a_j(f)\|^2 + \|a_j^*(f)\|^2 \tag{1.32}$$

for $f \in \operatorname{dom}(a_j)$ and $j = 1, \ldots, n$.

Now we consider the position operators $X_j(u) = x_j u$ and the momentum operators $D_j u = -i\frac{\partial u}{\partial x_j}$ on $L^2(\mathbb{R}^n)$. Recall that

$$\|X_j(u)\|^2 + \|D_j(u)\|^2 \geq \|u\|^2, \quad u \in \operatorname{dom}(X_j) \cap \operatorname{dom}(D_j), \tag{1.33}$$

which is a variant of the uncertainty principle; see [30].

We indicate that (1.33) follows from inequality (1.32). To show this, we use the Bargmann transform $\mathcal{B} : L^2(\mathbb{R}^n) \longrightarrow A^2(\mathbb{C}^n, e^{-|z|^2})$ given by

$$\mathcal{B}(f)(z) = \pi^{-n/4} \int_{\mathbb{R}^n} f(x) \exp[-z^2/2 + \sqrt{2}zx - x^2/2]\, dx, \quad f \in L^2(\mathbb{R}^n),$$

where $z^2 = \sum_{j=1}^n z_j^2$ and $xz = \sum_{j=1}^n x_j z_j$.

We claim that the Bargmann transform is a unitary operator between $L^2(\mathbb{R}^n)$ and $A^2(\mathbb{C}^n, e^{-|z|^2})$. To show this, we first mention that $\mathcal{B}(f)$ is an entire function and set again $n = 1$:

$$\mathcal{B}(f)(z) = \pi^{-1/4} \exp[-z^2/2] \sum_{k=0}^{\infty} \frac{z^k}{k!} \int_{\mathbb{R}} f(x)(\sqrt{2}x)^k \exp[-x^2/2]\, dx, \tag{1.34}$$

and

$$\begin{aligned}\left|\int_{\mathbb{R}} f(x)(\sqrt{2}x)^k \exp(-x^2/2)\, dx\right| &\le \left(\int_{\mathbb{R}} |f(x)|^2\, dx\right)^{1/2} \left(\int_{\mathbb{R}} x^{2k} e^{-x^2}\, dx\right)^{1/2} \\ &= \|f\|_{L^2(\mathbb{R})} \left(\int_0^{\infty} r^{k-1/2} e^{-r}\, dr\right)^{1/2} \\ &\le \|f\|_{L^2(\mathbb{R})} \sqrt{k!},\end{aligned}$$

which gives that $\mathcal{B}(f)$ is an entire function.

Let

$$\Phi_z(x) = \pi^{-n/4} \exp[-\bar{z}^2/2 + \sqrt{2}\bar{z}x - x^2/2]\, dx.$$

Then

$$\mathcal{B}f(z) = (f, \Phi_z)_{L^2(\mathbb{R}^n)}. \tag{1.35}$$

Lemma 1.17.

$$(\Phi_u, \Phi_v)_{L^2(\mathbb{R}^n)} = \exp\langle v, u\rangle$$

for all $u, v \in \mathbb{C}^n$.

Proof. $(\Phi_u, \Phi_v)_{L^2(\mathbb{R}^n)}$

$$\begin{aligned}&= \pi^{-n/2} \int_{\mathbb{R}^n} \exp[-\bar{u}^2/2 + \sqrt{2}\bar{u}x - x^2/2] \exp[-v^2/2 + \sqrt{2}vx - x^2/2]\, dx \\ &= \pi^{-n/2} \exp(v\bar{u}) \int_{\mathbb{R}^n} \exp\left[-\frac{1}{2}(v + \bar{u})^2 + \sqrt{2}x(v + \bar{u}) - x^2\right] dx \\ &= \pi^{-n/2} \exp(v\bar{u}) \int_{\mathbb{R}^n} \exp[-(x - (v + \bar{u})/\sqrt{2})^2]\, dx \\ &= (2\pi)^{-n/2} \exp(v\bar{u}) \int_{\mathbb{R}^n} \exp\left[-\frac{1}{2}y^2\right] dy \\ &= \exp\langle v, u\rangle.\end{aligned}$$

□

Using (1.26), we see now that

$$(\Phi_u, \Phi_v)_{L^2(\mathbb{R}^n)} = (\phi_u, \phi_v) \tag{1.36}$$

for all $u, v \in \mathbb{C}^n$.

The system $\{\Phi_z : z \in \mathbb{C}^n\}$ is total in $L^2(\mathbb{R}^n)$, i. e., $(f, \Phi_z)_{L^2(\mathbb{R}^n)} = 0$ for all $z \in \mathbb{C}^n$ implies $f = 0$: by (1.34) we have

$$\mathcal{B}f(-it/\sqrt{2}) = \pi^{-1/4} e^{-t^2/2} \int_{\mathbb{R}} f(x) e^{-itx} e^{-x^2/2}\, dx = 0$$

for all $t \in \mathbb{R}$. Hence the Fourier transform of the function $f(x)e^{-x^2/2}$ is 0, so $f = 0$ almost everywhere.

The linear combinations of vectors Φ_z are dense in $L^2(\mathbb{R}^n)$, which is a consequence of the Hahn–Banach theorem: we have to show that if a continuous linear functional L has the property that $L(\Phi_z) = 0$ for all $z \in \mathbb{C}^n$, then $L = 0$, which follows again from the fact that each continuous linear functional is given by a function f in $L^2(\mathbb{R}^n)$ and we have $L(\Phi_z) = (\Phi_z, f)_{L^2(\mathbb{R}^n)}$. The same is also true for the system $\{\phi_z : z \in \mathbb{C}^n\}$ in $A^2(\mathbb{C}^n, e^{-|z|^2})$, where we use the reproducing property (1.27).

Now we define the unitary operator $U : L^2(\mathbb{R}^n) \longrightarrow A^2(\mathbb{C}^n, e^{-|z|^2})$ by

$$U(\Phi_z) = \phi_z, \quad z \in \mathbb{C}^n.$$

We indicate that for linear combinations, we have by (1.26)

$$\|a_1\phi_{z_1} + \cdots + a_m\phi_{z_m}\|^2 = \|a_1\Phi_{z_1} + \cdots + a_m\Phi_{z_m}\|^2_{L^2(\mathbb{R}^n)},$$

and hence we can extend U by continuity and obtain

$$Uf(z) = (Uf, \phi_z) = (f, U^{-1}\phi_z)_{L^2(\mathbb{R}^n)} = (f, \Phi_z)_{L^2(\mathbb{R}^n)} = \mathcal{B}f(z)$$

by (1.35).

Concerning the operators a_j and a_j^*, we have

$$\mathcal{B}^* a_j \mathcal{B} = \frac{1}{\sqrt{2}}(X_j + iD_j) \quad \text{and} \quad \mathcal{B}^* a_j^* \mathcal{B} = \frac{1}{\sqrt{2}}(X_j - iD_j),$$

which follows from integration by parts and

$$\frac{\partial}{\partial x_j} \exp[-z^2/2 + \sqrt{2}zx - x^2/2] = (\sqrt{2}z_j - x_j) \exp[-z^2/2 + \sqrt{2}zx - x^2/2];$$

see also [30]. So we get

$$\|X_j(u)\|^2 = \frac{1}{2}\|\mathcal{B}^* a_j \mathcal{B}(u) + \mathcal{B}^* a_j^* \mathcal{B}(u)\|^2,$$

and

$$\|D_j(u)\|^2 = \frac{1}{2}\|\mathcal{B}^* a_j \mathcal{B}(u) - \mathcal{B}^* a_j^* \mathcal{B}(u)\|^2.$$

Hence

$$\begin{aligned}\|X_j(u)\|^2 + \|D_j(u)\|^2 &= \|\mathcal{B}^* a_j \mathcal{B}(u)\|^2 + \|\mathcal{B}^* a_j^* \mathcal{B}(u)\|^2 \\ &= \|a_j \mathcal{B}(u)\|^2 + \|a_j^* \mathcal{B}(u)\|^2 \\ &\geq \|\mathcal{B}(u)\|^2 \\ &= \|u\|^2,\end{aligned}$$

where we used (1.32) and that $\mathcal{B}$ is an isometry.

The last computation also shows that (1.33) implies (1.32).

1.6 Nonradial weights

We consider spaces of entire functions with nonradial weights and determine their Bergman kernels using properties of the Segal–Bargmann space (Section 1.4). For $0 \leq \beta \leq 1$, define

$$H_\beta = \left\{ f : \mathbb{C} \longrightarrow \mathbb{C} \text{ entire} : \|f\|^2 = (f, f) = \int_{\mathbb{C}} |f(z)|^2 \, e^{-((1+\beta)x^2 + (1-\beta)y^2)} \, d\lambda(z) < \infty \right\},$$

where $z = x + iy$. In general, we cannot expect that the monomials in z constitute a biorthogonal system for H_β. We define

$$\psi_k(z) = z^k \exp\left(\frac{\beta}{2} z^2\right)$$

for $k = 0, 1, \ldots$. Since

$$\left|\exp\left(\frac{\beta}{2} z^2\right)\right|^2 = \exp(\beta(x^2 - y^2)),$$

we have $\psi_k \in H_\beta$. Let $k \neq m$. For the inner product in H_β, we have

$$(\psi_k, \psi_m) = \int_{\mathbb{C}} z^k \exp\left(\frac{\beta}{2} z^2\right) \overline{\left(z^m \exp\left(\frac{\beta}{2} z^2\right)\right)} \, e^{-((1+\beta)x^2 + (1-\beta)y^2)} \, d\lambda(z).$$

The exponential terms deliver

$$\exp\left(\frac{\beta}{2}z^2\right)\left(\exp\left(\frac{\beta}{2}z^2\right)\right)^{-} e^{-((1+\beta)x^2+(1-\beta)y^2)}$$
$$= \exp(\beta(x^2-y^2))\, e^{-((1+\beta)x^2+(1-\beta)y^2)} = e^{-|z|^2}. \tag{1.37}$$

Hence

$$(\psi_k, \psi_m) = \int_{\mathbb{C}} z^k \bar{z}^m e^{-|z|^2}\, d\lambda(z) = 0.$$

Now let $f \in H_\beta$ be such that $(f, \psi_k) = 0$ for each $k = 0, 1, \ldots$. Then

$$(f, \psi_k) = \int_{\mathbb{C}} f(z) \exp\left(-\frac{\beta}{2}z^2\right) \exp\left(\frac{\beta}{2}z^2\right)\left(z^k \exp\left(\frac{\beta}{2}z^2\right)\right)^{-} e^{-((1+\beta)x^2+(1-\beta)y^2)}\, d\lambda(z);$$

using (1.37), we get

$$(f, \psi_k) = \int_{\mathbb{C}} f(z) \exp\left(-\frac{\beta}{2}z^2\right) \bar{z}^k e^{-|z|^2}\, d\lambda(z),$$

which implies that $f(z)\exp(-\frac{\beta}{2}z^2) = 0$, and hence $f = 0$. Therefore $(\psi_k)_k$ is a complete biorthogonal system in H_β.

The mapping $\Psi : H_\beta \longrightarrow A^2(\mathbb{C}, e^{-|z|^2})$ defined by

$$\Psi(f)(z) = f(z)\exp\left(-\frac{\beta}{2}z^2\right)$$

is an isometry:

$$\|\Psi(f)\|_{A^2(\mathbb{C}, e^{-|z|^2})} = \|f\|_{H_\beta}, \quad f \in H_\beta.$$

Hence the norms of ψ_k in H_β are $\|\psi_k\|^2 = \pi k!$, and the Bergman kernel of H_β has the form

$$K_\beta(z, w) = \frac{1}{\pi} \sum_{k=0}^{\infty} \frac{z^k \exp(\frac{\beta}{2}z^2)\,(w^k \exp(\frac{\beta}{2}w^2))^{-}}{k!}$$
$$= \frac{1}{\pi} \exp\left(\frac{\beta}{2}(z^2 + \bar{w}^2)\right) \exp(z\bar{w}).$$

Next, we compute the adjoint of $a(f) = \frac{df}{dz}$ on H_β.
For this purpose, we use the orthonormal basis

$$\chi_k(z) = \frac{z^k}{\sqrt{\pi k!}} \exp\left(\frac{\beta}{2}z^2\right), \quad k = 0, 1, \ldots,$$

and write $f = \sum_{k=0}^{\infty} f_k \chi_k$ and $g = \sum_{k=0}^{\infty} g_k \chi_k$ for $f \in \operatorname{dom}(a)$ and $g \in \operatorname{dom}(a^*)$, where $(f_k)_k, (g_k)_k \in l^2$.

Then we get

$$a(f) = f_1 \chi_0 + \sum_{k=1}^{\infty} (f_{k+1}\sqrt{k+1} + \beta f_{k-1}\sqrt{k})\chi_k.$$

Hence

$$\begin{aligned}(a(f), g)_{H_\beta} &= f_1 \bar{g}_0 + \sum_{k=1}^{\infty} (f_{k+1}\sqrt{k+1} + \beta f_{k-1}\sqrt{k})\bar{g}_k \\ &= f_0 \beta \bar{g}_1 + \sum_{k=1}^{\infty} f_k (\bar{g}_{k-1}\sqrt{k} + \beta \bar{g}_k \sqrt{k}) \\ &= (f, a^*(g))_{H_\beta}.\end{aligned}$$

This implies that

$$a^*(g) = \beta \frac{dg}{dz} + (1-\beta^2) z g.$$

For $\beta = 0$, we have the Segal–Bargmann space; for $\beta = 1$, we get $a^* = a$, so the operator a is self-adjoint on H_1.

In addition, we use Parseval's identity to compute the norms $\|a(f)\|_{H_\beta}^2$ and $\|a^*(f)\|_{H_\beta}^2$:

$$\|a(f)\|_{H_\beta}^2 = |f_1|^2 + \sum_{k=0}^{\infty} |\beta\sqrt{k+1} f_k + \sqrt{k+2} f_{k+2}|^2 \tag{1.38}$$

and

$$\|a^*(f)\|_{H_\beta}^2 = \beta^2 |f_1|^2 + \sum_{k=0}^{\infty} |\sqrt{k+1} f_k + \beta\sqrt{k+2} f_{k+2}|^2. \tag{1.39}$$

A simple computation shows that

$$\|a^*(f)\|_{H_\beta}^2 - \|a(f)\|_{H_\beta}^2 = (1-\beta^2)\|f\|_{H_\beta}^2;$$

it follows that $\operatorname{dom}(a) = \operatorname{dom}(a^*)$, and for $0 \le \beta < 1$, we obtain

$$\|f\|_{H_\beta}^2 \le \frac{1}{1-\beta^2} (\|a(f)\|_{H_\beta}^2 + \|a^*(f)\|_{H_\beta}^2),$$

compare with (1.32).

1.7 Notes

The basics on Hilbert spaces can be found, for instance, in [93]. There are numerous good texts on different aspects of Bergman spaces and their reproducing kernels; the reader may consult the books by Chen and Shaw [16], Krantz [69], and Jarnicki and Pflug [60]. The remark about the Riemann mapping and the Bergman kernel is contained in Bell's book [4]. Further results regarding the Bergman kernel and the solution to $\overline{\partial}$ will be discussed in the next chapters. Folland's book [30] yields an excellent approach to mathematical concepts of quantum mechanics, like creation and annihilation operators and the uncertainty principle. The fundamentals of the Fock (Segal–Bargmann) space can be found in Zhu's book [97]. We will come back to these notions at the end of this book; see Chapter 16. An even more general nonradial form of the weight is investigated in [80].

2 The canonical solution operator to $\overline{\partial}$

In this chapter, we will use properties of the Bergman kernel to solve the inhomogeneous Cauchy–Riemann equation

$$\frac{\partial u}{\partial \overline{z}} = g \quad \text{or} \quad \overline{\partial} u = g,$$

where

$$\frac{\partial}{\partial \overline{z}} = \frac{1}{2}\left(\frac{\partial}{\partial x} + i\frac{\partial}{\partial y}\right), \quad z = x + iy, \tag{2.1}$$

and $g \in A^2(\mathbb{D})$.

Before we proceed, we recall some basic facts from operator theory.

2.1 Compact operators on Hilbert spaces

Let H_1 and H_2 be separable Hilbert spaces, and let $A : H_1 \longrightarrow H_2$ be a bounded linear operator. The operator A is compact if the image $A(B_1)$ of the unit ball B_1 in H_1 is a relatively compact subset of H_2; since H_2 is complete, this is equivalent to the notion of a totally bounded set, i. e., for each $\epsilon > 0$, there exists a finite number of elements $v_1, \dots, v_m \in H_2$ such that

$$A(B_1) \subset \bigcup_{j=1}^{m} B(v_j, \epsilon),$$

where $B(v_j, \epsilon) = \{v \in H_2 : \|v - v_j\| < \epsilon\}$.

Another equivalent definition of compactness is as follows: for each bounded sequence $(u_k)_k$ in H_1, the image sequence $(A(u_k))_k$ has a convergent subsequence in H_2.

Let $\mathcal{L}(H_1, H_2)$ denote the space of all bounded linear operators from H_1 to H_2 endowed with the topology generated by the operator norm

$$\|A\| = \sup\{\|Au\| : \|u\| \leq 1\}.$$

In this way, $\mathcal{L}(H_1, H_2)$ becomes a Banach space. Let $\mathcal{K}(H_1, H_2)$ denote the subspace of all compact operators from H_1 to H_2.

The following characterization of compactness is useful for the special operators in the text; see, for instance, [23]:

Proposition 2.1. *Let H_1 and H_2 be Hilbert spaces, and let $S: H_1 \to H_2$ be a bounded linear operator. The following three statements are equivalent:*

https://doi.org/10.1515/9783111182926-002

- *S is compact.*
- *For every $\epsilon > 0$, there are $C = C_\epsilon > 0$ and a compact operator $T = T_\epsilon : H_1 \to H_2$ such that*

$$\|Sv\| \leq C\|Tv\| + \epsilon\|v\|. \tag{2.2}$$

- *For every $\epsilon > 0$, there are $C = C_\epsilon > 0$ and a compact operator $T = T_\epsilon : H_1 \to H_2$ such that*

$$\|Sv\|^2 \leq C\|Tv\|^2 + \epsilon\|v\|^2. \tag{2.3}$$

Proof. First, we show that (2.2) and (2.3) are equivalent.

Suppose that (2.3) holds. Write (2.3) with ϵ and C replaced by their squares to obtain

$$\|Sv\|^2 \leq C^2\|Tv\|^2 + \epsilon^2\|v\|^2 \leq (C\|Tv\| + \epsilon\|v\|)^2,$$

which implies (2.2).

Now suppose that (2.2) holds. Choose η such that $\epsilon = 2\eta^2$ and apply (2.2) with ϵ replaced by η to get

$$\|Sv\|^2 \leq C^2\|Tv\|^2 + 2\eta C\|v\|\|Tv\| + \eta^2\|v\|^2.$$

It is easily seen (small–large constant trick) that there is $C' > 0$ such that

$$2\eta C\|v\|\|Tv\| \leq \eta^2\|v\|^2 + C'\|Tv\|^2,$$

and hence

$$\|Sv\|^2 \leq (C^2 + C')\|Tv\|^2 + 2\eta^2\|v\|^2 = C''\|Tv\|^2 + \epsilon\|v\|^2.$$

To prove the proposition, it therefore suffices to show that (2.2) is equivalent to compactness. When S is known to be compact, we choose $T = S$ and $C = 1$, and then (2.2) holds for every positive ϵ.

For the converse, let $(v_n)_n$ be a bounded sequence in H_1. We want to extract a Cauchy subsequence from $(Sv_n)_n$. From (2.2) we have

$$\|Sv_n - Sv_m\| \leq C\|Tv_n - Tv_m\| + \epsilon\|v_n - v_m\|. \tag{2.4}$$

Given a positive integer N, we may choose ϵ sufficiently small in (2.4) so that the second term on the right-hand side is at most $1/(2N)$. The first term can be made smaller than $1/(2N)$ by extracting a subsequence of $(v_n)_n$ (still labeled the same) for which $(Tv_n)_n$ converges and then choosing n and m large enough.

Let $(v_n^{(0)})_n$ denote the original bounded sequence. The above argument shows that for each positive integer N, there is a sequence $(v_n^{(N)})_n$ satisfying: $(v_n^{(N)})_n$ is a subsequence of $(v_n^{(N-1)})_n$, and for any pair v and w in $(v_n^{(N)})_n$, we have $\|Sv - Sw\| \leq 1/N$.

Let $(w_k)_k$ be the diagonal sequence defined by $w_k = v_k^{(k)}$. Then $(w_k)_k$ is a subsequence of $(v_n^{(0)})_n$, and the image sequence under S of $(w_k)_k$ is a Cauchy sequence. Since H_2 is complete, the image sequence converges, and S is compact. □

Proposition 2.2. *$\mathcal{K}(H_1, H_2)$ is a closed subspace of $\mathcal{L}(H_1, H_2)$ endowed with the operator norm.*

Proof. Let $A \in \mathcal{L}(H_1, H_2)$. Suppose that for each $\epsilon > 0$, there is a compact operator A_ϵ such that $\|A - A_\epsilon\| \leq \epsilon$. Then for each $u \in H_1$, we have

$$\|Au - A_\epsilon u\| \leq \epsilon \|u\|.$$

Now we get

$$\begin{aligned}\|Au\| &= \|Au - A_\epsilon u + A_\epsilon u\| \\ &\leq \|Au - A_\epsilon u\| + \|A_\epsilon u\| \\ &\leq \epsilon\|u\| + \|A_\epsilon u\|.\end{aligned}$$

Proposition 2.1 implies that A is compact. □

Proposition 2.3. *Suppose that $A \in \mathcal{K}(H_1, H_2)$, $S \in \mathcal{L}(H_1, H_1)$, and $T \in \mathcal{L}(H_2, H_2)$. Then both AS and TA are compact.*

Proof. If $(u_k)_k$ is a bounded sequence in H_1, then $(S(u_k))_k$ is also bounded, because S is a bounded operator. Since A is compact, $(A(S(u_k)))_k$ has a convergent subsequence. Thus AS is compact.

To show that TA is compact, we use Proposition 2.1:

$$\begin{aligned}\|TAu\| &\leq \|T\|\,\|Au\| \leq \|T\|(\epsilon\|u\| + C\|Au\|) \\ &\leq \epsilon\,\|T\|\,\|u\| + C\,\|T\|\,\|Au\|.\end{aligned}$$

□

Corollary 2.4. *Let H be a Hilbert space. Then $\mathcal{K}(H, H)$ forms a two-sided closed ideal in $\mathcal{L}(H, H)$.*

Theorem 2.5. *Let $A : H_1 \longrightarrow H_2$ be a bounded linear operator.*

The following properties are equivalent:

(i) *A is compact;*
(ii) *the adjoint operator $A^* : H_2 \longrightarrow H_1$ is compact;*
(iii) *$A^*A : H_1 \longrightarrow H_1$ is compact.*

Proof. Suppose that A is compact. Then, by Proposition 2.3, AA^* is also compact. Given $\epsilon > 0$, it follows that there is a constant C such that

$$\begin{aligned}\|A^*u\|^2 = (A^*u, A^*u) = (u, AA^*u) \\ &\leq \|u\|\,\|AA^*u\| \\ &\leq \epsilon\|u\|^2 + C\|AA^*u\|^2.\end{aligned}$$

Therefore, by Proposition 2.1, A^* is compact.

Since $A^{**} = A$, (i) and (ii) are equivalent.

(i) implies (iii) (Proposition 2.3), so it remains to show that (iii) implies (i). Let A^*A be compact. Given $\epsilon > 0$, there is again a constant C such that

$$\|Au\|^2 = (Au, Au) = (u, A^*Au) \leq \epsilon\|u\|^2 + C\|A^*Au\|^2,$$

so Proposition 2.1 implies that A is compact. □

Proposition 2.6. *A bounded operator $A : H \longrightarrow H$ is compact if and only if there exists a sequence $(A_k)_k$ of linear operators with finite-dimensional range such that $\|A - A_k\| \to 0$ as $k \to \infty$.*

Proof. An operator with finite-dimensional range is compact, because any bounded sequence in a finite-dimensional Hilbert space has a convergent subsequence. The limit of compact operators in the operator norm is again compact, so we proved one direction. The converse will follow from the next two results. □

The following theorem is the spectral theorem for compact self-adjoint operators:

Theorem 2.7. *Let $A : H \longrightarrow H$ be a compact self-adjoint operator on a separable Hilbert space H. Then there exist a real zero sequence $(\mu_n)_n$ and an orthonormal system $(e_n)_n$ in H such that for $x \in H$,*

$$Ax = \sum_{n=0}^{\infty} \mu_n(x, e_n)e_n,$$

where the sum converges in the operator norm, i. e.,

$$\sup_{\|x\|\leq 1}\left\|Ax - \sum_{n=0}^{N} \mu_n(x, e_n)e_n\right\| \to 0$$

as $N \to \infty$.

Proof. First, we collect some elementary properties of self-adjoint operators:

Claim **(a)**: Let $A \in \mathcal{L}(H, H) = \mathcal{L}(H)$ and suppose that $(Ax, x) = 0$ for all $x \in H$. Then $A = 0$. We have

$$0 = (A(x+y), x+y) - (A(x-y), x-y) = 2[(Ax,y) + (Ay,x)],$$

and hence $(Ax,y) + (Ay,x) = 0$. Now replace x by ix. Then $i(Ax,y) - i(Ay,x) = 0$, and therefore $(Ax,y) = 0$ for all $x,y \in H$, which implies $A = 0$.

Claim **(b)**: $A = A^*$ if and only if $(Ax,x) \in \mathbb{R}$ for all $x \in H$.

The equality $A = A^*$ implies $(Ax,x)^- = (x,Ax) = (Ax,x) \in \mathbb{R}$. If $(Ax,x) = (Ax,x)^-$ for all $x \in H$, then $(Ax,x) = (x,A^*x)^- = (A^*x,x)$, and so $((A-A^*)x,x) = 0$ for all $x \in H$. By (a) we have $A = A^*$.

Now we show that for a self-adjoint operator $A \in \mathcal{L}(H)$, the norm of A is given by

$$\|A\| = \sup_{\|x\|=1} |(Ax,x)|. \tag{2.5}$$

For this aim, let $N_A = \sup_{\|x\|=1} |(Ax,x)|$. Then $N_A \le \|A\|$. We also have

$$2[(Ax,y) + (Ay,x)] = (A(x+y), x+y) - (A(x-y), x-y)$$

as $A^* = A$ and, by the parallelogram rule,

$$|4\Re(Ax,y)| \le N_A(\|x+y\|^2 + \|x-y\|^2) = 2N_A(\|x\|^2 + \|y\|^2).$$

There is $\theta \in \mathbb{R}$ such that $e^{-i\theta}(Ax,y) = |(Ax,y)|$. Now replace y by $e^{-i\theta}y$. Then

$$2|(Ax,y)| \le N_A(\|x\|^2 + \|y\|^2).$$

For $t > 0$, we replace x by tx and y by y/t. Then we obtain

$$2|(Ax,y)| \le N_A\left(t^2\|x\|^2 + \frac{1}{t^2}\|y\|^2\right).$$

We consider the right side of this inequality as a function in t, and after differentiation with respect to t, we get

$$2t\|x\|^2 - \frac{2}{t^3}\|y\|^2,$$

so the right side of the inequality is minimal if $t^2 = \|y\|/\|x\|$. Hence we obtain

$$2|(Ax,y)| \le 2N_A\|x\|\,\|y\|$$

and $\|A\| \le N_A$, which proves (2.5).

Next we show that if $A \ne 0$, then there exists an eigenvector $x_0 \in H$ of A such that $\|x_0\| = 1$ and

$$|(Ax_0,x_0)| = \|A\| = \sup_{\|x\|=1} |(Ax,x)|, \tag{2.6}$$

and we show that the corresponding eigenvalue λ_0 is real and satisfies $|\lambda_0| = \|A\|$.

By (2.5) we have $\|A\| = \sup_{\|x\|=1} |(Ax,x)|$. Hence there is a sequence $(x_n)_n$ in H such that $\|x_n\| = 1$ and $\lim_{n\to\infty} |(Ax_n, x_n)| = \|A\|$. The inner products (Ax_n, x_n) are real, so there exists a subsequence, again denoted by $(x_n)_n$, such that $\lim_{n\to\infty}(Ax_n, x_n) = \lambda_0$. Then we have $\lambda_0 = \|A\|$ or $\lambda_0 = -\|A\|$.

Now we get

$$\begin{aligned} 0 &\le \|Ax_n - \lambda_0 x_n\|^2 = \|Ax_n\|^2 - 2\lambda_0(Ax_n, x_n) + \lambda_0^2\|x_n\|^2 \\ &\le \|A\|^2 - 2\lambda_0(Ax_n, x_n) + \|A\|^2 \end{aligned}$$

and, letting $n \to \infty$,

$$\lim_{n\to\infty} \|Ax_n - \lambda_0 x_n\| = 0.$$

Since A is compact and $(x_n)_n$ is a bounded sequence, there exists a subsequence $(x_{n_k})_k$ such that $\lim_{k\to\infty} Ax_{n_k} = x$. Hence $\lim_{k\to\infty} \lambda_0 x_{n_k} = x$, and as $\lambda_0 \neq 0$, we also have $\lim_{k\to\infty} x_{n_k} = x_0$ with $\lambda_0 x_0 = x$. Since A is continuous, this implies

$$\lim_{k\to\infty} Ax_{n_k} = Ax_0,$$

and hence $x = \lambda_0 x_0 = Ax_0$. So x_0 is an eigenvector of A with eigenvalue λ_0. This proves (2.6).

Now each eigenvalue of A is real: $Ax = \lambda x$ and $x \neq 0$ imply $\lambda(x,x) = (Ax,x) = (x,Ax) = (x,\lambda x) = \overline{\lambda}(x,x)$. Therefore $\lambda \in \mathbb{R}$.

Let $H_\lambda = \{x \in H : Ax = \lambda x\}$ be the eigenspace of the eigenvalue λ. Since A is continuous, H_λ is closed. There exists an orthonormal basis in H_λ consisting of eigenvectors of A.

If $\lambda_1 \neq \lambda_2$ are eigenvalues, then $H_{\lambda_1} \perp H_{\lambda_2}$, because for $x \in H_{\lambda_1}$ and $y \in H_{\lambda_2}$, we have $\lambda_1(x,y) = (Ax,y) = (x,Ay) = (x,\lambda_2 y) = \lambda_2(x,y)$, and hence $(x,y) = 0$.

We claim that the eigenspace H_λ of an eigenvalue $\lambda \neq 0$ is always finite-dimensional: if H_λ is of infinite dimension, we would have an infinite orthonormal system $\{x_\alpha\}_{\alpha\in A}$ such that $Ax_\alpha = \lambda x_\alpha$, and since the restriction of A to H_λ is also compact, we could find a convergent subsequence of $(Ax_\alpha)_\alpha$, which contradicts to $\|\lambda x_{\alpha_1} - \lambda x_{\alpha_2}\|^2 = 2|\lambda|^2$ for $\alpha_1 \neq \alpha_2$.

Now we have two possible cases: (i) there are finitely many eigenvalues $\neq 0$; (ii) there are infinitely many eigenvalues $\neq 0$.

(i) In this case, we have

$$H = H_{\lambda_0} \oplus \cdots \oplus H_{\lambda_{k-1}} \oplus M,$$

where $M = (H_{\lambda_0} \oplus \cdots \oplus H_{\lambda_{k-1}})^\perp$. We claim that $A(M) \subseteq M$: for $y \in M$, we have $(Ay,x) = (y,Ax) = 0$ for all $x \in H_{\lambda_0} \oplus \cdots \oplus H_{\lambda_{k-1}}$, because $Ax \in H_{\lambda_0} \oplus \cdots \oplus H_{\lambda_{k-1}}$, and hence $Ay \in M$.

The restriction A_M of A to M is again compact and self-adjoint: for $x,y \in M$, we have $(A_M x,y) = (Ax,y) = (x,Ay) = (x,A_M y)$, because $A(M) \subseteq M$. We claim that $A_M = 0$.

Suppose $A_M \neq 0$. Then, by (2.5), there is $\lambda \neq 0$ such that $A_M x = \lambda x$ for some $x \neq 0$, and hence we would get $x \in H_\lambda$, which is a contradiction.

Let $P_j : H \longrightarrow H_{\lambda_j}$ be the orthogonal projection. Then we obtain

$$x = P_0 x + P_1 x + \cdots + P_{k-1} x + y,$$

where $Ay = 0$, and therefore

$$Ax = \lambda_0 P_0 x + \lambda_1 P_1 x + \cdots + \lambda_{k-1} P_{k-1} x.$$

(ii) For each $\epsilon > 0$, the set $\{\lambda_i : \lambda_i$ eigenvalue of A with $|\lambda_i| \geq \epsilon\}$ is finite; otherwise, we would get an infinite orthonormal system $\{x_i\}$ such that $Ax_i = \lambda_i x_i$, and the sequence $(Ax_i)_i$ would have to contain a convergent subsequence. However,

$$\|Ax_k - Ax_l\|^2 = \|\lambda_k x_k - \lambda_l x_l\|^2 = |\lambda_k|^2 + |\lambda_l|^2 \geq 2\epsilon^2,$$

which is a contradiction.

Let $\lambda_0, \ldots, \lambda_{k-1}$ be eigenvalues with $|\lambda_i| \geq \epsilon$, and let $M = (H_{\lambda_0} \oplus \cdots \oplus H_{\lambda_{k-1}})^\perp$. The restriction A_M of A to M is self-adjoint and compact; hence, by (2.5), there is an eigenvector x such that $A_M x = \lambda x$ and $|\lambda| = \|A_M\|$, and thus $Ax = \lambda x$. Therefore $x \in (H_{\lambda_i})^\perp$ and $|\lambda| = \|A_M\| < \epsilon$. So we have

$$Ax = Ax_1 + A_M x_2, \; x_1 \in M^\perp, \quad x_2 \in M.$$

Since $Ax_1 = \lambda_0 P_0 x + \cdots + \lambda_{k-1} P_{k-1} x$, we obtain

$$\|A_M x_2\| = \|Ax - \lambda_0 P_0 x - \cdots - \lambda_{k-1} P_{k-1} x\| < \epsilon \|x\|.$$

Letting $\epsilon \to 0$, we get

$$Ax = \sum_{j=0}^{\infty} \lambda_j P_j x.$$

Now let $\{\psi_1, \ldots, \psi_k\}$ be an orthonormal basis of H_{λ_i}. Then

$$AP_i(x) = \lambda_i P_i(x) = \lambda_i \sum_{j=1}^{k} (x, \psi_j) \psi_j.$$

The sequence of the eigenvalues can be ordered in the following way:

$$|\lambda_0| \geq |\lambda_1| \geq \cdots,$$

because for each $\epsilon > 0$, there are only finitely many λ_i with $|\lambda_i| \geq \epsilon$.

Finally, the orthogonality of the eigenvectors x_k and Bessel's inequality (1.6) give

$$\left\| Ax - \sum_{k=1}^{n} \lambda_k (x, x_k) x_k \right\|^2 = \sum_{k=n+1}^{\infty} |\lambda_k (x, x_k)|^2 \leq \left(\|x\| \sup_{k>n} |\lambda_k| \right)^2,$$

and since $(\lambda_k)_{k\geq 1}$ is a zero sequence, the above series converges to A in the operator norm. □

Now we drop the assumption of self-adjointness and obtain the following:

Proposition 2.8. *Let $A : H_1 \longrightarrow H_2$ be a compact operator. There exist a decreasing zero sequence $(s_n)_n$ in $\mathbb{R}^+$ and orthonormal systems $(e_n)_{n\geq 0}$ in H_1 and $(f_n)_{n\geq 0}$ in H_2 such that*

$$Ax = \sum_{n=0}^{\infty} s_n (x, e_n) f_n \quad \forall x \in H_1,$$

where the sum converges again in the operator norm.

Proof. To show this, we apply the spectral theorem for the self-adjoint compact operator $A^*A : H_1 \longrightarrow H_1$ and get

$$A^*Ax = \sum_{n=0}^{\infty} s_n^2 (x, e_n) e_n, \tag{2.7}$$

where s_n^2 are the eigenvalues of A^*A. If $s_n > 0$, then we set $f_n = s_n^{-1} A e_n$ and get

$$(f_n, f_m) = \frac{1}{s_n s_m} (Ae_n, Ae_m) = \frac{1}{s_n s_m} (A^*Ae_n, e_m) = \frac{s_n^2}{s_n s_m} (e_n, e_m) = \delta_{n,m}.$$

For $y \in H_1$ such that $y \perp e_n$ for each $n \in \mathbb{N}_0$, we have by (2.7) that

$$\|Ay\|^2 = (Ay, Ay) = (A^*Ay, y) = 0.$$

Hence we have

$$\begin{aligned} Ax &= A\left(x - \sum_{n=0}^{\infty} (x, e_n) e_n \right) + A\left(\sum_{n=0}^{\infty} (x, e_n) e_n \right) \\ &= \sum_{n=0}^{\infty} (x, e_n) A e_n = \sum_{n=0}^{\infty} s_n (x, e_n) f_n. \end{aligned}$$

Similarly to the last theorem, we get

$$\left\| Ax - \sum_{k=1}^{n} s_k (x, e_k) f_k \right\|^2 = \sum_{k=n+1}^{\infty} |s_k (x, e_k)|^2 \leq \left(\|x\| \sup_{k>n} |s_k| \right)^2,$$

which implies that the series converges in the operator norm. □

Remark 2.9. From the last statement we get the missing direction in the proof of Proposition 2.6.

The numbers s_n are called the s-numbers of A. They are uniquely determined by the operator A, since they are the square roots of the eigenvalues of A^*A.

Let $0 < p < \infty$. The operator A belongs to the Schatten class $\mathbf{S}_p$ if its sequence $(s_n)_n$ of s-numbers belongs to l^p. The elements of the Schatten class $\mathbf{S}_2$ are called Hilbert–Schmidt operators.

Proposition 2.10. *Let $A : H_1 \longrightarrow H_2$ be a bounded linear operator between Hilbert spaces. The following conditions are equivalent:*

(i) *there is an orthonormal basis $(e_i)_{i\in I}$ of H_1, such that $\sum_{i\in I}\|Ae_i\|^2 < \infty$;*
(ii) *for each orthonormal basis $(f_j)_{j\in J}$ of H_1, we have $\sum_{j\in J}\|Af_j\|^2 < \infty$;*
(iii) *A is a Hilbert–Schmidt operator.*

Proof. (i) implies (ii): Let $(e_i)_{i\in I}$ be as in (i), and let $(g_l)_{l\in L}$ be an orthonormal basis of H_2. Then by Parseval's equality (Proposition 1.10)

$$\sum_{l\in L}\|A^*g_l\|^2 = \sum_{l\in L}\sum_{i\in I}|(A^*g_l, e_i)|^2 = \sum_{i\in I}\sum_{l\in L}|(g_l, Ae_i)|^2 = \sum_{i\in I}\|Ae_i\|^2 < \infty.$$

If $(f_j)_{j\in J}$ is an arbitrary orthonormal basis of H_1, then we get again

$$\sum_{j\in J}\|Af_j\|^2 = \sum_{l\in L}\|A^*g_l\|^2 = \sum_{i\in I}\|Ae_i\|^2 < \infty.$$

(ii) implies (iii): Let $(e_i)_{i\in I}$ be an orthonormal basis of H_1, and let M be a finite subset of I. We set $P_M x = \sum_{i\in M}(x, e_i)e_i$. Then

$$\begin{aligned}\|(A - AP_M)x\| &= \|A(I - P_M)x\| = \Big\|\sum_{i\in I\setminus M}(x, e_i)Ae_i\Big\| \\ &\le \Big(\sum_{i\in I\setminus M}\|Ae_i\|^2\Big)^{1/2}\Big(\sum_{i\in I\setminus M}|(x, e_i)|^2\Big)^{1/2} \le \Big(\sum_{i\in I\setminus M}\|Ae_i\|^2\Big)^{1/2}\|x\|,\end{aligned}$$

where we used Bessel's inequality (1.6). By assumption, $\sum_{i\in I}\|Ae_i\|^2 < \infty$, and hence we can approximate A in the operator norm by operators of finite range. Therefore A is compact and, by Proposition 2.8, can be written as

$$Ax = \sum_{n=0}^{\infty} s_n(x, x_n)f_n.$$

By Remark 1.11 there exists an orthonormal basis $(\xi_j)_{j\in J}$ that contains the orthonormal system $(x_n)_{n\ge 0}$. Then we have

$$\sum_{j\in J}\|A\xi_j\|^2 = \sum_{j\in J}\left\|\sum_{n=0}^{\infty} s_n(\xi_j, x_n)f_n\right\|^2 = \sum_{n=0}^{\infty} s_n^2 < \infty,$$

so A is a Hilbert–Schmidt operator.

(iii) implies (i): If A is a Hilbert–Schmidt operator, then it can be written as above, and we obtain (i). □

Corollary 2.11. *Let $A : H_1 \longrightarrow H_2$ be a Hilbert–Schmidt operator. Then for each orthonormal basis $(e_i)_{i\in I}$ of H_1, we have*

$$\nu_2(A)^2 := \sum_{n=0}^{\infty} s_n^2 = \sum_{i\in I}\|Ae_i\|^2 \geq \|A\|^2.$$

In particular, ν_2 is a norm in the space $S_2(H_1, H_2)$ of all Hilbert–Schmidt operators between H_1 and H_2.

Proof. We have

$$\|Ax\| = \left\|\sum_{i\in I}(x, e_i)Ae_i\right\| \leq \left(\sum_{i\in I}\|Ae_i\|^2\right)^{1/2}\|x\|,$$

and hence $\nu_2(A) \geq \|A\|$. It is easily seen that $A \mapsto (\sum_{i\in I}\|Ae_i\|^2)^{1/2}$ is a norm. □

If H_1 and H_2 are separable Hilbert spaces, $A \in \mathcal{L}(H_1, H_2)$, $(e_n)_{n\geq 1}$ is an orthonormal basis of H_1, and $(f_n)_{n\geq 1}$ is an orthonormal basis of H_2, then for each $x \in H_1$, we have

$$Ax = \sum_{n=1}^{\infty}(x, e_n)Ae_n,$$

and hence for all $j \in \mathbb{N}$,

$$(Ax, f_j) = \sum_{n=1}^{\infty}(x, e_n)(Ae_n, f_j).$$

The coefficients of Ax with respect to $(f_j)_{j\geq 1}$ can be computed from the coefficients of x with respect to $(e_n)_{n\geq 1}$ by means of the infinite matrix

$$a_{j,n} = (Ae_n, f_j), \quad j, n \in \mathbb{N}.$$

Parseval's equation implies

$$\sum_{j,n=1}^{\infty}|a_{j,n}|^2 = \sum_{j,n=1}^{\infty}|(Ae_n, f_j)|^2 = \sum_{n=1}^{\infty}\|Ae_n\|^2,$$

and we obtain the following:

Corollary 2.12. *An operator $A \in \mathcal{L}(H_1, H_2)$ is a Hilbert–Schmidt operator if and only if the matrix $(a_{j,n})_{j,n\geq 1}$ of A with respect to arbitrary orthonormal bases satisfies*

$$\sum_{j,n=1}^{\infty} |a_{j,n}|^2 < \infty.$$

In this case,

$$\nu_2(A)^2 = \sum_{j,n=1}^{\infty} |a_{j,n}|^2.$$

Corollary 2.13. *A linear operator $A : l^2 \longrightarrow l^2$ is a Hilbert–Schmidt operator if and only if there exists a matrix $(a_{j,n})_{j,n\geq 1}$ with $\sum_{j,n=1}^{\infty} |a_{j,n}|^2 < \infty$ such that for all $x = (x_n)_{n\geq 1} \in l^2$, we have*

$$Ax = \left(\sum_{n=1}^{\infty} a_{j,n} x_n \right)_{j\geq 1}.$$

Proof. If A has the given form, we can apply the Cauchy–Schwarz inequality to get

$$\|Ax\|_2^2 \leq \left(\sum_{j,n=1}^{\infty} |a_{j,n}|^2 \right) \|x\|_2^2,$$

and hence everything follows from Corollary 2.12. □

Proposition 2.14. *Let $S \subseteq \mathbb{R}^n$ and $T \subseteq \mathbb{R}^m$ be open sets, and let $A : L^2(T) \longrightarrow L^2(S)$ be a bounded linear operator. Then A is a Hilbert–Schmidt operator if and only if there exists $K \in L^2(S \times T)$ such that*

$$Af(s) = \int_T K(s,t) f(t)\, dt, \quad f \in L^2(T).$$

In this case,

$$\nu_2(A) = \left(\int_{S\times T} |K(s,t)|^2\, ds\, dt \right)^{1/2}.$$

Proof. Let $(g_k)_{k\geq 1}$ and $(f_j)_{j\geq 1}$ be orthonormal bases of $L^2(T)$ and $L^2(S)$, respectively. Then $(h_{j,k})_{j,k\geq 1}$ defined by

$$h_{j,k}(s,t) := f_j(s)\overline{g_k(t)}, \quad (s,t) \in S \times T,$$

is an orthonormal basis of $L^2(S \times T)$. This can be shown as follows: if $F \in L^2(S \times T)$, then $t \mapsto F(s,t) f_j(s)$ is in $L^2(T)$ for almost every $s \in S$. If $F \perp h_{j,k}$ for all $j, k \in \mathbb{N}$, then it follows that the functions $t \mapsto F(s,t)f_j(s)$ vanish in $L^2(T)$ for all $j \in \mathbb{N}$ almost everywhere with respect to s. This implies $F = 0$ and $(h_{j,k})_{j,k\geq 1}$ is an orthonormal basis.

If A is a Hilbert–Schmidt operator and $(a_{j,k})_{j,k\geq 1}$ is its matrix with respect to $(g_k)_{k\geq 1}$ and $(f_j)_{j\geq 1}$, then, by Corollary 2.12,

$$K(s,t) := \sum_{j,k\geq 1} a_{j,k} h_{j,k}(s,t) = \sum_{j,k\geq 1} a_{j,k} f_j(s)\overline{g_k(t)}$$

belongs to $L^2(S \times T)$, and $\|K\|_2^2 = \sum_{j,k\geq 1} |a_{j,k}|^2 = \nu_2(A)^2$.

For $f \in L^2(T)$, we have

$$Af(s) = \sum_{j,k\geq 1} a_{j,k}(f, g_k) f_j(s) = \int_T \sum_{j,k\geq 1} a_{j,k} f_j(s)\overline{g_k(t)} f(t)\, dt = \int_T K(s,t) f(t)\, dt.$$

If A is given by the kernel function $K \in L^2(S \times T)$, then A is continuous, because

$$\|Af\|^2 \leq \left(\int_{S\times T} |K(s,t)|^2\, ds\, dt \right) \int_T |f(t)|^2\, dt.$$

For the matrix $(a_{j,k})_{j,k\geq 1}$ of A, we get

$$a_{j,k} = (Ag_k, f_j) = \int_S \left(\int_T K(s,t) g_k(t)\, dt \right) \overline{f_j(s)}\, ds = (K, h_{j,k})$$

for all $j, k \in \mathbb{N}$. Bessel's inequality and Corollary 2.12 imply that A is a Hilbert–Schmidt operator. □

2.2 The canonical solution operator to $\bar{\partial}$ restricted to $A^2(\mathbb{D})$

We return to the inhomogeneous Cauchy–Riemann equation on the disc $\mathbb{D}$ and use the notations of Chapter 1. Let

$$S(g)(z) = \int_{\mathbb{D}} K(z,w) g(w)(z-w)^{-} d\lambda(w). \tag{2.8}$$

Using the Bergman projection

$$P : L^2(\mathbb{D}) \longrightarrow A^2(\mathbb{D})$$

and (1.5), we get

$$S(g)(z) = \bar{z} g(z) - P(\tilde{g})(z),$$

where $\tilde{g}(w) = \overline{w}g(w)$. We claim that $S(g)$ is a solution of the inhomogeneous Cauchy–Riemann equation

$$\frac{\partial}{\partial \overline{z}}S(g)(z) = \frac{\partial \overline{z}}{\partial \overline{z}}g(z) + \overline{z}\frac{\partial g}{\partial \overline{z}} + \frac{\partial P(\tilde{g})}{\partial \overline{z}} = g(z),$$

because g and $P(\tilde{g})$ are holomorphic functions, and therefore $\overline{\partial}S(g) = g$. In addition, we have $S(g) \perp A^2(\mathbb{D})$, because for arbitrary $f \in A^2(\mathbb{D})$, we get

$$(Sg,f) = (\tilde{g} - P(\tilde{g}),f) = (\tilde{g},f) - (P(\tilde{g}),f) = (\tilde{g},f) - (\tilde{g},Pf) = (\tilde{g},f) - (\tilde{g},f) = 0.$$

The operator $S : A^2(\mathbb{D}) \longrightarrow L^2(\mathbb{D})$ is called the canonical solution operator to $\overline{\partial}$.

Now we want to show that S is a compact operator. For this purpose, we consider the adjoint operator S^* and prove that S^*S is compact, which implies that S is compact (Theorem 2.5).

For $g \in A^2(\mathbb{D})$ and $f \in L^2(\mathbb{D})$, we have

$$\begin{aligned}(Sg,f) &= \int_{\mathbb{D}}\Big(\int_{\mathbb{D}} K(z,w)g(w)(z-w)^{-}\,d\lambda(w)\Big)\overline{f(z)}\,d\lambda(z)\\ &= \int_{\mathbb{D}}\overline{\Big(\int_{\mathbb{D}} K(w,z)(z-w)f(z)\,d\lambda(z)\Big)}\,g(w)\,d\lambda(w) = (g,S^*f),\end{aligned}$$

and hence

$$S^*(f)(w) = \int_{\mathbb{D}} K(w,z)(z-w)f(z)\,d\lambda(z). \tag{2.9}$$

Now set

$$c_n^2 = \int_{\mathbb{D}} |z|^{2n}\,d\lambda(z) = \frac{\pi}{n+1}$$

and $\phi_n(z) = z^n/c_n$, $n \in \mathbb{N}_0$. Then the Bergman kernel $K(z,w)$ can be expressed in the form

$$K(z,w) = \sum_{k=0}^{\infty}\frac{z^k\overline{w}^k}{c_k^2}.$$

Next, we compute

$$P(\tilde{\phi}_n)(z) = \int_{\mathbb{D}}\sum_{k=0}^{\infty}\frac{z^k\overline{w}^k}{c_k^2}\,\overline{w}\,\frac{w^n}{c_n}\,d\lambda(w) = \sum_{k=1}^{\infty}\frac{z^{k-1}}{c_{k-1}^2}\int_{\mathbb{D}}\frac{\overline{w}^k w^n}{c_n}\,d\lambda(w) = \frac{c_n z^{n-1}}{c_{n-1}^2},$$

and hence we have

$$S(\phi_n)(z) = \overline{z}\,\phi_n(z) - \frac{c_n z^{n-1}}{c_{n-1}^2}, \qquad n \in \mathbb{N}.$$

Now we apply S^* and get

$$S^* S(\phi_n)(w) = \int_{\mathbb{D}} \sum_{k=0}^{\infty} \frac{w^k \overline{z}^k}{c_k^2}(z - w)\left(\frac{\overline{z} z^n}{c_n} - \frac{c_n z^{n-1}}{c_{n-1}^2}\right) d\lambda(z).$$

The last integral is computed in two steps: first, by the multiplication by z,

$$\begin{aligned}
\int_{\mathbb{D}} \sum_{k=0}^{\infty} \frac{w^k \overline{z}^k}{c_k^2}\left(\frac{\overline{z} z^{n+1}}{c_n} - \frac{c_n z^n}{c_{n-1}^2}\right) d\lambda(z) &= \int_{\mathbb{D}} \frac{z^{n+1}}{c_n} \sum_{k=0}^{\infty} \frac{w^k \overline{z}^{k+1}}{c_k^2}\, d\lambda(z) \\
&\quad - \frac{c_n}{c_{n-1}^2} \int_{\mathbb{D}} z^n \sum_{k=0}^{\infty} \frac{w^k \overline{z}^k}{c_k^2}\, d\lambda(z) \\
&= \frac{w^n}{c_n^3} \int_{\mathbb{D}} |z|^{2n+2}\, d\lambda(z) - \frac{w^n}{c_{n-1}^2 c_n} \int_{\mathbb{D}} |z|^{2n}\, d\lambda(z) \\
&= \left(\frac{c_{n+1}^2}{c_n^3} - \frac{c_n}{c_{n-1}^2}\right) w^n.
\end{aligned}$$

Next, by the multiplication by w,

$$\begin{aligned}
w \int_{\mathbb{D}} \sum_{k=0}^{\infty} \frac{w^k \overline{z}^k}{c_k^2}\left(\frac{\overline{z} z^n}{c_n} - \frac{c_n z^{n-1}}{c_{n-1}^2}\right) d\lambda(z) &= w \int_{\mathbb{D}} \frac{z^n}{c_n} \sum_{k=0}^{\infty} \frac{w^k \overline{z}^{k+1}}{c_k^2}\, d\lambda(z) \\
&\quad - w \int_{\mathbb{D}} \frac{c_n z^{n-1}}{c_{n-1}^2} \sum_{k=0}^{\infty} \frac{w^k \overline{z}^k}{c_k^2}\, d\lambda(z) \\
&= w\left(\frac{c_n w^{n-1}}{c_{n-1}^2} - \frac{c_n w^{n-1}}{c_{n-1}^2}\right) \\
&= 0.
\end{aligned}$$

From this it follows that

$$S^* S(\phi_n)(w) = \left(\frac{c_{n+1}^2}{c_n^2} - \frac{c_n^2}{c_{n-1}^2}\right)\phi_n(w), \qquad n = 1, 2, \ldots;$$

for $n = 0$, an analogous computation shows

$$S^* S(\phi_0)(w) = \frac{c_1^2}{c_0^2}\,\phi_0(w).$$

Finally, we get the following:

Proposition 2.15. *Let $S : A^2(\mathbb{D}) \longrightarrow L^2(\mathbb{D})$ be the canonical solution operator for $\bar{\partial}$, and let $(\phi_k)_k$ be the normalized monomials. Then*

$$S^* S\phi = \frac{c_1^2}{c_0^2}(\phi, \phi_0)\phi_0 + \sum_{n=1}^{\infty}\left(\frac{c_{n+1}^2}{c_n^2} - \frac{c_n^2}{c_{n-1}^2}\right)(\phi, \phi_n)\phi_n \tag{2.10}$$

for each $\phi \in A^2(\mathbb{D})$.

Since

$$\frac{c_{n+1}^2}{c_n^2} - \frac{c_n^2}{c_{n-1}^2} = \frac{1}{(n+2)(n+1)} \to 0 \quad \text{as } n \to \infty,$$

it follows that $S^* S$ and S are compact.

We have also shown that the s-numbers of S are $(\frac{c_{n+1}^2}{c_n^2} - \frac{c_n^2}{c_{n-1}^2})^{1/2}$, and since

$$\sum_{n=1}^{\infty}\left(\frac{c_{n+1}^2}{c_n^2} - \frac{c_n^2}{c_{n-1}^2}\right) < \infty,$$

it follows that S is Hilbert–Schmidt.

This can also be shown directly using Proposition 2.14. For this purpose, we claim that the function $(z, w) \mapsto K(z, w)(z - w)^-$ belongs to $L^2(\mathbb{D} \times \mathbb{D})$.

We have to prove that

$$\int_{\mathbb{D}}\int_{\mathbb{D}} \frac{|\overline{z} - \overline{w}|^2}{|1 - z\overline{w}|^4}\, d\lambda(z)\, d\lambda(w) < \infty.$$

An easy estimate gives $|z - w| \le |1 - z\overline{w}|$ for $z, w \in \mathbb{D}$. Hence

$$\int_{\mathbb{D}}\int_{\mathbb{D}} \frac{|\overline{z} - \overline{w}|^2}{|1 - z\overline{w}|^4}\, d\lambda(z)\, d\lambda(w) \le \int_{\mathbb{D}}\int_{\mathbb{D}} \frac{1}{|1 - z\overline{w}|^2}\, d\lambda(z)\, d\lambda(w).$$

Introducing the polar coordinates $z = r\,e^{i\theta}$ and $w = s\,e^{i\phi}$, we can write the last integral in the form

$$\begin{aligned}\int_{\mathbb{D}}\int_{\mathbb{D}} \frac{1}{|1 - z\overline{w}|^2}\, d\lambda(z)\, d\lambda(w) &= \int_0^1\int_0^1\int_0^{2\pi}\int_0^{2\pi} \frac{r\,s\, d\theta\, d\phi\, dr\, ds}{1 - 2r\,s\,\cos(\theta - \phi) + r^2 s^2} \\ &= \int_0^1\int_0^1\int_0^{2\pi}\int_0^{2\pi} \frac{1 - r^2 s^2}{1 - 2r\,s\,\cos(\theta - \phi) + r^2 s^2}\,\frac{r\,s}{1 - r^2 s^2}\, d\theta\, d\phi\, dr\, ds.\end{aligned}$$

Integration of the Poisson kernel with respect to θ yields

$$\int_0^{2\pi} \frac{1-\rho^2}{1-2\rho\cos(\theta-\phi)+\rho^2}\, d\theta = 2\pi, \quad 0<\rho<1.$$

Therefore we have

$$\int_0^1\int_0^1\int_0^{2\pi}\int_0^{2\pi} \frac{1-r^2s^2}{1-2rs\cos(\theta-\phi)+r^2s^2}\,\frac{rs}{1-r^2s^2}\, d\theta\, d\phi\, dr\, ds$$

$$= (2\pi)^2 \int_0^1\int_0^1 \frac{rs}{1-r^2s^2}\, dr\, ds = -(2\pi)^2 \int_0^1 \frac{\log(1-s^2)}{2s}\, ds < \infty.$$

In the following, we consider the weighted spaces of entire functions

$$A^2(\mathbb{C}, e^{-|z|^m}) = \left\{ f : \mathbb{C} \longrightarrow \mathbb{C} : \|f\|_m^2 := \int_{\mathbb{C}} |f(z)|^2 e^{-|z|^m}\, d\lambda(z) < \infty \right\},$$

where $m > 0$. Let

$$c_k^2 = \int_{\mathbb{C}} |z|^{2k} e^{-|z|^m}\, d\lambda(z).$$

Then

$$K_m(z,w) = \sum_{k=0}^{\infty} \frac{z^k \overline{w}^k}{c_k^2}$$

is the reproducing kernel for $A^2(\mathbb{C}, e^{-|z|^m})$.

In the sequel the expression

$$\frac{c_{k+1}^2}{c_k^2} - \frac{c_k^2}{c_{k-1}^2}$$

will be important. Using the integral representation of the Γ-function, we easily see that the above expression is equal to

$$\frac{\Gamma(\frac{2k+4}{m})}{\Gamma(\frac{2k+2}{m})} - \frac{\Gamma(\frac{2k+2}{m})}{\Gamma(\frac{2k}{m})}.$$

For $m = 2$, this expression equals 1 for each $k = 1, 2, \ldots$. We will be interested in the limit behavior as $k \to \infty$. By Stirling's formula the limit behavior is equivalent to the limit behavior of the expression

$$\left(\frac{2k+2}{m}\right)^{2/m} - \left(\frac{2k}{m}\right)^{2/m}$$

as $k \to \infty$. Hence we have shown the following:

Lemma 2.16. *The expression*

$$\frac{\Gamma(\frac{2k+4}{m})}{\Gamma(\frac{2k+2}{m})} - \frac{\Gamma(\frac{2k+2}{m})}{\Gamma(\frac{2k}{m})}$$

tends to ∞ for $0 < m < 2$, is equal to 1 for $m = 2$, and tends to zero for $m > 2$ as k tends to ∞.

For $0 < \rho < 1$, define $f_\rho(z) := f(\rho z)$ and $\tilde{f}_\rho(z) = \overline{z} f_\rho(z)$ for $f \in A^2(\mathbb{C}, e^{-|z|^m})$. Then we easily see that $\tilde{f}_\rho \in L^2(\mathbb{C}, e^{-|z|^m})$ but there are functions $g \in A^2(\mathbb{C}, e^{-|z|^m})$ such that $\overline{z} g \notin L^2(\mathbb{C}, e^{-|z|^m})$.

Let $P_m : L^2(\mathbb{C}, e^{-|z|^m}) \longrightarrow A^2(\mathbb{C}, e^{-|z|^m})$ denote the orthogonal projection. Then P_m can be written in the form

$$P_m(f)(z) = \int_{\mathbb{C}} K_m(z,w) f(w) e^{-|w|^m}\, d\lambda(w), \quad f \in L^2(\mathbb{C}, e^{-|z|^m}).$$

Proposition 2.17. *Let $m \geq 2$. Then there is a constant $C_m > 0$, depending only on m, such that*

$$\int_{\mathbb{C}} |\tilde{f}_\rho(z) - P_m(\tilde{f}_\rho)(z)|^2 e^{-|z|^m}\, d\lambda(z) \leq C_m \int_{\mathbb{C}} |f(z)|^2 e^{-|z|^m}\, d\lambda(z)$$

for each $0 < \rho < 1$ and for each $f \in A^2(\mathbb{C}, e^{-|z|^m})$.

Proof. First, we observe that for the Taylor expansion of $f(z) = \sum_{k=0}^\infty a_k z^k$, we have

$$\begin{aligned} P_m(\tilde{f}_\rho)(z) &= \int_{\mathbb{C}} \sum_{k=0}^\infty \frac{z^k \overline{w}^k}{c_k^2} \left(\overline{w} \sum_{j=0}^\infty a_j \rho^j w^j \right) e^{-|w|^m}\, d\lambda(w) \\ &= \sum_{k=1}^\infty a_k \frac{c_k^2}{c_{k-1}^2} \rho^k z^{k-1}. \end{aligned}$$

Now we obtain

$$\begin{aligned} \int_{\mathbb{C}} |\tilde{f}_\rho(z) - P_m(\tilde{f}_\rho)(z)|^2 e^{-|z|^m}\, d\lambda(z) &= \int_{\mathbb{C}} \left(\overline{z} \sum_{k=0}^\infty a_k \rho^k z^k - \sum_{k=1}^\infty a_k \frac{c_k^2}{c_{k-1}^2} \rho^k z^{k-1} \right) \\ &\quad \times \left(z \sum_{k=0}^\infty \overline{a_k} \rho^k \overline{z}^k - \sum_{k=1}^\infty \overline{a_k} \frac{c_k^2}{c_{k-1}^2} \rho^k \overline{z}^{k-1} \right) e^{-|z|^m}\, d\lambda(z) \\ &= \int_{\mathbb{C}} \left(\sum_{k=0}^\infty |a_k|^2 \rho^{2k} |z|^{2k+2} - 2 \sum_{k=1}^\infty |a_k|^2 \frac{c_k^2}{c_{k-1}^2} \rho^{2k} |z|^{2k} \right. \end{aligned}$$

$$+\sum_{k=1}^{\infty}|a_k|^2\frac{c_k^4}{c_{k-1}^4}\rho^{2k}|z|^{2k-2}\Bigg)e^{-|z|^m}\,d\lambda(z)$$

$$|a_0|^2c_1^2+\sum_{k=1}^{\infty}|a_k|^2c_k^2\rho^{2k}\left(\frac{c_{k+1}^2}{c_k^2}-\frac{c_k^2}{c_{k-1}^2}\right).$$

Now the result follows from the facts that

$$\int_{\mathbb{C}}|f(z)|^2e^{-|z|^m}\,d\lambda(z)=\sum_{k=0}^{\infty}|a_k|^2c_k^2$$

and that the sequence $(\frac{c_{k+1}^2}{c_k^2}-\frac{c_k^2}{c_{k-1}^2})_k$ is bounded. □

Remark 2.18. Already in the last proposition, the sequence $(\frac{c_{k+1}^2}{c_k^2}-\frac{c_k^2}{c_{k-1}^2})_k$ plays an important role, and it will turn out that this sequence is the sequence of eigenvalues of the operator $S_m^*S_m$ (see below).

Proposition 2.19. *Let $m\geq 2$ and consider an entire function $f\in A^2(\mathbb{C},e^{-|z|^m})$ with Taylor series expansion $f(z)=\sum_{k=0}^{\infty}a_kz^k$. Let*

$$F(z):=\overline{z}\sum_{k=0}^{\infty}a_kz^k-\sum_{k=1}^{\infty}a_k\frac{c_k^2}{c_{k-1}^2}z^{k-1}$$

and define $S_m(f):=F$. Then $S_m:A^2(\mathbb{C},e^{-|z|^m})\longrightarrow L^2(\mathbb{C},e^{-|z|^m})$ is a continuous linear operator representing the canonical solution operator to $\overline{\partial}$ restricted to $A^2(\mathbb{C},e^{-|z|^m})$, i. e., $\overline{\partial}S_m(f)=f$ and $S_m(f)\perp A^2(\mathbb{C},e^{-|z|^m})$.

Proof. By the proof of Proposition 2.17 and by Abel's and Fatou's theorems (see, for instance, [26]) we have

$$\begin{aligned}\int_{\mathbb{C}}|F(z)|^2e^{-|z|^m}\,d\lambda(z)&=\int_{\mathbb{C}}\lim_{\rho\to1}|\tilde{f}_\rho(z)-P_m(\tilde{f}_\rho)(z)|^2e^{-|z|^m}\,d\lambda(z)\\&\leq\sup_{0<\rho<1}\int_{\mathbb{C}}|\tilde{f}_\rho(z)-P_m(\tilde{f}_\rho)(z)|^2e^{-|z|^m}\,d\lambda(z)\\&\leq C_m\int_{\mathbb{C}}|f(z)|^2e^{-|z|^m}\,d\lambda(z),\end{aligned}$$

and hence the function

$$F(z):=\overline{z}\sum_{k=0}^{\infty}a_kz^k-\sum_{k=1}^{\infty}a_k\frac{c_k^2}{c_{k-1}^2}z^{k-1}$$

belongs to $L^2(\mathbb{C},e^{-|z|^m})$ and satisfies

$$\int_{\mathbb{C}} |F(z)|^2 e^{-|z|^m}\, d\lambda(z) \le C_m \int_{\mathbb{C}} |f(z)|^2 e^{-|z|^m}\, d\lambda(z). \tag{2.11}$$

The above computation also shows that $\lim_{\rho\to 1} \|\tilde f_\rho - P_m(\tilde f_\rho)\|_m = \|F\|_m$ and, by a standard argument for L^p-spaces (see, for instance, [26]),

$$\lim_{\rho\to 1} \|\tilde f_\rho - P_m(\tilde f_\rho) - F\|_m = 0.$$

A similar computation as in the case $A^2(\mathbb{D})$ shows that the function F defined above satisfies $\bar\partial F = f$. Let $S_m(f) := F$. Then, by the last remarks, $S_m : A^2(\mathbb{C}, e^{-|z|^m}) \longrightarrow L^2(\mathbb{C}, e^{-|z|^m})$ is a continuous linear solution operator for $\bar\partial$. For arbitrary $h \in A^2(\mathbb{C}, e^{-|z|^m})$, we have

$$(h, S_m(f))_m = (h, F)_m = \lim_{\rho\to 1}(h, \tilde f_\rho - P_m(\tilde f_\rho))_m = \lim_{\rho\to 1}(h - P_m(h), \tilde f_\rho)_m = 0,$$

where $(\cdot,\cdot)_m$ denotes the inner product in $L^2(\mathbb{C}, e^{-|z|^m})$. Hence S_m is the canonical solution operator for $\bar\partial$ restricted to $A^2(\mathbb{C}, e^{-|z|^m})$. □

Remark 2.20. Let

$$f(z) = \sum_{k=0}^{\infty} \frac{z^k}{\sqrt{(k+1)!}\,\sqrt{k+1}}.$$

Then $f \in A^2(\mathbb{C}, e^{-|z|^2})$, since

$$\|f\|_2^2 = 2\pi \sum_{k=0}^{\infty} \frac{k!}{(k+1)!(k+1)} = 2\pi \sum_{k=0}^{\infty} \frac{1}{(k+1)^2} < \infty.$$

However,

$$\|zf\|_2^2 = 2\pi \sum_{k=0}^{\infty} \frac{(k+1)!}{(k+1)!(k+1)} = 2\pi \sum_{k=0}^{\infty} \frac{1}{(k+1)} = \infty,$$

and hence $zf \notin L^2(\mathbb{C}, e^{-|z|^2})$.

The expression for the function F in the last theorem corresponds formally to the expression $\bar z f - P_m(\bar z f)$; in general, $\bar z f \notin L^2(\mathbb{C}, e^{-|z|^m})$ for $f \in A^2(\mathbb{C}, e^{-|z|^m})$, but $f \mapsto F$ defines a bounded linear operator from $A^2(\mathbb{C}, e^{-|z|^m})$ to $L^2(\mathbb{C}, e^{-|z|^m})$.

Theorem 2.21. *The canonical solution operator to $\bar\partial$ restricted to the space $A^2(\mathbb{C}, e^{-|z|^m})$ is compact if and only if*

$$\lim_{k\to\infty}\left(\frac{c_{k+1}^2}{c_k^2} - \frac{c_k^2}{c_{k-1}^2}\right) = 0.$$

Proof. For a complex polynomial p, the canonical solution operator S_m can be written in the form

$$S_m(p)(z) = \int_{\mathbb{C}} K_m(z,w)p(w)(\overline{z} - \overline{w})e^{-|w|^m}\, d\lambda(w),$$

and therefore we can express the conjugate S_m^* in the form

$$S_m^*(q)(w) = \int_{\mathbb{C}} K_m(w,z)q(z)(z - w)e^{-|z|^m}\, d\lambda(z)$$

if q is a finite linear combination of the terms $\overline{z}^k z^l$. This follows by considering the inner product $(S_m(p), q)_m = (p, S_m^*(q))_m$.

Now we claim that

$$S_m^* S_m(u_n)(w) = \left(\frac{c_{n+1}^2}{c_n^2} - \frac{c_n^2}{c_{n-1}^2}\right)u_n(w), \quad n = 1, 2, \ldots,$$

and

$$S_m^* S_m(u_0)(w) = \frac{c_1^2}{c_0^2}\, u_0(w),$$

where $\{u_n(z) = z^n/c_n, n = 0, 1, \ldots\}$ is the standard orthonormal basis of $A^2(\mathbb{C}, e^{-|z|^m})$.

In a similar way as before, for the case of $A^2(\mathbb{D})$, we see that

$$S_m(u_n)(z) = \overline{z}u_n(z) - \frac{c_n z^{n-1}}{c_{n-1}^2}, \quad n = 1, 2, \ldots.$$

Hence

$$\begin{aligned} S_m^* S_m(u_n)(w) &= \int_{\mathbb{C}} K_m(w,z)(z - w)\left(\frac{\overline{z}z^n}{c_n} - \frac{c_n z^{n-1}}{c_{n-1}^2}\right)e^{-|z|^m}\, d\lambda(z) \\ &= \int_{\mathbb{C}} \sum_{k=0}^{\infty} \frac{w^k \overline{z}^k}{c_k^2}(z - w)\left(\frac{\overline{z}z^n}{c_n} - \frac{c_n z^{n-1}}{c_{n-1}^2}\right)e^{-|z|^m}\, d\lambda(z). \end{aligned}$$

As before, we get

$$\int_{\mathbb{C}} \sum_{k=0}^{\infty} \frac{w^k \overline{z}^k}{c_k^2}\left(\frac{\overline{z}z^{n+1}}{c_n} - \frac{c_n z^n}{c_{n-1}^2}\right)e^{-|z|^m}\, d\lambda(z) = \left(\frac{c_{n+1}^2}{c_n^3} - \frac{c_n}{c_{n-1}^2}\right)w^n$$

and

$$w \int_{\mathbb{C}} \sum_{k=0}^{\infty} \frac{w^k \overline{z}^k}{c_k^2} \left(\frac{\overline{z} z^n}{c_n} - \frac{c_n z^{n-1}}{c_{n-1}^2} \right) e^{-|z|^m} \, d\lambda(z) = w \left(\frac{c_n w^{n-1}}{c_{n-1}^2} - \frac{c_n w^{n-1}}{c_{n-1}^2} \right) = 0,$$

which implies that

$$S_m^* S_m(u_n)(w) = \left(\frac{c_{n+1}^2}{c_n^2} - \frac{c_n^2}{c_{n-1}^2} \right) u_n(w), \quad n = 1, 2, \ldots.$$

The case $n = 0$ follows from an analogous computation.

The last statement says that $S_m^* S_m$ is a diagonal operator with respect to the orthonormal basis $\{u_n(z) = z^n / c_n\}$ of $A^2(\mathbb{C}, e^{-|z|^m})$. Therefore we easily see that $S_m^* S_m$ is compact if and only if

$$\lim_{n \to \infty} \left(\frac{c_{n+1}^2}{c_n^2} - \frac{c_n^2}{c_{n-1}^2} \right) = 0. \qquad \square$$

Theorem 2.22. *The canonical solution operator for $\overline{\partial}$ restricted to the space $A^2(\mathbb{C}, e^{-|z|^m})$ is compact if $m > 2$. The canonical solution operator for $\overline{\partial}$ as an operator from $L^2(\mathbb{C}, e^{-|z|^2})$ into itself is not compact.*

Proof. The first statement follows immediately from Theorem 2.21 and Lemma 2.16. For the second statement, we use (2.11) to show that the canonical solution operator is continuous as operator from $A^2(\mathbb{C}, e^{-|z|^2})$ to $L^2(\mathbb{C}, e^{-|z|^2})$.

By Hörmander's L^2-estimate for the solution of the $\overline{\partial}$-equation (see Chapter 8, Theorem 8.4), for each $g \in L^2(\mathbb{C}, e^{-|z|^2})$, there is a function $f \in L^2(\mathbb{C}, e^{-|z|^2})$ such that $\overline{\partial} f = g$ and

$$\int_{\mathbb{C}} |f(z)|^2 e^{-|z|^2} \, d\lambda(z) \leq \int_{\mathbb{C}} |g(z)|^2 e^{-|z|^2} \, d\lambda(z).$$

Hence the canonical solution operator for $\overline{\partial}$ as an operator from $L^2(\mathbb{C}, e^{-|z|^2})$ into itself is continuous, and its restriction to the closed subspace $A^2(\mathbb{C}, e^{-|z|^2})$ fails to be compact by Theorem 2.21. By the definition of compactness this implies that the canonical solution operator is not compact as an operator from $L^2(\mathbb{C}, e^{-|z|^2})$ into itself. $\square$

Remark 2.23. In the case of the Fock space $A^2(\mathbb{C}, e^{-|z|^2})$ the composition $S_2^* S_2$ equals the identity on $A^2(\mathbb{C}, e^{-|z|^2})$, which follows from the proof of Theorem 2.21.

Theorem 2.24. *Let $m \geq 2$. The canonical solution operator for $\overline{\partial}$ restricted to $A^2(\mathbb{C}, e^{-|z|^m})$ fails to be Hilbert–Schmidt.*

Proof. By Proposition 2.19 we know that the canonical solution operator is continuous, and we can use the techniques from before to get

$$
\begin{aligned}
\|S_m(u_n)\|_m^2 &= \frac{1}{c_n^2}\int_{\mathbb{C}}\left|\overline{z}\,z^n - \frac{c_n^2}{c_{n-1}^2}\,z^{n-1}\right|^2 e^{-|z|^m}\,d\lambda(z)\\
&= \frac{1}{c_n^2}\int_{\mathbb{C}} |z|^{2n-2}\left(|z|^4 - \frac{2c_n^2|z|^2}{c_{n-1}^2} + \frac{c_n^4}{c_{n-1}^4}\right)e^{-|z|^m}\,d\lambda(z)\\
&= \frac{1}{c_n^2}\int_{\mathbb{C}} |z|^{2n+2}\,e^{-|z|^m}\,d\lambda(z) - \frac{2}{c_{n-1}^2}\int_{\mathbb{C}} |z|^{2n}\,e^{-|z|^m}\,d\lambda(z)\\
&\quad + \frac{c_n^2}{c_{n-1}^4}\int_{\mathbb{C}} |z|^{2n-2}\,e^{-|z|^m}\,d\lambda(z)\\
&= \frac{c_{n+1}^2}{c_n^2} - \frac{c_n^2}{c_{n-1}^2}.
\end{aligned}
$$

Hence

$$
\sum_{n=0}^{\infty} \|S_m(u_n)\|_m^2 < \infty
$$

if and only if

$$
\lim_{n\to\infty} \frac{c_{n+1}^2}{c_n^2} < \infty.
$$

By Proposition 2.10, S_m is a Hilbert–Schmidt operator if and only if

$$
\sum_{n=0}^{\infty} \|S_m(u_n)\|_m^2 < \infty.
$$

In our case, we have

$$
\frac{c_{n+1}^2}{c_n^2} = \Gamma\left(\frac{2n+4}{m}\right)/\Gamma\left(\frac{2n+2}{m}\right),
$$

which, by Stirling's formula, implies that the corresponding canonical solution operator to $\overline{\partial}$ fails to be Hilbert–Schmidt. □

In the case of several variables the corresponding operator S^*S is more complicated; nevertheless, using a suitable orthogonal decomposition, we can generalize the above results.

2.3 Notes

Most of the general material about compact operators on Hilbert spaces can be found in [77] or [93]. The results about the canonical solution operator to $\overline{\partial}$ restricted to holomorphic functions were taken from [38] and [39]. This operator can be viewed as a Hankel operator with antiholomorphic symbol $\overline{z}$. Generalizations of the results on weighted spaces can be found in [20]. A recent book by Zhu [97] is devoted to background information and different topics of analysis in weighted spaces of entire functions.

3 Spectral properties of the canonical solution operator to $\bar{\partial}$

In this chapter, we concentrate on several complex variables and generalize the results of Chapter 2. We discuss compactness and Schatten class membership of the canonical solution operator to $\bar{\partial}$ in several complex variables.

3.1 Complex differential forms

For this purpose, we introduce the notion of complex differential forms. Let $\Omega \subseteq \mathbb{C}^n$ be an open subset, and let $f : \Omega \longrightarrow \mathbb{C}$ be a $\mathcal{C}^1$-function. We write $z_j = x_j + iy_j$ and for $P \in \Omega$, consider the differential

$$df_P = \sum_{j=1}^{n} \left(\frac{\partial f}{\partial x_j}(P)\, dx_j + \frac{\partial f}{\partial y_j}(P)\, dy_j \right).$$

We use the complex differentials

$$dz_j = dx_j + idy_j, \quad d\bar{z}_j = dx_j - idy_j$$

and the derivatives

$$\frac{\partial}{\partial z_j} = \frac{1}{2}\left(\frac{\partial}{\partial x_j} - i\frac{\partial}{\partial y_j} \right), \quad \frac{\partial}{\partial \bar{z}_j} = \frac{1}{2}\left(\frac{\partial}{\partial x_j} + i\frac{\partial}{\partial y_j} \right)$$

and rewrite the differential df_p in the form

$$df_P = \sum_{j=1}^{n} \left(\frac{\partial f}{\partial z_j}(P)\, dz_j + \frac{\partial f}{\partial \bar{z}_j}(P)\, d\bar{z}_j \right) = \partial f_P + \bar{\partial} f_P.$$

A general differential form is given by

$$\omega = \sum_{|J|=p,|K|=q}{}' a_{J,K}\, dz_J \wedge d\bar{z}_K,$$

where $\sum'_{|J|=p,|K|=q}$ denotes the sum taken only over all increasing multi-indices $J = (j_1, \dots, j_p)$ and $K = (k_1, \dots, k_q)$, and

$$dz_J = dz_{j_1} \wedge \cdots \wedge dz_{j_p}, \quad d\bar{z}_K = d\bar{z}_{k_1} \wedge \cdots \wedge d\bar{z}_{k_q}.$$

The derivative $d\omega$ of ω is defined by

$$d\omega = \sum_{|J|=p,|K|=q}{}' da_{J,K} \wedge dz_J \wedge d\bar{z}_K = \sum_{|J|=p,|K|=q}{}' (\partial a_{J,K} + \bar{\partial} a_{J,K}) \wedge dz_J \wedge d\bar{z}_K,$$

https://doi.org/10.1515/9783111182926-003

and we set

$$\partial\omega = \sum_{|J|=p,|K|=q}{}' \partial a_{J,K} \wedge dz_J \wedge d\overline{z}_K \quad \text{and} \quad \overline{\partial}\omega = \sum_{|J|=p,|K|=q}{}' \overline{\partial} a_{J,K} \wedge dz_J \wedge d\overline{z}_K.$$

We have $d = \partial + \overline{\partial}$, and since $d^2 = 0$, it follows that

$$0 = (\partial + \overline{\partial}) \circ (\partial + \overline{\partial})\omega = (\partial \circ \partial)\omega + (\partial \circ \overline{\partial} + \overline{\partial} \circ \partial)\omega + (\overline{\partial} \circ \overline{\partial})\omega,$$

which implies $\partial^2 = 0, \overline{\partial}^2 = 0$, and $\partial \circ \overline{\partial} + \overline{\partial} \circ \partial = 0$ by comparing the types of the differential forms involved.

3.2 (0, 1)-forms with holomorphic coefficients

Let Ω be a bounded domain in $\mathbb{C}^n$, and let $A^2_{(0,1)}(\Omega)$ denote the space of all $(0,1)$-forms with holomorphic coefficients belonging to $L^2(\Omega)$. With the same proof as in Chapter 2, we show that the canonical solution operator

$$S : A^2_{(0,1)}(\Omega) \longrightarrow L^2(\Omega)$$

has the form

$$S(g)(z) = \int_\Omega K(z,w)\langle g(w), z - w\rangle \, d\lambda(w), \tag{3.1}$$

where K denotes the Bergman kernel of Ω,

$$g(z) = \sum_{j=1}^n g_j(z)\, d\overline{z}_j \in A^2_{(0,1)}(\Omega),$$

and

$$\langle g(w), z - w\rangle = \sum_{j=1}^n g_j(w)(\overline{z}_j - \overline{w}_j)$$

for $z = (z_1, \dots, z_n)$ and $w = (w_1, \dots, w_n)$.

Let $v(z) = \sum_{j=1}^n \overline{z}_j g_j(z)$. Then it follows that

$$\overline{\partial} v = \sum_{j=1}^n \frac{\partial v}{\partial \overline{z}_j} d\overline{z}_j = \sum_{j=1}^n g_j d\overline{z}_j = g.$$

Hence the canonical solution operator S can be written in the form $S(g) = v - P(v)$, where $P : L^2(\Omega) \longrightarrow A^2(\Omega)$ is the Bergman projection. If $\tilde{v}$ is another solution to $\overline{\partial} u = g$, then $v - \tilde{v} \in A^2(\Omega)$, and hence $v = \tilde{v} + h$, where $h \in A^2(\Omega)$. Therefore

$$v - P(v) = \tilde{v} + h - P(\tilde{v}) - P(h) = \tilde{v} - P(\tilde{v}).$$

Since $g_j \in A^2(\Omega), j = 1, \dots, n$, we have

$$g_j(z) = \int_\Omega K(z,w) g_j(w)\, d\lambda(w).$$

Now we get

$$\begin{aligned} S(g)(z) &= \sum_{j=1}^n \overline{z}_j g_j(z) - \int_\Omega K(z,w)\Big(\sum_{j=1}^n \overline{w}_j g_j(w) \Big)\, d\lambda(w) \\ &= \int_\Omega \Big[\Big(\sum_{j=1}^n \overline{z}_j g_j(w) \Big) K(z,w) - \Big(\sum_{j=1}^n \overline{w}_j g_j(w) \Big) K(z,w) \Big]\, d\lambda(w) \\ &= \int_\Omega K(z,w) \langle g(w), z - w \rangle\, d\lambda(w), \end{aligned}$$

and (3.1) is proved.

Remark 3.1. It is pointed out that a (0, 1)-form $g = \sum_{j=1}^n g_j\, d\overline{z}_j$ with holomorphic coefficients is not invariant under the pull back by a holomorphic map $F = (F_1, \dots, F_n) : \Omega_1 \longrightarrow \Omega$. Then

$$F^* g = \sum_{l=1}^n g_l\, d\overline{F}_l = \sum_{j=1}^n \Big(\sum_{l=1}^n g_l \frac{\partial \overline{F}_l}{\partial \overline{z}_j} \Big)\, d\overline{z}_j,$$

where we used the fact that

$$d\overline{F}_l = \partial \overline{F}_l + \overline{\partial}\, \overline{F}_l = \sum_{j=1}^n \frac{\partial \overline{F}_l}{\partial z_j}\, dz_j + \sum_{j=1}^n \frac{\partial \overline{F}_l}{\partial \overline{z}_j}\, d\overline{z}_j = \sum_{j=1}^n \frac{\partial \overline{F}_l}{\partial \overline{z}_j}\, d\overline{z}_j.$$

The expressions $\frac{\partial \overline{F}_l}{\partial \overline{z}_j}$ are not holomorphic.

Nevertheless, it is true that $\overline{\partial} u = g$ implies $\overline{\partial}(u \circ F) = F^* g$, which follows from the fact that for a general differential form ω and a holomorphic map F, we have

$$\partial(F^*\omega) = F^*(\partial\omega) \quad \text{and} \quad \overline{\partial}(F^*\omega) = F^*(\overline{\partial}\omega).$$

For holomorphic (n, n)-forms, we have the following result

Proposition 3.2. *Let ω be a holomorphic (n, n)-form, i. e.,*

$$\omega = \tilde{\omega}\, dz_1 \wedge \cdots \wedge dz_n \wedge d\overline{z}_1 \wedge \cdots \wedge d\overline{z}_n,$$

where $\tilde{\omega} \in A^2(\Omega)$. Let u be the following $(n, n-1)$-form

$$u = \sum_{j=1}^{n} u_j \, dz_1 \wedge \cdots \wedge dz_n \wedge d\overline{z}_1 \wedge \cdots \wedge [d\overline{z}_j] \wedge \cdots \wedge d\overline{z}_n,$$

where

$$u_j(z) = \frac{(-1)^{n+j-1}}{n} \int_{\Omega} (\overline{z}_j - \overline{w}_j) K(z,w) \tilde{\omega}(w) \, d\lambda(w),$$

and $[d\overline{z}_j]$ *means that* $d\overline{z}_j$ *is omitted.*

Then $u_j \perp A^2(\Omega)$, $j = 1, \ldots, n$, *and* $\bar\partial u = \omega$.

Proof. It follows that

$$u_j(z) = \frac{(-1)^{n+j-1}}{n} (\overline{z}_j \tilde{\omega}(z) - P(\overline{w}_j \tilde{\omega})(z)),$$

from which we obtain

$$\frac{\partial u_j}{\partial \overline{z}_k} = \frac{(-1)^{n+j-1}}{n} \left(\frac{\partial \overline{z}_j}{\partial \overline{z}_k} \tilde{\omega} + \overline{z}_j \frac{\partial \tilde{\omega}}{\partial \overline{z}_k} \right) = \frac{(-1)^{n+j-1}}{n} \delta_{jk} \, \tilde{\omega},$$

where δ_{jk} is the Kronecker delta symbol. Hence

$$\begin{aligned} \bar\partial u &= \sum_{k=1}^{n} \sum_{j=1}^{n} \frac{\partial u_j}{\partial \overline{z}_k} d\overline{z}_k \wedge dz_1 \wedge \cdots \wedge dz_n \wedge d\overline{z}_1 \wedge \cdots \wedge [d\overline{z}_j] \wedge \cdots \wedge d\overline{z}_n \\ &= \sum_{k=1}^{n} \sum_{j=1}^{n} ((-1)^{n+j-1}/n) \, \delta_{jk} \, \tilde{\omega} \, d\overline{z}_k \\ &\quad \wedge dz_1 \wedge \cdots \wedge dz_n \wedge d\overline{z}_1 \wedge \cdots \wedge [d\overline{z}_j] \wedge \cdots \wedge d\overline{z}_n \\ &= \tilde{\omega} \, dz_1 \wedge \cdots \wedge dz_n \wedge d\overline{z}_1 \wedge \cdots \wedge d\overline{z}_n. \end{aligned}$$

□

Remark 3.3. The pull -back by a holomorphic map F has in this case the form

$$F^* \omega = \left| \det \frac{\partial F_j}{\partial z_k} \right|^2 \tilde{\omega} \, dz_1 \wedge \cdots \wedge dz_n \wedge d\overline{z}_1 \wedge \cdots \wedge d\overline{z}_n.$$

3.3 Compactness and Schatten class membership

Now we will study boundedness, compactness, and Schatten class membership of the canonical solution operator to $\bar\partial$, restricted to $(0,1)$-forms with holomorphic coefficients, on $L^2(d\mu)$, where μ is a Borel measure on $\mathbb{C}^n$ with the property that the monomials form an orthogonal family in $L^2(d\mu)$. The characterizations are formulated in terms of moment properties of μ.

This situation covers a number of basic examples:

- The Lebesgue measure on bounded domains in $\mathbb{C}^n$ that are invariant under the torus action

$$(\theta_1, \ldots, \theta_n)(z_1, \ldots, z_n) \mapsto (e^{i\theta_1} z_1, \ldots, e^{i\theta_n} z_n)$$

(i. e., Reinhardt domains).
- Weighted L^2 spaces with radially symmetric weights (e. g., generalized Fock spaces).
- Weighted L^2 spaces with decoupled radial weights, that is,

$$d\mu = e^{\sum_j \varphi_j(|z_j|^2)} d\lambda,$$

where $\varphi_j \colon \mathbb{R} \to \mathbb{R}$ is a weight function.

We denote by

$$A^2(d\mu) = \overline{\{z^\alpha : \alpha \in \mathbb{N}^n\}}$$

the closure of the monomials in $L^2(d\mu)$ and write

$$m_\alpha = c_\alpha^{-1} = \int |z^\alpha|^2 d\mu.$$

We will give necessary and sufficient conditions in terms of these multimoments of the measure μ for the canonical solution operator to $\bar\partial$ – when restricted to $(0,1)$-forms with coefficients in $A^2(d\mu)$ – to be bounded, compact, and belonging to the Schatten class $\mathbf{S}_p$. This is accomplished by presenting a complete diagonalization of the solution operator by orthonormal bases with corresponding estimates.

In the following, we will investigate the *canonical solution operator S* to $\bar\partial$ that to each

$$\omega = \sum_{j=0}^{n} f_j d\bar z_j, \quad f_j \in A^2(d\mu),$$

assigns the solution to the equation $\bar\partial u = \omega$, which is orthogonal to $A^2(d\mu)$.

We will frequently encounter multi-indices γ that may have one (but not more than one) entry equal to -1: in that case, we define $c_\gamma = 0$. We will denote the set of these multi-indices by Γ. We let $e_j = (0, \ldots, 1, \ldots, 0)$ be the multi-index with 1 in the jth spot and 0 elsewhere.

Theorem 3.4. *$S \colon A^2_{(0,1)}(d\mu) \to L^2(d\mu)$ is bounded if and only if there exists a constant C such that*

$$\frac{c_{\gamma+e_j}}{c_{\gamma+2e_j}} - \frac{c_\gamma}{c_{\gamma+e_j}} < C$$

for all multi-indices $\gamma \in \Gamma$ *and all* $j = 1, \ldots, n$.

We have a similar criterion for compactness.

Theorem 3.5. $S\colon A^2_{(0,1)}(d\mu) \to L^2(d\mu)$ *is compact if and only if*

$$\lim_{\gamma}\left(\frac{c_{\gamma+e_j}}{c_{\gamma+2e_j}} - \frac{c_\gamma}{c_{\gamma+e_j}}\right) = 0 \tag{3.2}$$

for all $j = 1, \ldots, n$.

In particular, the only *if* implication of Theorem 3.5 implies several known noncompactness statements for S (see, e. g., [63, 88]), as well as the noncompactness of S on the polydisc; see Chapter 15. The main interest in these noncompactness statements is that if S fails to be compact, so does the $\bar\partial$-Neumann operator N; see Chapter 4.

The multimoments also lend themselves to characterizing the finer spectral property of being in the Schatten class $\mathbf{S}_p$. Let us recall that an operator $T : H_1 \to H_2$ belongs to the Schatten class $\mathbf{S}_p$ if the self-adjoint operator T^*T has a sequence of eigenvalues belonging to ℓ^p.

Theorem 3.6. *Let* $p > 0$. *Then* $S\colon A^2_{(0,1)}(d\mu) \to L^2(d\mu)$ *is in the Schatten-p-class* $\mathbf{S}_p$ *if and only if*

$$\sum_{\gamma\in\Gamma}\left(\sum_j\left(\frac{c_{\gamma+e_j}}{c_{\gamma+2e_j}} - \frac{c_\gamma}{c_{\gamma+e_j}}\right)\right)^{\frac{p}{2}} < \infty. \tag{3.3}$$

Condition (3.3) is substantially easier to check if $p = 2$ (we will show that then the sum is in fact a telescoping sum), i. e., for the case of the Hilbert–Schmidt class; we state this as a theorem.

Theorem 3.7. *The canonical solution operator* S *is in the Hilbert–Schmidt class if and only if*

$$\lim_{k\to\infty}\sum_{\substack{\gamma\in\mathbb{N}^n,|\gamma|=k\\1\le j\le n}}\frac{c_\gamma}{c_{\gamma+e_j}} < \infty. \tag{3.4}$$

Before we prove the theorems, we will derive some consequences.

Let us apply Theorem 3.4 to the case of decoupled weights or, more generally, of product measures $d\mu = d\mu_1 \times \cdots \times d\mu_n$, where each $d\mu_k$ is a (circle-invariant) measure on $\mathbb{C}$. Note that for such measures, there is *definitely* no compactness by Theorem 3.5. If we denote

$$c_j^k = \left(\int_{\mathbb{C}} |z|^{2k}\,d\mu_j\right)^{-1},$$

then we have that

$$c_{(\gamma_1,\dots,\gamma_n)} = \prod_{j=1}^{n} c_j^{\gamma_j}.$$

We thus obtain the following corollary.

Corollary 3.8. *For a product measure $d\mu = d\mu_1 \times \cdots \times d\mu_n$ as above, the canonical solution operator $S: A^2_{(0,1)}(d\mu) \to L^2(d\mu)$ is bounded if and only if there exists a constant C such that*

$$\frac{c_j^{k+1}}{c_j^{k+2}} - \frac{c_j^k}{c_j^{k+1}} < C$$

for all $k \in \mathbb{N}_0$ and all $j = 1,\dots,n$. Equivalently, S is bounded if and only if the canonical solution operator $S_j: A^2(d\mu_j) \to L^2(d\mu_j)$ is bounded for every $j = 1,\dots,n$.

To see that (3.2) is not satisfied for product measures, consider multi-indices $\gamma \in \Gamma$ such that $\gamma_j = -1$: then $c_\gamma = 0$ (by definition) and $c_{\gamma+e_j} \neq 0$, and therefore

$$\left(\frac{c_{\gamma+e_j}}{c_{\gamma+2e_j}} - \frac{c_\gamma}{c_{\gamma+e_j}}\right) = \frac{c_j^0}{c_j^1} - 0 > \delta > 0$$

for all multi-indices γ with $\gamma_j = -1$.

In the case of a rotation-invariant measure μ, we write

$$m_d = \int_{\mathbb{C}^n} |z|^{2d}\, d\mu$$

and claim that

$$c_\gamma = \frac{(n-1+|\gamma|)!}{(n-1)!\,\gamma!\,m_{|\gamma|}}, \tag{3.5}$$

where $|\gamma| = \gamma_1 + \cdots + \gamma_n$ and $\gamma! = \gamma_1!\dots\gamma_n!$.

To show this, let $\mathcal{U}$ be the unitary group consisting of all $n \times n$ unitary matrices, and let dU denote the Haar probability measure on $\mathcal{U}$. Let σ be the rotation-invariant probability measure on the unit sphere $\mathbb{S}$ in $\mathbb{C}^n$. For a multi-index $\alpha = (\alpha_1,\dots,\alpha_n)$, we define $z^\alpha = z_1^{\alpha_1}\dots z_n^{\alpha_n}$, $\alpha! = \alpha_1!\dots\alpha_n!$, and $|\alpha| = \alpha_1 + \cdots + \alpha_n$.

Due to the invariance of μ, it follows by Fubini's theorem that

$$\begin{aligned}\int_{\mathbb{C}^n} z^\alpha\, \bar{z}^\beta\, d\mu(z) &= \int_{\mathcal{U}}\int_{\mathbb{C}^n} (Uz)^\alpha\, (\overline{Uz})^\beta\, d\mu(z)\, dU \\ &= \int_{\mathbb{C}^n}\int_{\mathcal{U}} (Uz)^\alpha\, (\overline{Uz})^\beta\, dU\, d\mu(z)\end{aligned}$$

$$= \int_{\mathbb{C}^n} |z|^{|\alpha|+|\beta|} \int_{\mathbb{S}} \zeta^\alpha \overline{\zeta}^\beta \, d\sigma(\zeta)\, d\mu(z), \tag{3.6}$$

where we used the fact that for a continuous function $f \in \mathcal{C}(\mathbb{S})$, we have

$$\int_{\mathbb{S}} f(\zeta)\, d\sigma(\zeta) = \int_{\mathcal{U}} f(U\eta)\, dU$$

for all $\eta \in \mathbb{S}$ (see [84, Proposition 1.4.7]).

It is clear that for $\alpha \neq \beta$, we have

$$\int_{\mathbb{S}} \zeta^\alpha \overline{\zeta}^\beta \, d\sigma(\zeta) = 0.$$

Next, we claim that for any multi-index γ,

$$\int_{\mathbb{S}} |\zeta^\gamma|^2 \, d\sigma(\zeta) = \frac{(n-1)!\, \gamma!}{(n-1+|\gamma|)!}. \tag{3.7}$$

To prove (3.7), we use the integral

$$I = \int_{\mathbb{C}^n} |z^\gamma|^2 \exp(-|z|^2)\, d\lambda_{2n}(z) = \prod_{j=1}^n \int_{\mathbb{C}} |w|^{2\gamma_j} \exp(-|w|^2)\, d\lambda_2(w),$$

where λ_{2n} is the Lebesgue measure on $\mathbb{R}^{2n}$. It follows easily that $I = \pi^n \gamma!$. Now applying integration in polar coordinates to I, we get

$$\pi^n \gamma! = 2n\, c_n \int_0^\infty r^{2|\gamma|+2n-1} e^{-r^2}\, dr \int_{\mathbb{S}} |\zeta^\gamma|^2\, d\sigma(\zeta),$$

where c_n is the volume of the unit ball in $\mathbb{C}^n$.

Hence

$$\int_{\mathbb{S}} |\zeta^\gamma|^2\, d\sigma(\zeta) = \frac{\pi^n \gamma!}{(n-1+|\gamma|)!\, n c_n},$$

and taking $\gamma = 0$, we get $c_n = \pi^n/n!$, which proves (3.7).

For $d \in \mathbb{N}$, we set

$$m_d = \int_{\mathbb{C}^n} |z|^{2d}\, d\mu \quad \text{and} \quad c_\gamma^{-1} = \int_{\mathbb{C}^n} |z^\gamma|^2\, d\mu$$

and from (3.6) and (3.7) obtain

$$c_\gamma = \frac{(n-1+|\gamma|)!}{(n-1)!\,\gamma!\,m_{|\gamma|}}.$$

To express the conditions of our theorems, we compute (setting $d = |\gamma| + 1$)

$$\sum_j \left(\frac{c_{\gamma+e_j}}{c_{\gamma+2e_j}} - \frac{c_\gamma}{c_{\gamma+e_j}} \right) = \begin{cases} \frac{d+2n-1}{d+n}\frac{m_{d+1}}{m_d} - \frac{m_d}{m_{d-1}}, & \gamma_j \neq -1 \text{ for all } j, \\ \frac{1}{d+n}\frac{m_{d+1}}{m_d} & \text{else.} \end{cases} \tag{3.8}$$

Note that the Cauchy–Schwarz inequality implies that the first case in (3.8) always dominates the second case for $n \geq 2$; for $n = 1$, we observe that the second case in (3.8) reduces to $\frac{m_1}{m_0}$; compare with Proposition 2.15.

Using this observation and some trivial inequalities, we get the following corollaries.

Corollary 3.9. *Let μ be a rotation-invariant measure on $\mathbb{C}^n$. Then the canonical solution operator to $\bar\partial$ is bounded on $A^2_{(0,1)}(d\mu)$ if and only if*

$$\sup_{d\in\mathbb{N}} \left(\frac{(2n+d-1)m_{d+1}}{(n+d)m_d} - \frac{m_d}{m_{d-1}} \right) < \infty. \tag{3.9}$$

Corollary 3.10. *Let μ be a rotation-invariant measure on $\mathbb{C}^n$. Then the canonical solution operator to $\bar\partial$ is compact on $A^2_{(0,1)}(d\mu)$ if and only if*

$$\lim_{d\to\infty} \left(\frac{(2n+d-1)m_{d+1}}{(n+d)m_d} - \frac{m_d}{m_{d-1}} \right) = 0. \tag{3.10}$$

Corollary 3.11. *Let μ be a rotation-invariant measure on $\mathbb{C}^n$. Then the canonical solution operator to $\bar\partial$ is a Hilbert–Schmidt operator on $A^2_{(0,1)}(d\mu)$ if and only if*

$$\lim_{d\to\infty} \binom{n+d-2}{n-1} \frac{m_{d+1}}{m_d} < \infty. \tag{3.11}$$

Remark 3.12. It follows that the canonical solution operator to $\bar\partial$ is a Hilbert–Schmidt operator on $A^2(\mathbb{D})$ but fails to be Hilbert–Schmidt on $A^2(\mathbb{B}^n)$, where $\mathbb{B}^n$ is the unit ball in $\mathbb{C}^n$ for $n \geq 2$.

Corollary 3.13. *Let μ be a rotation-invariant measure on $\mathbb{C}^n$, and let $p > 0$. Then the canonical solution operator to $\bar\partial$ is in the Schatten class $\mathbf{S}_p$ as an operator from $A^2_{(0,1)}(d\mu)$ to $L^2(d\mu)$ if and only if*

$$\sum_{d=1}^{\infty} \binom{n+d-2}{n-1} \left(\frac{(2n+d-1)m_{d+1}}{(n+d)m_d} - \frac{m_d}{m_{d-1}} \right)^{\frac{p}{2}} < \infty. \tag{3.12}$$

Now we prove the theorems of this section. In what follows, we denote by

$$u_\alpha = \sqrt{c_\alpha} z^\alpha$$

the orthonormal basis of monomials for the space $A^2(d\mu)$ and by

$$U_{\alpha,j} = u_\alpha d\bar z_j \tag{3.13}$$

the corresponding basis of $A^2_{(0,1)}(d\mu)$. We first note that it is always possible to solve the $\bar\partial$-equation for the elements of this basis; indeed, $\bar\partial \bar z_j u_\alpha = U_{\alpha,j}$. The canonical solution operator is also easily determined for forms with monomial coefficients:

Lemma 3.14. *The canonical solution $Sz^\alpha d\bar z_j$ for monomial forms is given by*

$$Sz^\alpha d\bar z_j = \bar z_j z^\alpha - \frac{c_{\alpha-e_j}}{c_\alpha} z^{\alpha-e_j}, \quad \alpha \in \mathbb{N}_0^n. \tag{3.14}$$

Proof. We have $\langle \bar z_j z^\alpha, z^\beta\rangle = \langle z^\alpha, z^{\beta+e_j}\rangle$; so this expression is nonzero only if $\beta = \alpha - e_j$ (in particular, this implies (3.14) for multi-indices α with $\alpha_j = 0$; recall our convention that $c_\gamma = 0$ if one of the entries of γ is negative). Thus $Sz^\alpha d\bar z_j = \bar z_j z^\alpha + cz^{\alpha-e_j}$, and c is computed by

$$0 = \langle \bar z_j z^\alpha + cz^{\alpha-e_j}, z^{\alpha-e_j}\rangle = c_\alpha^{-1} + cc_{\alpha-e_j}^{-1},$$

which gives $c = -c_{\alpha-e_j}/c_\alpha$. □

We are going to introduce an orthogonal decomposition

$$A^2_{(0,1)}(d\mu) = \bigoplus_{\gamma\in\Gamma} E_\gamma$$

of $A^2_{(0,1)}(d\mu)$ into at most n-dimensional subspaces E_γ indexed by multi-indices $\gamma \in \Gamma$ (we will describe the index set below) and a corresponding sequence of mutually orthogonal finite-dimensional subspaces $F_\gamma \subset L^2(d\mu)$, which diagonalizes S (by this we mean that $SE_\gamma = F_\gamma$). To motivate the definition of E_γ, note that

$$\langle Sz^\alpha d\bar z_k, Sz^\beta d\bar z_\ell\rangle = \begin{cases} 0, & \beta \neq \alpha + e_\ell - e_k, \\ \frac{1}{c_\alpha}\left(\frac{c_\alpha}{c_{\alpha+e_\ell}} - \frac{c_{\alpha-e_k}}{c_{\alpha+e_\ell-e_k}}\right), & \beta = \alpha + e_\ell - e_k, \end{cases} \tag{3.15}$$

so that $\langle Sz^\alpha d\bar z_k, Sz^\beta d\bar z_\ell\rangle \neq 0$ if and only if there exists a multi-index γ such that $\alpha = \gamma + e_k$ and $\beta = \gamma + e_\ell$. Recall (3.13) and define

$$E_\gamma = \operatorname{span}\{U_{\gamma+e_j,j} : 1 \le j \le n\} = \operatorname{span}\{z^{\gamma+e_j} d\bar z_j : 1 \le j \le n\},$$

and likewise $F_\gamma = SE_\gamma$. Recall that Γ is defined to be the set of all multi-indices whose entries are greater than or equal to -1 and has at most one negative entry. Note that E_γ

is one-dimensional if exactly one entry in γ equals -1 and n-dimensional otherwise. We have already observed that F_γ are mutually orthogonal subspaces of $L^2(d\mu)$ (see (3.15)).

Whenever we use multi-indices γ and integers $p \in \{1, \ldots, n\}$ as indices, we use the convention that p runs over all p such that $\gamma + e_p \geq 0$; that is, for a fixed multi-index $\gamma \in \Gamma$, either the indices are all $p \in \{1, \ldots, n\}$, or there is exactly one p such that $\gamma_p = -1$, in which case the index is exactly this one p.

We next observe that we can find an orthonormal basis of E_γ and an orthonormal basis of F_γ such that $S_\gamma = S|_{E_\gamma} : E_\gamma \to F_\gamma$ acts diagonally in these bases. First, note that it suffices to do this if $\dim E_\gamma = n$ (since an operator between one-dimensional spaces is automatically diagonal). Fixing γ, the functions $U_j := U_{\gamma+e_j,j}$ are an orthonormal basis of E_γ. The operator S_γ is clearly nonsingular on this space, so the functions $SU_j = \Psi_j$ constitute a basis of F_γ. For a basis B of vectors $v^j = (v_1^j, \ldots, v_n^j)$, $j = 1, \ldots, n$, of $\mathbb{C}^n$, we consider the new basis

$$V_k = \sum_{j=1}^{n} v_k^j U_j.$$

Since the basis given by the U_j is orthonormal, the basis given by the V_k is also orthonormal, provided that the vectors $v_k = (v_k^1, \ldots, v_k^n)$ constitute an orthonormal basis for $\mathbb{C}^n$ with the standard Hermitian product. Let us write

$$\Phi_k = SV_k = \sum_j v_k^j SU_j.$$

The inner product $\langle \Phi_p, \Phi_q \rangle$ is then given by $\sum_{j,k} v_p^j \bar{v}_q^k \langle SU_j, SU_k \rangle$. We therefore have

$$\begin{aligned}
&\begin{pmatrix} \langle \Phi_1, \Phi_1 \rangle & \cdots & \langle \Phi_1, \Phi_n \rangle \\ \vdots & & \vdots \\ \langle \Phi_n, \Phi_1 \rangle & \cdots & \langle \Phi_n, \Phi_n \rangle \end{pmatrix} \\
&= \begin{pmatrix} v_1^1 & \cdots & v_1^n \\ \vdots & & \vdots \\ v_n^1 & \cdots & v_n^n \end{pmatrix}
\begin{pmatrix} \langle \Psi_1, \Psi_1 \rangle & \cdots & \langle \Psi_1, \Psi_n \rangle \\ \vdots & & \vdots \\ \langle \Psi_n, \Psi_1 \rangle & \cdots & \langle \Psi_n, \Psi_n \rangle \end{pmatrix}
\begin{pmatrix} \bar{v}_1^1 & \cdots & \bar{v}_n^1 \\ \vdots & & \vdots \\ \bar{v}_1^n & \cdots & \bar{v}_n^n \end{pmatrix}.
\end{aligned} \tag{3.16}$$

Since the matrix $(\langle \Psi_j, \Psi_k \rangle)_{j,k}$ is Hermitian, we can unitarily diagonalize it; that is, we can choose an orthonormal basis B of $\mathbb{C}^n$ such that with this choice of B, the vectors

$$\varphi_{\gamma,k} = V_k = \sum_j v_k^j U_{\gamma+e_j,j} \tag{3.17}$$

of E_γ are orthonormal, and their images $\Phi_k = SV_k$ are orthogonal in F_γ. Therefore $\Phi_k / \|\Phi_k\|$ is an orthonormal basis of F_γ such that $S_\gamma : E_\gamma \to F_\gamma$ is diagonal when expressed in terms of the bases $\{V_1, \ldots, V_n\} \subset E_\gamma$ and $\{\Phi_1, \ldots, \Phi_n\} \subset F_\gamma$ with entries $\|\Phi_k\|$.

Furthermore, the $\|\Phi_k\|$ are exactly the square roots of the eigenvalues of the matrix $(\langle\Psi_p,\Psi_q\rangle)_{p,q}$, which by (3.15) is given by

$$\begin{aligned}\langle\Psi_p,\Psi_q\rangle &= \langle SU_{\gamma+e_p,p}, SU_{\gamma+e_q,q}\rangle\\ &= \sqrt{c_{\gamma+e_p}}\sqrt{c_{\gamma+e_q}}\langle S\, z^{\gamma+e_p}\, d\bar z_p, S\, z^{\gamma+e_q}\, d\bar z_q\rangle\\ &= \sqrt{c_{\gamma+e_p}c_{\gamma+e_q}}\frac{1}{c_{\gamma+e_p}}\left(\frac{c_{\gamma+e_p}}{c_{\gamma+e_p+e_q}}-\frac{c_\gamma}{c_{\gamma+e_q}}\right)\\ &= \frac{c_{\gamma+e_p}c_{\gamma+e_q}-c_\gamma c_{\gamma+e_p+e_q}}{c_{\gamma+e_p+e_q}\sqrt{c_{\gamma+e_p}c_{\gamma+e_q}}}.\end{aligned}\tag{3.18}$$

Summarizing, we have the following proposition.

Proposition 3.15. *With μ as above, the canonical solution operator*

$$S\colon A^2_{(0,1)}(d\mu)\to L^2_{(0,1)}(d\mu)$$

admits a diagonalization by orthonormal bases. In fact, we have a decomposition $A^2_{(0,1)} = \bigoplus_\gamma E_\gamma$ into mutually orthogonal finite-dimensional subspaces E_γ indexed by the multi-indices γ with at most one negative entry (equal to -1), which are of dimension 1 or n, and orthonormal bases $\varphi_{\gamma,j}$ of E_γ such that $S\varphi_{\gamma,j}$ is a set of mutually orthogonal vectors in $L^2(d\mu)$. For fixed γ, the norms $\|S\varphi_{\gamma,j}\|$ are the square roots of the eigenvalues of the matrix $C_\gamma = (C_{\gamma,p,q})_{p,q}$ given by

$$C_{\gamma,p,q} = \frac{c_{\gamma+e_p}c_{\gamma+e_q}-c_\gamma c_{\gamma+e_p+e_q}}{c_{\gamma+e_p+e_q}\sqrt{c_{\gamma+e_p}c_{\gamma+e_q}}}.\tag{3.19}$$

In particular, we have that

$$\sum_{j=1}^n \|S\varphi_{\gamma,j}\|^2 = \operatorname{tr}(C_{\gamma,p,q})_{p,q} = \sum_{p=1}^n\left(\frac{c_{\gamma+e_p}}{c_{\gamma+2e_p}}-\frac{c_\gamma}{c_{\gamma+e_p}}\right).\tag{3.20}$$

To prove Theorem 3.4, we use Proposition 3.15. We have seen that we have an orthonormal basis $\varphi_{\gamma,j}$, $\gamma\in\Gamma$, $j\in\{1,\dots,n\}$, such that the images $S\varphi_{\gamma,j}$ are mutually orthogonal. Thus S is bounded if and only if there exists a constant C such that

$$\|S\varphi_{\gamma,j}\|^2 \le C$$

for all $\gamma\in\Gamma$ and $j\in\{1,\dots,\dim E_\gamma\}$. If $\dim E_\gamma = 1$, then γ has exactly one entry (say the jth one) equal to -1; in that case, let us write $\varphi_\gamma = U_{\gamma+e_j}d\bar z_j$. We have $S\varphi_\gamma = \sqrt{c_{\gamma+e_j}}\bar z_j z^{\gamma+e_j}$, and so

$$\|S\varphi_\gamma\|^2 = \frac{c_{\gamma+e_j}}{c_{\gamma+2e_j}}.$$

On the other hand, if $\dim E_\gamma = n$, then we argue as follows. Writing $\|S\varphi_{\gamma,j}\|^2 = \lambda_{\gamma,j}^2$ with $\lambda_{\gamma,j} > 0$, from (3.20) we find that

$$\sum_{j=1}^{n} \lambda_{\gamma,j}^2 = \sum_{j=1}^{n} \left(\frac{c_{\gamma+e_j}}{c_{\gamma+2e_j}} - \frac{c_\gamma}{c_{\gamma+e_j}} \right).$$

The last two equations complete the proof of Theorem 3.4.

To prove Theorem 3.5, we use Proposition 2.1.

Proof of Theorem 3.5. We first show that (3.2) implies compactness. We will use the notation already used in the proof of Theorem 3.4; that is, we write $\|S\varphi_{\gamma,j}\|^2 = \lambda_{\gamma,j}^2$. Let $\varepsilon > 0$. There exists a finite set A_ε of multi-indices $\gamma \in \Gamma$ such that for all $\gamma \notin A_\varepsilon$,

$$\sum_{j=1}^{n} \lambda_{\gamma,j}^2 = \sum_{j=1}^{n} \left(\frac{c_{\gamma+e_j}}{c_{\gamma+2e_j}} - \frac{c_\gamma}{c_{\gamma+e_j}} \right) < \varepsilon.$$

Hence, if we consider the finite-dimensional (and thus compact) operator T_ε defined by

$$T_\varepsilon\Big(\sum a_{\gamma,j}\varphi_{\gamma,j}\Big) = \sum_{\gamma \in A_\varepsilon} a_{\gamma,j} S\varphi_{\gamma,j}$$

for every $v = \sum a_{\gamma,j}\varphi_{\gamma,j} \in A^2_{(0,1)}(d\mu)$, then we obtain

$$\begin{aligned}
\|Sv\|^2 &= \|T_\varepsilon v\|^2 + \Big\| S \sum_{\gamma \notin A_\varepsilon} a_{\gamma,j}\varphi_{\gamma,j} \Big\|^2 \\
&= \|T_\varepsilon v\|^2 + \sum_{\gamma \notin A_\varepsilon} |a_{\gamma,j}|^2 \|S\varphi_{\gamma,j}\|^2 \\
&= \|T_\varepsilon v\|^2 + \sum_{\gamma \notin A_\varepsilon} |a_{\gamma,j}|^2 \lambda_{\gamma,j}^2 \\
&\le \|T_\varepsilon v\|^2 + \varepsilon \sum_{\gamma \notin A_\varepsilon} |a_{\gamma,j}|^2 \\
&\le \|T_\varepsilon v\|^2 + \varepsilon \|v\|^2.
\end{aligned}$$

Hence (2.2) holds, and we have proved the first implication in Theorem 3.5.

We now turn to the other direction. Assume that (3.2) is not satisfied. Then there exist $K > 0$ and an infinite family A of multi-indices γ such that for all $\gamma \in A$,

$$\sum_{j=1}^{n} \lambda_{\gamma,j}^2 = \sum_{j=1}^{n} \left(\frac{c_{\gamma+e_j}}{c_{\gamma+2e_j}} - \frac{c_\gamma}{c_{\gamma+e_j}} \right) > nK.$$

In particular, for each $\gamma \in A$, there exists j_γ such that $\lambda_{\gamma,j_\gamma}^2 > K$. Thus we have an infinite orthonormal family $\{\varphi_{\gamma,j_\gamma} : \gamma \in A\}$ of vectors such that their images $S\varphi_{\gamma,j_\gamma}$ are orthogonal and have the norms bounded from below by $\sqrt{K}$, which contradicts compactness. □

In the following, we will also need to introduce the usual grading on the index set Γ, that is, we write

$$\Gamma_k = \{\gamma \in \Gamma : |\gamma| = k\}, \quad k \geq -1. \tag{3.21}$$

To study the membership in the Schatten class, we need the following elementary lemma.

Lemma 3.16. *Assume that p and q are continuous real-valued functions on $\mathbb{R}^N$ that are homogeneous of degree 1 (i. e., $p(tx) = tp(x)$ and $q(tx) = tq(x)$ for $t \in \mathbb{R}$) and that $q(x) = 0$ and $p(x) = 0$ implies $x = 0$. Then there exists a constant C such that*

$$\frac{1}{C}|q(x)| \leq |p(x)| \leq C|q(x)|. \tag{3.22}$$

Proof. Note that the set $B_q = \{x : |q(x)| = 1\}$ is compact: it is closed since q is continuous, and since $|q|$ is bounded from below on the unit sphere of $\mathbb{R}^N$ by some $m > 0$, it is necessarily contained in the closed ball of radius $1/m$. Now the function $|p|$ is bounded on the compact set B_q, say, by $1/C$ from below and C from above. Thus for all $x \in \mathbb{R}^N$,

$$\frac{1}{C} \leq \left|p\left(\frac{x}{q(x)}\right)\right| \leq C,$$

which proves (3.22). □

Proof of Theorem 3.6. Note that S is in the Schatten class $\mathbf{S}_p$ if and only if

$$\sum_{\gamma \in \Gamma, j} \lambda_{\gamma,j}^p < \infty. \tag{3.23}$$

We rewrite this sum as

$$\sum_{\gamma \in \Gamma}\left(\sum_j \lambda_{\gamma,j}^p\right) =: M \in \mathbb{R} \cup \{\infty\}.$$

Lemma 3.16 implies that there exists a constant C such that for every $\gamma \in \Gamma$,

$$\frac{1}{C}\left(\sum_j \lambda_{\gamma,j}^2\right)^{p/2} \leq \sum_j \lambda_{\gamma,j}^p \leq C\left(\sum_j \lambda_{\gamma,j}^2\right)^{p/2}.$$

Hence $M < \infty$ if and only if

$$\sum_\gamma \left(\sum_j \lambda_{\gamma,j}^2\right)^{p/2} < \infty,$$

which after applying (3.20) becomes condition (3.6) claimed in Theorem 3.6. □

Proof of Theorem 3.7. S is in the Hilbert–Schmidt class if and only if

$$\sum_{\gamma\in\Gamma,j} \lambda_{\gamma,j}^2 < \infty. \tag{3.24}$$

We will prove that

$$\sum_{\ell=-1}^{k} \sum_{\gamma\in\Gamma_\ell,j} \lambda_{\gamma,j}^2 = \sum_{\substack{\alpha\in\mathbb{N}^n,|\alpha|=k+1\\ 1\le p\le n}} \frac{c_\alpha}{c_{\alpha+e_p}}, \tag{3.25}$$

which immediately implies Theorem 3.7. The proof is by induction over k. For $k = -1$, the left-hand side of (3.25) is

$$\sum_{j=1}^{n} \lambda_{-e_j,j}^2 = \sum_{j=1}^{n} \|z_j\|^2 c_0 = \sum_{j=1}^{n} \frac{c_0}{c_{e_p}},$$

which is equal to the right-hand side of (3.25). Now assume that (3.25) holds for $k = K-1$. We will show that this implies that it holds for $k = K$. We write

$$\begin{aligned}\sum_{\ell=-1}^{K} \sum_{\gamma\in\Gamma_\ell,j} \lambda_{\gamma,j}^2 &= \sum_{\substack{\alpha\in\mathbb{N}^n,|\alpha|=K-1\\ 1\le p\le n}} \frac{c_\alpha}{c_{\alpha+e_p}} + \sum_{\gamma\in\Gamma_K,j} \left(\frac{c_{\gamma+e_j}}{c_{\gamma+2e_j}} - \frac{c_\gamma}{c_{\gamma+e_j}}\right)\\ &= \sum_{\substack{\alpha\in\mathbb{N}^n,|\alpha|=K\\ 1\le p\le n}} \frac{c_\alpha}{c_{\alpha+e_p}}.\end{aligned}$$

This finishes the proof of Theorem 3.7. □

3.4 Notes

Most of the material in this chapter is taken from [47], which can be viewed as a generalization of results from [38, 39] and [71]. Corollary 3.13 improves Theorem C of [71] in the sense that it also covers the case $0 < p < 2$. We would like to note that our techniques can be adapted to the setting of [71] by considering the canonical solution operator on a Hilbert space $\mathcal{H}$ of holomorphic functions endowed with a norm comparable to the L^2-norm on each subspace generated by monomials of a fixed degree d if in addition to the requirements in [71], we also assume that the monomials belong to $\mathcal{H}$; this introduces the additional weights found by [71] in the formulas, as the reader can check. In our setting the formulas are somewhat “cleaner” by working with $A^2(d\mu)$ (in particular, Corollary 3.11 only holds in this setting).

The restriction of the canonical solution operator to $\bar\partial$ can be seen as a Hankel operator. Additional contributions were made by Knirsch and Schneider [63, 88]. Later, in

Chapter 12, we will investigate more general properties of this restriction, which are related to commutators between multiplication operators and the Bergman projection. In Chapter 13, we will study the canonical solution operator to $\overline{\partial}$ on L^2-spaces without restriction.

4 The $\overline{\partial}$-complex

Our main task will be to solve the inhomogeneous Cauchy–Riemann equation $\overline{\partial} u = f$, where the right-hand side f is given and satisfies the necessary condition $\overline{\partial} f = 0$. For $n > 1$, this is an overdetermined system of partial differential equations, which will be reduced to a system with equal numbers of unknowns and equations. We also need some background from the theory of unbounded operators on Hilbert spaces and the basics of the theory of distributions. Afterwards, we introduce the complex Laplacian (box operator). We show that under suitable assumptions this operator has a bounded inverse, the $\overline{\partial}$-Neumann operator, and we discuss important properties of the latter.

4.1 Unbounded operators on Hilbert spaces

In the sequel, we develop elements of unbounded self-adjoint operators, which are used for the $\overline{\partial}$-complex.

Definition 4.1. Let H_1, H_2 be Hilbert spaces, and let $T : \operatorname{dom}(T) \longrightarrow H_2$ be a densely defined linear operator, i. e., $\operatorname{dom}(T)$ is a dense linear subspace of H_1. Let $\operatorname{dom}(T^*)$ be the space of all $y \in H_2$ such that $x \mapsto (Tx, y)_2$ defines a continuous linear functional on $\operatorname{dom}(T)$. Since $\operatorname{dom}(T)$ is dense in H_1, there exists a uniquely determined element $T^*y \in H_1$ such that $(Tx, y)_2 = (x, T^*y)_1$ (Riesz representation Theorem 1.6). The map $y \mapsto T^*y$ is linear, and $T^* : \operatorname{dom}(T^*) \longrightarrow H_1$ is the adjoint operator to T.

An operator T is said to be closed if the graph

$$\mathcal{G}(T) = \{(f, Tf) \in H_1 \times H_2 : f \in \operatorname{dom}(T)\}$$

is a closed subspace of $H_1 \times H_2$.

The inner product in $H_1 \times H_2$ is

$$((x,y),(u,v)) = (x,u)_1 + (y,v)_2.$$

If $\tilde{V}$ is a linear subspace of H_1 containing $\operatorname{dom}(T)$ and an operator $\tilde{T}$ defined on $\tilde{V}$ is such that $\tilde{T}x = Tx$ for all $x \in \operatorname{dom}(T)$, then we say that $\tilde{T}$ is an extension of T.

An operator T with domain $\operatorname{dom}(T)$ is said to be closable if it has a closed extension $\tilde{T}$.

Lemma 4.2. *Let T be a densely defined closable operator. Then there is a closed extension $\overline{T}$, called its closure, whose domain is the smallest among all closed extensions.*

Proof. Let $\mathcal{V}$ be the set of $x \in H_1$ for which there exist $x_k \in \operatorname{dom}(T)$ and $y \in H_2$ such that $\lim_{k\to\infty} x_k = x$ and $\lim_{k\to\infty} Tx_k = y$. Since $\tilde{T}$ is a closed extension of T, it follows that $x \in \operatorname{dom}(\tilde{T})$ and $\tilde{T}x = y$. Therefore y is uniquely determined by x. We define $\overline{T}x = y$ with $\operatorname{dom}(\overline{T}) = \mathcal{V}$. Then $\overline{T}$ is an extension of T, and every closed extension of T is also

https://doi.org/10.1515/9783111182926-004

an extension of $\overline{T}$. The graph of $\overline{T}$ is the closure of the graph of T in $H_1 \times H_2$. Hence $\overline{T}$ is a closed operator. □

Lemma 4.3. *Let $T_1 : \operatorname{dom}(T_1) \longrightarrow H_2$ be a densely defined operator, and let $T_2 : H_2 \longrightarrow H_3$ be a bounded operator. Then $(T_2\, T_1)^* = T_1^*\, T_2^*$, which includes that $\operatorname{dom}((T_2\, T_1)^*) = \operatorname{dom}(T_1^*\, T_2^*)$.*

Proof. Note that

$$\operatorname{dom}(T_1^*\, T_2^*) = \{f \in \operatorname{dom}(T_2^*) : T_2^*(f) \in \operatorname{dom}(T_1^*)\}.$$

Let $f \in \operatorname{dom}(T_1^*\, T_2^*)$ and $g \in \operatorname{dom}(T_2\, T_1)$. Then

$$(T_1^*\, T_2^* f, g) = (T_2^* f, T_1 g) = (f, T_2\, T_1 g),$$

and hence $\operatorname{dom}(T_1^*\, T_2^*) \subseteq \operatorname{dom}((T_2\, T_1)^*)$.

Now let $f \in \operatorname{dom}((T_2\, T_1)^*)$. As T_2^* is bounded and everywhere defined on H_3, we have for all $g \in \operatorname{dom}(T_2\, T_1) = \operatorname{dom}(T_1)$ that

$$((T_2\, T_1)^* f, g) = (f, T_2\, T_1 g) = (T_2^* f, T_1 g).$$

Hence $T_2^* f \in \operatorname{dom}(T_1^*)$ and $f \in \operatorname{dom}(T_1^*\, T_2^*)$. □

Lemma 4.4. *Let T be a densely defined operator on H, and let S be a bounded operator on H. Then $(T + S)^* = T^* + S^*$.*

Proof. Let $f \in \operatorname{dom}(T^* + S^*) = \operatorname{dom}(T^*)$. Then for all $g \in \operatorname{dom}(T + S) = \operatorname{dom}(T)$, we have

$$((T^* + S^*)f, g) = (T^* f, g) + (S^* f, g) = (f, Tg) + (f, Sg) = (f, (T + S)g),$$

and hence $f \in \operatorname{dom}((T + S)^*)$ and $(T + S)^* f = T^* f + S^* f$.

If $f \in \operatorname{dom}((T + S)^*)$, then for all $g \in \operatorname{dom}(T + S) = \operatorname{dom}(T)$, we have

$$([(T + S)^* - S^*]f, g) = (f, (T + S)g) - (f, Sg) = (f, Tg),$$

and therefore $f \in \operatorname{dom}(T^*)$ and $\operatorname{dom}((T + S)^*) = \operatorname{dom}(T^* + S^*) = \operatorname{dom}(T^*)$. □

Lemma 4.5. *Let $T : \operatorname{dom}(T) \longrightarrow H_2$ be a densely defined linear operator and define $V : H_1 \times H_2 \longrightarrow H_2 \times H_1$ by $V((x, y)) = (y, -x)$. Then*

$$\mathcal{G}(T^*) = [V(\mathcal{G}(T))]^{\perp} = V(\mathcal{G}(T)^{\perp});$$

in particular, T^* is always closed.

Proof. $(y, z) \in \mathcal{G}(T^*) \Leftrightarrow (Tx, y)_2 = (x, z)_1$ for each $x \in \operatorname{dom}(T) \Leftrightarrow ((x, Tx), (-z, y)) = 0$ for each $x \in \operatorname{dom}(T) \Leftrightarrow V^{-1}((y, z)) = (-z, y) \in \mathcal{G}(T)^{\perp}$. Hence $\mathcal{G}(T^*) = V(\mathcal{G}(T)^{\perp})$, and since V is unitary, we have $V^* = V^{-1}$ and $[V(\mathcal{G}(T))]^{\perp} = V(\mathcal{G}(T)^{\perp})$. □

Lemma 4.6. *Let $T : \mathrm{dom}(T) \longrightarrow H_2$ be a densely defined closed linear operator. Then*

$$H_2 \times H_1 = V(\mathcal{G}(T)) \oplus \mathcal{G}(T^*).$$

Proof. Since $\mathcal{G}(T)$ is closed, $\mathcal{G}(T^*)^\perp = V(\mathcal{G}(T))$ by Lemma 4.5. □

Lemma 4.7. *Let $T : \mathrm{dom}(T) \longrightarrow H_2$ be a densely defined closed linear operator. Then $\mathrm{dom}(T^*)$ is dense in H_2 and $T^{**} = T$.*

Proof. Let $z \perp \mathrm{dom}(T^*)$. Hence $(z, y)_2 = 0$ for each $y \in \mathrm{dom}(T^*)$. We have

$$V^{-1} : H_2 \times H_1 \longrightarrow H_1 \times H_2,$$

where $V^{-1}((y, x)) = (-x, y)$, and $V^{-1}V = \mathrm{Id}$. Now by Lemma 4.6 we have

$$H_1 \times H_2 \cong V^{-1}(H_2 \times H_1) = V^{-1}(V(\mathcal{G}(T)) \oplus \mathcal{G}(T^*)) \cong \mathcal{G}(T) \oplus V^{-1}(\mathcal{G}(T^*)).$$

Hence $(z, y)_2 = 0 \Leftrightarrow ((0, z), (-T^*y, y)) = 0$ for each $y \in \mathrm{dom}(T^*)$. This implies $(0, z) \in \mathcal{G}(T)$, and therefore $z = T(0) = 0$, which means that $\mathrm{dom}(T^*)$ is dense in H_2.

Since T and T^* are densely defined and closed, by Lemma 4.5 we have

$$\mathcal{G}(T) = \mathcal{G}(T)^{\perp\perp} = [V^{-1}\mathcal{G}(T^*)]^\perp = \mathcal{G}(T^{**}),$$

where $-V^{-1}$ corresponds to V in considering operators from H_2 to H_1. □

Lemma 4.8. *Let $T : \mathrm{dom}(T) \longrightarrow H_2$ be a densely defined linear operator. Then* $\ker T^* = (\mathrm{im}\, T)^\perp$, *which means that* $\ker T^*$ *is closed.*

Proof. Let $v \in \ker T^*$ and $y \in \mathrm{im}\, T$, which means that there exists $u \in \mathrm{dom}(T)$ such that $Tu = y$. Hence

$$(v, y)_2 = (v, Tu)_2 = (T^*v, u)_1 = 0,$$

and $\ker T^* \subseteq (\mathrm{im}\, T)^\perp$.

If $y \in (\mathrm{im}\, T)^\perp$, then $(y, Tu)_2 = 0$ for each $u \in \mathrm{dom}(T)$, which implies that $y \in \mathrm{dom}(T^*)$ and $(y, Tu)_2 = (T^*y, u)_1$ for each $u \in \mathrm{dom}(T)$. Since $\mathrm{dom}(T)$ is dense in H_1, we obtain $T^*y = 0$ and $(\mathrm{im}\, T)^\perp \subseteq \ker T^*$. □

Lemma 4.9. *Let $T : \mathrm{dom}(T) \longrightarrow H_2$ be a densely defined closed linear operator. Then* $\ker T$ *is a closed linear subspace of H_1.*

Proof. We use Lemma 4.8 for T^* and get $\ker T^{**} = (\mathrm{im}\, T^*)^\perp$. Since, by Lemma 4.7, $T^{**} = T$, we obtain that $\ker T = (\mathrm{im}\, T^*)^\perp$ and $\ker T$ is a closed linear subspace of H_1. □

For our applications to the $\overline{\partial}$-equation, it will be important to know whether the differential operators involved have a closed range or are even surjective. In the following propositions, we will explain the functional analysis background and show how inequalities correspond to the closed range property.

First, we collect some corresponding results about bounded operators and begin with the Baire category theorem and the open-mapping theorem.

Theorem 4.10. *Let X be a complete metric space.*
(i) *If $\{U_n\}_n$ is a sequence of open dense subsets of X, then $\bigcap_{n=1}^{\infty} U_n$ is dense in X.*
(ii) *X is not a countable union of nowhere dense sets, i. e., of sets E_n such that $\overline{E}_n$ has an empty interior.*

Proof. (i) Let W be a nonempty open set in X. We have to show that

$$\left(\bigcap_{n=1}^{\infty} U_n\right) \cap W \neq \emptyset.$$

Since $U_1 \cap W$ is open and nonempty, it contains a ball $B(r_1, x_1)$, and we can assume that $0 < r_1 < 1$. We choose sequences $(x_n)_n$ in X and $(r_n)_n$ in $\mathbb{R}_+$ inductively as follows: having chosen x_j and r_j for $j < n$, we observe that $U_n \cap B(r_{n-1}, x_{n-1})$ is open and nonempty, so we can choose x_n and r_n such that $0 < r_n < 2^{-n}$ and $\overline{B(r_n, x_n)} \subset U_n \cap B(r_{n-1}, x_{n-1})$. Then, for $n, m \geq N$, we see that $x_n, x_m \in B(r_N, x_N)$, and since $r_n \to 0$, the sequence $(x_n)_n$ is a Cauchy sequence. As X is complete, the limit $x = \lim_{n\to\infty} x_n$ exists. Since $x_n \in B(r_N, x_N)$ for all $n \geq N$, we have

$$x \in \overline{B(r_N, x_N)} \subset U_N \cap B(r_1, x_1) \subset U_N \cap W$$

for all N, which proves (i).

(ii) If $\{E_n\}_n$ is a sequence of nowhere dense sets, then $\{(\overline{E}_n)^c\}_n$ is a sequence of open dense sets. Since, by (i), $\bigcap_{n=1}^{\infty}(\overline{E}_n)^c \neq \emptyset$, we have

$$\bigcup_{n=1}^{\infty} E_n \subset \bigcup_{n=1}^{\infty} \overline{E}_n \neq X. \qquad \square$$

A set $E \subset X$ is called meager (or of the first category) if it is a countable union of nowhere dense sets; otherwise, E is of the second category. Now we apply the Baire category theorem to linear maps between Banach spaces to prove the open-mapping the closed graph theorems.

Theorem 4.11. *Let X and Y be Banach spaces, and let $T : X \longrightarrow Y$ be a bounded linear operator that is surjective. Then T is open.*

Proof. Let B_r denote the open ball of radius r and center 0 in X. We have to show that $T(B_r)$ contains an open ball about 0 in Y. Since $X = \bigcup_{n=1}^{\infty} B_n$ and T is surjective, we have $Y = \bigcup_{n=1}^{\infty} T(B_n)$. As Y is complete and the map $y \mapsto ny$ is a homeomorphism of Y that maps $T(B_1)$ to $T(B_n)$, applying Theorem 4.10, we obtain that $T(B_1)$ cannot be nowhere dense in Y. Hence there exist $y_0 \in Y$ and $r > 0$ such that the ball $B(4r, y_0) = \{y \in Y : \|y - y_0\| < 4r\}$ is contained in $\overline{T(B_1)}$. Pick $y_1 = Tx_1 \in T(B_1)$ such that $\|y_1 - y_0\| < 2r$. Then $B(2r, y_1) \subset \overline{T(B_1)}$, so if $\|y\| < 2r$, then we have

$$y = Tx_1 + (y - y_1) \in \overline{T(x_1 + B_1)} \subset \overline{T(B_2)}.$$

Dividing both sides by 2, we obtain that there exists $r > 0$ such that if $\|y\| < r$, and then $y \in \overline{T(B_1)}$. If we could replace $\overline{T(B_1)}$ by $T(B_1)$, the proof would be complete.

Dilation invariance implies that if $\|y\| < r2^{-n}$, then $y \in \overline{T(B_{2^{-n}})}$. Suppose that $\|y\| < r/2$; we can find $x_1 \in B_{1/2}$ such that $\|y - Tx_1\| < r/4$, and proceeding inductively, we can find $x_n \in B_{2^{-n}}$ such that

$$\left\| y - \sum_{j=1}^{n} Tx_j \right\| < r2^{-n-1}. \tag{4.1}$$

We have $\sum_{n=1}^{\infty} \|x_n\| < \sum_{n=1}^{\infty} 2^{-n} = 1$, and since X is complete, the absolutely convergent series $\sum_{n=1}^{\infty} x_n$ is also convergent, say to $x = \sum_{n=1}^{\infty} x_n \in X$, and by (4.1) we have $y = Tx$. This means that $T(B_1)$ contains all y with $\|y\| < r/2$. □

Corollary 4.12. *Let $T : X \longrightarrow Y$ be a bijective bounded linear operator between Banach spaces. Then T^{-1} is also bounded.*

Proof. By Theorem 4.11, T is open, and hence T^{-1} is bounded. □

Now it is easy to prove the closed graph theorem.

Theorem 4.13. *Let X and Y be Banach spaces. Let $T : X \longrightarrow Y$ be a closed operator with* $\operatorname{dom}(T) = X$. *Then T is bounded.*

Proof. Let π_1 and π_2 be the projections of $\mathcal{G}(T)$ onto X and Y. They are both continuous. Since $X \times Y$ is also complete, $\mathcal{G}(T)$ is a closed subspace of $X \times Y$ and therefore also complete. Since π_1 is a bijection from $\mathcal{G}(T)$ to X, π_1^{-1} is bounded by Corollary 4.12. Therefore $T = \pi_2 \circ \pi_1^{-1}$ is bounded. □

Before we proceed in the theory for unbounded operators on Hilbert spaces, we still prove the uniform boundedness principle again as an application of the Baire category theorem.

Theorem 4.14. *Suppose X and Y are normed spaces and $\mathcal{A}$ is a subset of $\mathcal{L}(X, Y)$.*

(i) *If $\sup_{T \in \mathcal{A}} \|Tx\| < \infty$ for all x in a nonmeager subset of X, then*

$$\sup_{T \in \mathcal{A}} \|T\| < \infty.$$

(ii) *If X is a Banach space and $\sup_{T \in \mathcal{A}} \|Tx\| < \infty$ for all $x \in X$, then*

$$\sup_{T \in \mathcal{A}} \|T\| < \infty.$$

Proof. Let

$$E_n = \left\{x \in X : \sup_{T\in\mathcal{A}} \|Tx\| \le n\right\} = \bigcap_{T\in\mathcal{A}} \{x \in X : \|Tx\| \le n\}.$$

Then the sets E_n are closed, so under the hypothesis of (i), some E_n must contain a non-trivial closed ball $\overline{B(r, x_0)}$. Then $E_{2n} \supset \overline{B(r, 0)}$, for if $\|x\| \le r$, then $x - x_0 \in E_n$ and

$$\|Tx\| \le \|T(x - x_0)\| + \|Tx_0\| \le 2n.$$

In other words, $\|Tx\| \le 2n$ whenever $T \in \mathcal{A}$ and $\|x\| \le r$, so

$$\sup_{T\in\mathcal{A}} \|T\| \le 2n/r.$$

(ii) follows from (i) by Theorem 4.10. □

The following characterization of compactness uses the uniform boundedness principle and the notion of weak convergence.

Definition 4.15. A sequence $(x_k)_k$ in a Hilbert space H is a weak null-sequence if $(x_k, x) \to 0$ for each $x \in H$. A sequence $(x_k)_k$ converges weakly to x_0 if $(x_k - x_0)_k$ is a weak null sequence.

Remark 4.16. A weakly convergent sequence $(x_k)_k$ in a Hilbert space is always bounded: we have

$$\sup_k |(x_k - x_0, x)| < \infty$$

for all $x \in H$. Then, by Theorem 4.14,

$$\sup_k \|x_k - x_0\| = \sup_k \sup_{\|x\|\le 1} |(x_k - x_0, x)| < \infty,$$

and therefore $\|x_k\| \le \|x_k - x_0\| + \|x_0\| < \infty$ for all $k \in \mathbb{N}$.

In the same way, we can show that each weak Cauchy sequence is bounded.

If $A \in \mathcal{L}(H_1, H_2)$ and $(x_k)_k$ is a weakly convergent sequence in H_1, then $(Ax_k)_k$ converges weakly in H_2, which follows from

$$(Ax_k - Ax_0, y)_2 = (x_k - x_0, A^* y)_1,$$

where $y \in H_2$.

Proposition 4.17. *Let $A \in \mathcal{L}(H_1, H_2)$ be a bounded operator between Hilbert spaces. Then A is compact if and only if $(Ax_k)_k$ converges to 0 in H_2 for each weak null sequence $(x_k)_k$ in H_1.*

Proof. Let A be compact, and let $(x_k)_k$ be a weak null sequence in H_1. Then $(Ax_k)_k$ is a weak null sequence in H_2. Suppose that $(\|Ax_k\|)_k$ does not converge to 0. Then there exist $\epsilon > 0$ and a subsequence $(y_k)_k$ of $(x_k)_k$ such that $\|Ay_k\| \geq \epsilon$ for each $k \in \mathbb{N}$. By Remark 4.16 the sequence $(\|y_k\|)_k$ is bounded. As A is compact, there exists a subsequence $(z_k)_k$ of $(y_k)_k$ such that $(Az_k)_k$ is convergent in H_2. Since $(Az_k)_k$ converges weakly to 0, we would have

$$\|Az_k\| = \sup_{\|y\|\leq 1} |(Az_k, y)| \to 0,$$

contradicting $\|Ay_k\| \geq \epsilon$.

For the opposite direction, we need a number of prerequisites. First, we claim the following statement: if $(x_k)_k$ is a bounded sequence in a Hilbert space H and if $((x_k, y))_k$ is a Cauchy sequence in $\mathbb{C}$ for each y in a dense subset M of H, then $(x_k)_k$ is a weak Cauchy sequence in H. To show this, let $x \in H$ be an arbitrary element and choose $y \in M$ such that $\|x - y\| < \epsilon$. Then

$$|(x_k - x_m, x)| \leq |(x_k, x - y)| + |(x_k - x_m, y)| + |(x_m, y - x)|.$$

Since $(x_k)_k$ is a bounded sequence, we get that $((x_k, x))_k$ is a Cauchy sequence in $\mathbb{C}$ for each $x \in H$. Next, we claim that each weak Cauchy sequence in H is also weakly convergent: a weak Cauchy sequence is bounded, and therefore we have $|(x_k, x)| \leq C\|x\|$ for some constant $C > 0$ and for each $k \in \mathbb{N}$. Hence

$$F(x) := \lim_{k\to\infty} (x, x_k)$$

defines a continuous linear functional for which there exists $x_0 \in H$ such that

$$(x, x_0) = \lim_{k\to\infty} (x, x_k),$$

which means that $(x_k)_k$ converges weakly to x_0.

Next, we show that each bounded sequence $(x_k)_k$ contains a weakly convergent subsequence: the sequence $((x_k, x))_k$ is bounded in $\mathbb{C}$, and hence we can find subsequences $(x_k^{(m)})_k$ of $(x_k^{(m-1)})_k$ for $m \in \mathbb{N}$ with $(x_k)_k = (x_k^{(0)})_k$ such that $((x_k^{(m)}, x_m))_k$ converges. The diagonal sequence $(x_k^{(k)})_k$ has the property that the sequence $((x_k^{(k)}, x_m))_k$ is convergent for each $m \in \mathbb{N}$. This implies that $((x_k^{(k)}, x))_k$ converges for each x in the linear span L of the elements $\{x_\ell : \ell \in \mathbb{N}\}$. Since the sequence $(x_k^{(k)})_k$ is bounded, we have from above that $((x_k^{(k)}, x))_k$ is convergent for each $x \in \overline{L}$. It is clear that $((x_k^{(k)}, x)) = 0$ for each $x \in L^\perp$, and therefore we get that $(x_k^{(k)})_k$ is a weak Cauchy sequence, which converges weakly by our prerequisites.

Now we can finish the proof of the opposite direction: let $(x_k)_k$ be a bounded sequence in H_1. We know that it contains a weakly convergent subsequence $x_{k_\ell} \to x_0$, and hence the sequence $(x_{k_\ell} - x_0)_\ell$ is a weak null sequence. By assumption we obtain that $Ax_{k_\ell} \to Ax_0$ in H_2. So A is a compact operator. □

Now we continue to derive further important properties of unbounded operators on Hilbert spaces, which will later be used for the $\overline{\partial}$-operator.

Lemma 4.18. *Let $T : H_1 \longrightarrow H_2$ be a bounded linear operator. Then $T(H_1)$ is closed if and only if $T|_{(\ker T)^\perp}$ is bounded from below, i. e.,*

$$\|Tf\| \geq C\|f\| \quad \forall f \in (\ker T)^\perp.$$

Proof. If $T(H_1)$ is closed, then the mapping

$$T : (\ker T)^\perp \longrightarrow T(H_1)$$

is bijective and continuous and, by the open-mapping Theorem 4.11, also open. This implies the desired inequality.

To prove the other direction, let $(f_n)_n$ be a sequence in H_1 such that $Tf_n \to y$ in H_2. We have to show that there exists $h \in H_1$ such that $Th = y$. Decompose $f_n = g_n + h_n$, where $g_n \in \ker T$ and $h_n \in (\ker T)^\perp$. By assumption we have

$$\|h_n - h_m\| \leq C\|Th_n - Th_m\| = C\|Tf_n - Tf_m\| < \epsilon$$

for all sufficiently large n and m. Hence $(h_n)_n$ is a Cauchy sequence. Let $h = \lim_{n\to\infty} h_n$. Then we have

$$\|Tf_n - Th\| = \|Th_n - Th\| \leq \|T\| \, \|h_n - h\|,$$

and therefore

$$y = \lim_{n\to\infty} Tf_n = Th.$$ □

Lemma 4.19. *Let T be as before. Then $T(H_1)$ is closed if and only if $T^*(H_2)$ is closed.*

Proof. Since $T^{**} = T$, it suffices to show one direction. We will show that the closedness of $T(H_1)$ implies that $(\ker T)^\perp = \operatorname{im} T^*$; since $(\ker T)^\perp$ is closed, we will be finished.

Let $x \in \operatorname{im} T^*$. Then there exists $y \in H_2$ such that $x = T^*y$. Now we get for $x' \in \ker T$ that

$$(x, x') = (T^*y, x') = (y, Tx') = 0,$$

and hence $\operatorname{im} T^* \subseteq (\ker T)^\perp$.

For $x' \in (\ker T)^\perp$, we define the linear functional

$$\lambda(Tx) = (x, x')$$

on the closed subspace $T(H_1)$ of H_2. We remark that λ is well defined, since $Tx = T\tilde{x}$ implies that $x - \tilde{x} \in \ker T$, and hence $(x - \tilde{x}, x') = 0$ and $(x, x') = (\tilde{x}, x')$. The operator $T : H_1 \longrightarrow T(H_1)$ is continuous and surjective. Since $T(H_1)$ is closed, the open-mapping theorem implies $\|v\| \leq C\|Tv\|$ for all $v \in (\ker T)^\perp$, where $C > 0$ is a constant. Set $y = Tx$ and write $x = u + v$, where $u \in \ker T$ and $v \in (\ker T)^\perp$. Then we obtain

$$\begin{aligned} |\lambda(y)| = |(x, x')| &= |(v, x')| \\ &\leq \|v\|\|x'\| \\ &\leq C\|Tv\|\|x'\| \\ &= C\|Tx\|\|x'\| \\ &= C\|y\|\|x'\|. \end{aligned}$$

Hence λ is continuous on $\operatorname{im} T$. By Theorem 1.6 there exists a uniquely determined element $z \in \operatorname{im} T$ such that

$$\lambda(y) = (y, z)_2 = (x, x')_1.$$

This implies $(y, z)_2 = (Tx, z)_2 = (x, T^*z)_1 = (x, x')_1$ for all $x \in H_1$, and hence $x' = T^*z \in \operatorname{im} T^*$. ☐

Lemma 4.20. *Let $T : H_1 \longrightarrow H_2$ be a densely defined closed operator. Then $\operatorname{im} T$ is closed in H_2 if and only if $T|_{\operatorname{dom}(T)\cap(\ker T)^\perp}$ is bounded from below, i. e.,*

$$\|Tf\| \geq C\|f\| \quad \forall f \in \operatorname{dom}(T) \cap (\ker T)^\perp.$$

Proof. On the graph $\mathcal{G}(T)$, we define the operator $\tilde{T}((f, Tf)) = Tf$ and get a bounded linear operator

$$\tilde{T} : \mathcal{G}(T) \longrightarrow H_2,$$

since

$$\|\tilde{T}((f, Tf))\| = \|Tf\| \leq (\|f\|^2 + \|Tf\|^2)^{1/2} = \|(f, Tf)\|,$$

and $\operatorname{im} \tilde{T} = \operatorname{im} T$.

By Lemma 4.18, $\operatorname{im} T$ is closed if and only if $\tilde{T}|_{(\ker \tilde{T})^\perp}$ is bounded from below.

We have $\ker \tilde{T} = \ker T \oplus \{0\}$, and it remains to show that $\tilde{T}|_{(\ker \tilde{T})^\perp}$ is bounded from below if and only if $T|_{\operatorname{dom}(T)\cap(\ker T)^\perp}$ is bounded from below. This follows from

$$\|\tilde{T}((f, Tf))\| = \|Tf\| \geq C(\|f\|^2 + \|Tf\|^2)^{1/2},$$

and hence, for $0 < C < 1$,

$$\|Tf\|^2 \geq \frac{C^2}{1-C^2}\,\|f\|^2. \qquad \square$$

Lemma 4.21. *Let $P, Q : H \longrightarrow H$ be orthogonal projections on the Hilbert space H. Then the following statements are equivalent:*
(i) $\operatorname{im}(PQ)$ *is closed;*
(ii) $\operatorname{im}(QP)$ *is closed;*
(iii) $\operatorname{im}(I-P)(I-Q)$ *is closed;*
(iv) $P(H) + (I-Q)(H)$ *is closed.*

Proof. (i) and (ii) are equivalent by Lemma 4.19 and the fact that $QP = Q^*P^* = (PQ)^*$.

Suppose (ii) holds and let $(f_n)_n$ and $(g_n)_n$ be sequences in H such that $Pf_n + (I-Q)g_n \to h$. Then

$$Q(Pf_n + (I-Q)g_n) = QPf_n \to Qh.$$

By assumption, $\operatorname{im}(QP)$ is closed, and hence there exists $f \in H$ such that $QPf = Qh$; it follows that $Qh = Pf - (I-Q)(Pf)$ and

$$\begin{aligned} h = Qh + (I-Q)h &= Pf - (I-Q)(Pf) + (I-Q)h \\ &= Pf + (I-Q)(h-Pf) \in P(H) + (I-Q)(H), \end{aligned}$$

which yields (iv).

If (iv) holds and $(f_n)_n$ is a sequence in H such that $QPf_n \to h$, then we get

$$QPf_n = Pf_n - (I-Q)Pf_n \in P(H) + (I-Q)(H).$$

Hence there exist $f, g \in H$ such that $h = Pf + (I-Q)g$, and it follows that

$$Qh = Q\Big(\lim_{n\to\infty} QPf_n\Big) = \lim_{n\to\infty} Q^2Pf_n = h$$

and

$$h = Pf + (I-Q)g = Qh = QPf \in \operatorname{im}(QP),$$

and therefore (ii) holds.

Finally, replace P by $I-P$ and Q by $I-Q$. Then, using the statements proved so far, we obtain the equivalence

$$\operatorname{im}(I-P)(I-Q) \text{ closed} \Leftrightarrow (I-P)(H) + Q(H) \text{ closed},$$

which proves the remaining statements. $\square$

At this point, we are able to prove Lemma 4.19 for densely defined closed operators.

Proposition 4.22. *Let $T : H_1 \longrightarrow H_2$ be a densely defined closed operator. Then* $\operatorname{im} T$ *is closed if and only if* $\operatorname{im} T^*$ *is closed.*

Proof. Let $P : H_1 \times H_2 \longrightarrow \mathcal{G}(T)$ be the orthogonal projection of $H_1 \times H_2$ onto the closed subspace $\mathcal{G}(T)$ of $H_1 \times H_2$, and let $Q : H_1 \times H_2 \longrightarrow \{0\} \times H_2$ be the canonical orthogonal projection. Then $\operatorname{im} T \cong \operatorname{im} QP$, and since $I - Q : H_1 \times H_2 \longrightarrow H_1 \times \{0\}$ and

$$I - P : H_1 \times H_2 \longrightarrow \mathcal{G}(T)^\perp = V(\dot{\mathcal{G}}(T^*)) \cong \mathcal{G}(T^*),$$

we obtain the desired result from Lemma 4.21. □

Proposition 4.23. *Let $T : H_1 \longrightarrow H_2$ be a densely defined closed operator, and let G be a closed subspace of H_2 such that $G \supseteq \operatorname{im} T$. Suppose that $T^*|_{\operatorname{dom}(T^*)\cap G}$ is bounded from below, i. e., $\|f\| \leq C\|T^*f\|$ for all $f \in \operatorname{dom}(T^*) \cap G$, where $C > 0$ is a constant. Then $G = \operatorname{im} T$.*

Proof. We have $\ker T^* = (\operatorname{im} T)^\perp$. Since $\operatorname{im} T \subseteq G$, it follows that $\ker T^* \supseteq G^\perp$. If $G^\perp$ is a proper subspace of $\ker T^*$, then $G \cap \ker T^* \neq \{0\}$, which is a contradiction to the assumption that $T^*|_{\operatorname{dom}(T^*)\cap G}$ is bounded from below. Hence $\ker T^* = G^\perp$, and

$$G = G^{\perp\perp} = (\ker T^*)^\perp = \operatorname{im} T^{\perp\perp} = \overline{(\operatorname{im} T)}.$$

In addition, we have

$$T^*|_{\operatorname{dom}(T^*)\cap G} = T^*|_{\operatorname{dom}(T^*)\cap(\ker T^*)^\perp},$$

and by Lemma 4.20 we obtain that $\operatorname{im} T^*$ is closed. By Proposition 4.22, $\operatorname{im} T$ is also closed, and we get that $G = \operatorname{im} T$. □

Remark 4.24. The last proposition also holds in the other direction: if $T : H_1 \longrightarrow H_2$ is a densely defined closed operator and G is a closed subspace of H_2 with $G = \operatorname{im} T$, then $T^*|_{\operatorname{dom}(T^*)\cap G}$ is bounded from below. Since in this case $G = \operatorname{im} T$, we have that $\operatorname{im} T$ is closed, and hence, by Lemma 4.22, $\operatorname{im} T^*$ is also closed. Therefore Lemma 4.20 and the fact that $G = (\ker T^*)^\perp$ imply that $T^*|_{\operatorname{dom}(T^*)\cap G}$ is bounded from below.

Proposition 4.25. *Let $T : H_1 \longrightarrow H_2$ be a densely defined closed operator, and let G be a closed subspace of H_2 such that $G \supseteq \operatorname{im} T$. Suppose that $T^*|_{\operatorname{dom}(T^*)\cap G}$ is bounded from below. Then for each $v \in H_1$ such that $v \perp \ker T$, there exists $f \in \operatorname{dom}(T^*) \cap G$ such that $T^*f = v$ and $\|f\| \leq C\|v\|$.*

Proof. We have $\ker T = (\operatorname{im} T^*)^\perp$, and hence $v \in (\ker T)^\perp = \overline{\operatorname{im} T^*}$. In addition, $G^\perp \subseteq (\operatorname{im} T)^\perp = \ker T^*$, and therefore

$$\operatorname{im} T^*|_{\operatorname{dom}(T^*)\cap G} = \operatorname{im} T^*,$$

which means that $\operatorname{im} T^*$ is closed and that for $v \in (\ker T)^\perp = \operatorname{im} T^*$, there exists $f \in \operatorname{dom}(T^*) \cap G$ such that $T^*f = v$. The desired norm-inequality follows from the assumption that $T^*|_{\operatorname{dom}(T^*)\cap G}$ is bounded from below. □

In the following, we introduce the fundamental concept of an unbounded self-adjoint operator, which will be crucial for both spectral theory and its applications to complex analysis.

Definition 4.26. A densely defined linear operator $T : \operatorname{dom}(T) \longrightarrow H$ is symmetric if $(Tx, y) = (x, Ty)$ for all $x, y \in \operatorname{dom}(T)$. We say that T is self-adjoint if T is symmetric and $\operatorname{dom}(T) = \operatorname{dom}(T^*)$. This is equivalent to requiring that $T = T^*$ and implies that T is closed. We say that T is essentially self-adjoint if it is symmetric and its closure $\overline{T}$ is self-adjoint.

Remark 4.27.
(a) If T is a symmetric operator, then there are two natural closed extensions. We have $\operatorname{dom}(T) \subseteq \operatorname{dom}(T^*)$ and $T^* = T$ on $\operatorname{dom}(T)$. Since T^* is closed (Lemma 4.5), T^* is a closed extension of T, and it is the maximal closed extension. Since T is also closable by Lemma 4.2, $\overline{T}$ exists and is the minimal closed extension.
(b) If T is essentially self-adjoint, then its self-adjoint extension is unique. To prove this, let S be a self-adjoint extension of T. Then S is closed, and, being an extension of T, it is also an extension of its smallest extension $\overline{T}$. Hence

$$\overline{T} \subset S = S^* \subset (\overline{T})^* = \overline{T},$$

and $S = \overline{T}$.

Lemma 4.28. *Let T be a densely defined symmetric operator.*
(i) *If $\operatorname{dom}(T) = H$, then T is self-adjoint and bounded.*
(ii) *If T is self-adjoint and injective, then $\operatorname{im} T$ is dense in H, and T^{-1} is self-adjoint.*
(iii) *If $\operatorname{im} T$ is dense in H, then T is injective.*
(iv) *If $\operatorname{im} T = H$, then T is self-adjoint, and T^{-1} is bounded.*

Proof. (i) By assumption $\operatorname{dom}(T) \subseteq \operatorname{dom}(T^*)$. If $\operatorname{dom}(T) = H$, then it follows that T is self-adjoint and therefore also closed (Lemma 4.5) and continuous by the closed graph theorem.

(ii) Suppose $y \perp \operatorname{im} T$. Then $x \mapsto (Tx, y) = 0$ is continuous on $\operatorname{dom}(T)$, and hence $y \in \operatorname{dom}(T^*) = \operatorname{dom}(T)$, and $(x, Ty) = (Tx, y) = 0$ for all $x \in \operatorname{dom}(T)$. Thus $Ty = 0$, and since T is assumed to be injective, it follows that $y = 0$. This proves that $\operatorname{im} T$ is dense in H.

Therefore T^{-1} is densely defined with $\operatorname{dom}(T^{-1}) = \operatorname{im} T$, and $(T^{-1})^*$ exists. Now let $U : H \times H \longrightarrow H \times H$ be defined by $U((x, y)) = (-y, x)$. It easily follows that $U^2 = -I$ and $U^2(M) = M$ for any subspace M of $H \times H$, and we get $\mathcal{G}(T^{-1}) = U(\mathcal{G}(-T))$ and

$U(\mathcal{G}(T^{-1})) = \mathcal{G}(-T)$. Being self-adjoint, T is closed; hence $-T$ is closed, and T^{-1} is closed. By Lemma 4.6 applied to T^{-1} and to $-T$ we get the orthogonal decompositions

$$H \times H = U(\mathcal{G}(T^{-1})) \oplus \mathcal{G}((T^{-1})^*)$$

and

$$H \times H = U(\mathcal{G}(-T)) \oplus \mathcal{G}(-T) = \mathcal{G}(T^{-1}) \oplus U(\mathcal{G}(T^{-1})).$$

Consequently,

$$\mathcal{G}((T^{-1})^*) = [U(\mathcal{G}(T^{-1}))]^{\perp} = \mathcal{G}(T^{-1}),$$

which shows that $(T^{-1})^* = T^{-1}$.

(iii) Suppose $Tx = 0$. Then $(x, Ty) = (Tx, y) = 0$ for each $y \in \operatorname{dom}(T)$. Thus $x \perp \operatorname{im}(T)$, and therefore $x = 0$.

(iv) Since $\operatorname{im}(T) = H$, (iii) implies that T is injective and $\operatorname{dom}(T^{-1}) = H$. If $x, y \in H$, then $x = Tz$ and $y = Tw$ for some $z \in \operatorname{dom}(T)$ and $w \in \operatorname{dom}(T)$, so that

$$(T^{-1}x, y) = (z, Tw) = (Tz, w) = (x, T^{-1}y).$$

Hence T^{-1} is symmetric. (i) implies that T^{-1} is self-adjoint (and bounded), and now it follows from (ii) that $T = (T^{-1})^{-1}$ is also self-adjoint. □

Lemma 4.29. *Let T be a densely defined closed operator,* $\operatorname{dom}(T) \subseteq H_1$, *and* $T : \operatorname{dom}(T) \longrightarrow H_2$. *Then $B = (I + T^*T)^{-1}$ and $C = T(I + T^*T)^{-1}$ are everywhere defined and bounded with $\|B\| \le 1$ and $\|C\| \le 1$; in addition, B is self-adjoint and positive.*

Proof. Let $h \in H_1$ be an arbitrary element and consider $(h, 0) \in H_1 \times H_2$. From the proof of Lemma 4.7 we get

$$H_1 \times H_2 = \mathcal{G}(T) \oplus V^{-1}(\mathcal{G}(T^*)), \tag{4.2}$$

which implies that $(h, 0)$ can be written in a unique way as

$$(h, 0) = (f, Tf) + (-T^*(-g), -g)$$

for $f \in \operatorname{dom}(T)$ and $g \in \operatorname{dom}(T^*)$, which gives $h = f + T^*g$ and $0 = Tf - g$. We set $Bh := f$ and $Ch := g$. In this way, we get two linear operators B and C everywhere defined on H_1. The two equations from above can now be written as

$$I = B + T^*C, \quad 0 = TB - C,$$

which gives

$$C = TB \quad \text{and} \quad I = B + T^*TB = (I + T^*T)B. \tag{4.3}$$

Decomposition in (4.2) is orthogonal, and therefore we obtain

$$\|h\|^2 = \|(h,0)\|^2 = \|(f,Tf)\|^2 + \|(T^*g,-g)\|^2 = \|f\|^2 + \|Tf\|^2 + \|T^*g\|^2 + \|g\|^2,$$

and hence

$$\|Bh\|^2 + \|Ch\|^2 = \|f\|^2 + \|g\|^2 \leq \|h\|^2,$$

which implies $\|B\| \leq 1$ and $\|C\| \leq 1$.

For each $u \in \operatorname{dom}(T^*T)$, we get

$$((I + T^*T)u, u) = (u,u) + (Tu,Tu) \geq (u,u),$$

and hence if $(I + T^*T)u = 0$, then $u = 0$. Therefore $(I + T^*T)^{-1}$ exists, and (4.3) implies that $(I + T^*T)^{-1}$ is defined everywhere and $B = (I + T^*T)^{-1}$. Finally, let $u, v \in H_1$. Then

$$\begin{aligned}(Bu, v) &= (Bu, (I + T^*T)Bv) = (Bu, Bv) + (Bu, T^*TBv)\\ &= (Bu, Bv) + (T^*TBu, Bv) = ((I + T^*T)Bu, Bv) = (u, Bv),\end{aligned}$$

and

$$(Bu, u) = (Bu, (I + T^*T)Bu) = (Bu, Bu) + (TBu, TBu) \geq 0,$$

which proves the lemma. □

At this point, we can describe the notion of the core of an operator, which will be very useful later for spectral analysis.

Definition 4.30. Let T be a closable operator with domain $\operatorname{dom}(T)$. A subspace $D \subset \operatorname{dom}(T)$ is called a core of the operator T if the closure of the restriction $T|_D$ is an extension of T.

Remark 4.31. If T is a closed operator, then $\overline{T|_D} = T$.

Lemma 4.32. *Let T be a densely defined closed operator, $\operatorname{dom}(T) \subseteq H_1$, and $T : \operatorname{dom}(T) \longrightarrow H_2$. Then $\operatorname{dom}(T^*T)$ is a core of the operator T.*

Proof. We have to show that $\mathcal{G}(T) = \overline{\mathcal{G}(T|_{\operatorname{dom}(T^*T)})}$. For this purpose, we consider elements (x, Tx) in the graph of T. We suppose that $(x, Tx) \perp (y, Ty)$ for each $y \in \operatorname{dom}(T^*T)$. Then

$$(x, (I + T^*T)y) = (x,y) + (Tx,Ty) = ((x,Tx),(y,Ty)) = 0,$$

and as $\operatorname{im}(I + T^*T) = H_1$ (Lemma 4.29), we conclude that $x = 0$, which means that $\mathcal{G}(T|_{\operatorname{dom}(T^*T)})$ is dense in $\mathcal{G}(T)$. □

Finally, we describe a general method to construct self-adjoint operators associated with Hermitian sesquilinear forms. This leads to a self-adjoint extension of an unbounded operator, which is known as the Friedrichs extension.

Definition 4.33. Let $(\mathcal{V}, \|\cdot\|_{\mathcal{V}})$ and $(H, \|\cdot\|_H)$ be Hilbert spaces such that

$$\mathcal{V} \subset H, \tag{4.4}$$

and suppose that there exists a constant $C > 0$ such that

$$\|u\|_H \leq C\,\|u\|_{\mathcal{V}} \tag{4.5}$$

for all $u \in \mathcal{V}$. We also assume that $\mathcal{V}$ is dense in H.

In this situation the space H can be imbedded into the dual space $\mathcal{V}'$: for $h \in H$, the mapping

$$L(u) = (u, h)_H, \quad u \in \mathcal{V},$$

is continuous on $\mathcal{V}$, which follows from (4.5):

$$|L(u)| \leq \|u\|_H \|h\|_H \leq C\|h\|_H \|u\|_{\mathcal{V}}.$$

Hence there exists a uniquely determined $v_h \in \mathcal{V}'$ such that

$$v_h(u) = (u, h)_H, \quad u \in \mathcal{V},$$

and the mapping $h \mapsto v_h$ is injective, as $\mathcal{V}$ is dense in H.

Definition 4.34. A form $a : \mathcal{V} \times \mathcal{V} \longrightarrow \mathbb{C}$ is sesquilinear if it is linear in the first component and antilinear in the second component. The form a is continuous if there exists a constant $C > 0$ such that

$$|a(u, v)| \leq C\|u\|_{\mathcal{V}}\,\|v\|_{\mathcal{V}} \tag{4.6}$$

for all $u, v \in \mathcal{V}$, and it is Hermitian if

$$a(u, v) = \overline{a(v, u)}$$

for all $u, v \in \mathcal{V}$.

The form a is called $\mathcal{V}$-elliptic if there exists a constant $\alpha > 0$ such that

$$|a(u, u)| \geq \alpha\|u\|_{\mathcal{V}}^2 \tag{4.7}$$

for all $u \in \mathcal{V}$.

Proposition 4.35. *Let a be a continuous $\mathcal{V}$-elliptic form on $\mathcal{V}\times\mathcal{V}$. Using* (4.6) *and the Riesz representation Theorem* 1.6, *we can define a linear operator*

$$A : \mathcal{V} \longrightarrow \mathcal{V}$$

such that

$$a(u, v) = (Au, v)_{\mathcal{V}}. \tag{4.8}$$

This operator A is a topological isomorphism from $\mathcal{V}$ onto $\mathcal{V}$.

Proof. First, we show that A is injective: (4.8) and (4.7) imply that for $u \in \mathcal{V}$, we have

$$\|Au\|_{\mathcal{V}}\|u\|_{\mathcal{V}} \geq |(Au, u)_{\mathcal{V}}| \geq \alpha\|u\|_{\mathcal{V}}^2,$$

and hence

$$\|Au\|_{\mathcal{V}} \geq \alpha\|u\|_{\mathcal{V}}, \tag{4.9}$$

which implies that A is injective.

Now we claim that $A(\mathcal{V})$ is dense in $\mathcal{V}$. Let $u \in \mathcal{V}$ be such that $(Av, u)_{\mathcal{V}} = 0$ for each $v \in \mathcal{V}$. Taking $v = u$, we get $a(u, u) = 0$, and, by (4.7), $u = 0$, which proves the claim.

Next, we observe that (4.8) implies $a(u, Au) = \|Au\|_{\mathcal{V}}^2$. Therefore, using (4.6), we obtain $\|A(u)\|_{\mathcal{V}} \leq C\|u\|_{\mathcal{V}}$, and hence $A \in \mathcal{L}(\mathcal{V})$. If $(v_n)_n$ is a Cauchy sequence in $A(\mathcal{V})$ and $Au_n = v_n$, then we derive from (4.9) that $(u_n)_n$ is also a Cauchy sequence. Let $u = \lim_{n\to\infty} u_n$. We know already that A is continuous, and therefore $\lim_{n\to\infty} Au_n = Au$, which shows that $\lim_{n\to\infty} v_n = v = Au$ and $A(\mathcal{V})$ is closed. As we have already shown that $A(\mathcal{V})$ is dense in $\mathcal{V}$, we conclude that A is surjective.

Finally, (4.9) yields that A^{-1} is continuous. □

Proposition 4.36. *Let a be a Hermitian continuous $\mathcal{V}$-elliptic form on $\mathcal{V}\times\mathcal{V}$ and suppose that* (4.4) *and* (4.5) *hold. Let* dom(*S*) *be the set of all $u \in \mathcal{V}$ such that the mapping $v \mapsto a(u, v)$ is continuous on $\mathcal{V}$ for the topology induced by H. For each $u \in$* dom(*S*), *there exists a uniquely determined element $Su \in H$ such that*

$$a(u, v) = (Su, v)_H \tag{4.10}$$

for each $v \in \mathcal{V}$ (by the Riesz representation Theorem 1.6*).*

Then S : dom(*S*) *$\longrightarrow H$ is a bijective densely defined self-adjoint operator, and $S^{-1} \in \mathcal{L}(H)$. Moreover,* dom(*S*) *is also dense in $\mathcal{V}$.*

Proof. First, we show that S is injective. For each $u \in \operatorname{dom}(S)$, we get from (4.7) and (4.5) that

$$\begin{aligned}\alpha\|u\|_H^2 &\le C\alpha\|u\|_{\mathcal{V}}^2 \le C|a(u,u)| \\ &= C|(Su,u)_H| \le C\|Su\|_H\|u\|_H,\end{aligned}$$

which implies that

$$\alpha\|u\|_H \le C\|Su\|_H \tag{4.11}$$

for all $u \in \mathrm{dom}(S)$, and therefore S is injective.

Now let $h \in H$ and consider the mapping $v \mapsto (h,v)_H$ for $v \in \mathcal{V}$. Then by (4.5) we obtain

$$|(h,v)_H| \le \|h\|_H\|v\|_H \le C\|h\|_H\|v\|_{\mathcal{V}},$$

which implies that there exists a uniquely determined $w \in \mathcal{V}$ such that $(h,v)_H = (w,v)_{\mathcal{V}}$ for all $v \in \mathcal{V}$. Now we apply Proposition 4.35 and get from (4.8) that $a(u,v) = (w,v)_{\mathcal{V}}$, where $u = A^{-1}w$. Since $a(u,v) = (h,v)_H$ for each $v \in \mathcal{V}$, we conclude that $u \in \mathrm{dom}(S)$ and that $Su = h$, which shows that S is surjective.

Suppose that $(u,h)_H = 0$ for each $u \in \mathrm{dom}(S)$. As S is surjective, there is $v \in \mathrm{dom}(S)$ such that $Sv = h$, and we get that $(u,Sv)_H = 0$ for each $u \in \mathrm{dom}(S)$. Using the $\mathcal{V}$-ellipticity (4.7), we get for $u = v$ that

$$0 = (Sv,v)_H = a(v,v) \ge \alpha\|v\|_{\mathcal{V}}^2,$$

which implies that $v = 0$ and consequently $h = 0$. Therefore we have shown that $\mathrm{dom}(S)$ is dense in H.

As $a(u,v)$ is Hermitian, we get for $u,v \in \mathrm{dom}(S)$ that

$$(Su,v)_H = a(u,v) = \overline{a(v,u)} = \overline{(Sv,u)_H} = (u,Sv)_H.$$

Hence S is symmetric, and $\mathrm{dom}(S) \subset \mathrm{dom}(S^*)$. Let $v \in \mathrm{dom}(S^*)$. Since S is surjective, there exists $v_0 \in \mathrm{dom}(S)$ such that $Sv_0 = S^*v$. This implies

$$(Su,v_0)_H = (u,Sv_0)_H = (u,S^*v)_H = (Su,v)_H$$

for all $u \in \mathrm{dom}(S)$. Using again the surjectivity of S, we derive that $v = v_0 \in \mathrm{dom}(S)$. This implies that $\mathrm{dom}(S) = \mathrm{dom}(S^*)$ and that S is self-adjoint.

Finally, we show that $\mathrm{dom}(S)$ is dense in $\mathcal{V}$. Let $h \in \mathcal{V}$ be such that $(u,h)_{\mathcal{V}} = 0$ for all $u \in \mathrm{dom}(S)$. By Proposition 4.35 there exists $f \in \mathcal{V}$ such that $Af = h$. Then

$$\begin{aligned}0 = (u,h)_{\mathcal{V}} &= (u,Af)_{\mathcal{V}} = \overline{(Af,u)_{\mathcal{V}}} \\ &= \overline{a(f,u)} = a(u,f) = (Su,f)_H.\end{aligned}$$

Since S is surjective, we obtain $f = 0$ and $h = Af = 0$. □

4.2 Distributions

In our context the unbounded operator T will mainly be the $\bar{\partial}$-operator. To achieve an appropriate closed extension, we will have to consider the derivatives $\frac{\partial}{\partial \bar{z}_k}$ in the sense of distributions. Therefore we will now briefly summarize elementary definitions and results of distribution theory.

Definition 4.37. Let $\Omega \subseteq \mathbb{R}^n$ be an open subset, and let $\mathcal{D}(\Omega) = \mathcal{C}_0^\infty(\Omega)$ be the space of $\mathcal{C}^\infty$-functions with compact support (test functions).

A sequence $(\phi_j)_j$ tends to 0 in $\mathcal{D}(\Omega)$ if there exists a compact set $K \subset \Omega$ such that $\operatorname{supp}(\phi_j) \subset K$ for every j and

$$\frac{\partial^{|\alpha|}\phi_j}{\partial x_1^{\alpha_1} \dots \partial x_n^{\alpha_n}} \to 0$$

uniformly on K for each $\alpha = (\alpha_1, \dots, \alpha_n)$.

A distribution is a linear functional u on $\mathcal{D}(\Omega)$ such that for every compact subset $K \subset \Omega$, there exist $k \in \mathbb{N}_0 = \mathbb{N} \cup \{0\}$ and a constant $C > 0$ such that

$$|u(\phi)| \leq C \sum_{|\alpha| \leq k} \sup_{x \in K} \left| \frac{\partial^{|\alpha|}\phi(x)}{\partial x_1^{\alpha_1} \dots \partial x_n^{\alpha_n}} \right|$$

for each $\phi \in \mathcal{D}(\Omega)$ with support in K. We denote the space of distributions on Ω by $\mathcal{D}'(\Omega)$.

It is easily seen that $u \in \mathcal{D}'(\Omega)$ if and only if $u(\phi_j) \to 0$ for every sequence $(\phi_j)_j$ in $\mathcal{D}(\Omega)$ converging to 0 in $\mathcal{D}(\Omega)$.

Example 4.38.

(1) Let $f \in L^1_{loc}(\Omega)$, where

$$L^1_{loc}(\Omega) = \{f : \Omega \longrightarrow \mathbb{C} \text{ measurable} : f|_K \in L^1(K)\ \forall K \subset \Omega,\ K \text{ compact}\}.$$

The mapping $T_f(\phi) = \int_\Omega f(x)\phi(x)\, d\lambda(x)$, $\phi \in \mathcal{D}(\Omega)$, is a distribution.

(2) Let $a \in \Omega$ and $\delta_a(\phi) := \phi(a)$, which is the point evaluation in a. The distribution δ_a is called the Dirac delta distribution.

In the sequel, certain operations for ordinary functions, such as multiplication of functions and differentiation, are generalized to distributions.

Definition 4.39. Let $f \in \mathcal{C}^\infty(\Omega)$ and $u \in \mathcal{D}'(\Omega)$. The multiplication of u with f is defined by $(fu)(\phi) := u(f\phi)$ for $\phi \in \mathcal{D}(\Omega)$. Notice that $f\phi \in \mathcal{D}(\Omega)$.

For $u \in \mathcal{D}'(\mathbb{R}^n)$ and $f \in \mathcal{D}(\mathbb{R}^n)$, the convolution of u and f is defined by

$$(u * f)(x) := u(y \mapsto f(x - y)).$$

If $u = T_g$ for some locally integrable function g, then it is the usual convolution of functions

$$(T_g * f)(x) = \int_{\Omega} g(y) f(x-y)\, d\lambda(y) = (g * f)(x).$$

Let

$$D_k = \frac{\partial}{\partial x_k} \quad \text{and} \quad D^\alpha = \frac{\partial^{|\alpha|}}{\partial x_1^{\alpha_1} \dots \partial x_n^{\alpha_n}},$$

where $\alpha = (\alpha_1, \dots, \alpha_n)$ is a multi-index. The partial derivative of a distribution $u \in \mathcal{D}'(\Omega)$ is defined by

$$(D_k u)(\phi) := -u(D_k \phi), \quad \phi \in \mathcal{D}(\Omega);$$

higher-order mixed derivatives are defined as

$$(D^\alpha u)(\phi) := (-1)^{|\alpha|} u(D^\alpha \phi), \quad \phi \in \mathcal{D}(\Omega).$$

This definition stems from integrating by parts:

$$\int_{\Omega} (D_k f) \phi \, d\lambda = -\int_{\Omega} f (D_k \phi)\, d\lambda,$$

where $f \in \mathcal{C}^1(\Omega)$ and $\phi \in \mathcal{D}(\Omega)$.

Definition 4.40. Let $\Omega_1 \subseteq \Omega$ be an open subset, and let $u \in \mathcal{D}'(\Omega)$. The restriction of u to Ω_1 is the distribution $u|_{\Omega_1}$ defined by

$$u|_{\Omega_1}(\phi) = u(\phi) \quad \text{for } \phi \in \mathcal{D}(\Omega_1) \subseteq \mathcal{D}(\Omega).$$

The support $\operatorname{supp}(u)$ of $u \in \mathcal{D}'(\Omega)$ consists of $x \in \Omega$ such that there is no open neighborhood U of x with $u|_U = 0$. We denote the space of distributions with compact support by $\mathcal{E}'(\Omega)$.

A fundamental solution of a differential operator $P(D)$ is a distribution $E \in \mathcal{E}'(\mathbb{R}^n)$ with

$$P(D)E = \delta_0.$$

In what follows, we will find a fundamental solution for $\overline{\partial}$, and we will investigate regularity properties of the $\overline{\partial}$-operator. For this purpose, we recall Stokes' theorem in the following way: let $\Omega \subset \mathbb{C}$ be a bounded domain with piecewise $\mathcal{C}^1$-boundary γ (positively oriented), let $f \in \mathcal{C}^1(\overline{\Omega})$ and $\zeta \in \Omega$; for $0 < \epsilon < \operatorname{dist}(\zeta, \Omega^c)$, let $\Omega_\epsilon := \{z \in \Omega : |z - \zeta| > \epsilon\}$, where the boundary of Ω_ϵ consists of γ and the negatively oriented circle $\rho_\epsilon = \{z : |z-\zeta| = \epsilon\}$. Consider the differential $\omega = \frac{f(z)}{z-\zeta}\,dz$. Then by Stokes' theorem

$$\int_{\Omega_\epsilon} d\omega = \int_\gamma \omega + \int_{\rho_\epsilon} \omega = \int_\gamma \frac{f(z)}{z-\zeta}\,dz - \int_0^{2\pi} if(\zeta + \epsilon e^{i\theta})\,d\theta. \tag{4.12}$$

We have

$$d\omega = \frac{\partial}{\partial \bar{z}}\left(\frac{f(z)}{z-\zeta}\right) d\bar{z} \wedge dz = -\frac{\partial f}{\partial \bar{z}}\frac{1}{z-\zeta}\,dz \wedge d\bar{z},$$

and by continuity $\int_0^{2\pi} if(\zeta + \epsilon e^{i\theta})\,d\theta \to 2\pi i f(\zeta)$ as $\epsilon \to 0$. Hence we obtain

$$f(\zeta) = \frac{1}{2\pi i}\int_\gamma \frac{f(z)}{z-\zeta}\,dz + \frac{1}{2\pi i}\int_\Omega \frac{\partial f}{\partial \bar{z}}\frac{1}{z-\zeta}\,dz \wedge d\bar{z}. \tag{4.13}$$

In particular, if $\phi \in \mathcal{D}(\Omega)$, then

$$\phi(\zeta) = -\frac{1}{\pi}\int_\Omega \frac{\partial \phi}{\partial \bar{z}}\frac{1}{z-\zeta}\,d\lambda(z), \tag{4.14}$$

where we used the fact that $dz \wedge d\bar{z} = (-2i)\,dx \wedge dy = (-2i)\,d\lambda(z)$.

This can be interpreted in the sense of distributions as

$$\delta_\zeta(\phi) = \phi(\zeta) = -\frac{1}{\pi}\int_\Omega \frac{\partial \phi}{\partial \bar{z}}\frac{1}{z-\zeta}\,d\lambda(z) = \int_\Omega \frac{\partial}{\partial \bar{z}}\left(z \mapsto \frac{1}{\pi(z-\zeta)}\right)\phi(z)\,d\lambda(z), \tag{4.15}$$

which means that $z \mapsto \frac{1}{\pi z}$ is a fundamental solution to $\bar{\partial}$.

Proposition 4.41. *Let $\Omega \subseteq \mathbb{C}$ be a domain, and let $u \in \mathcal{D}'(\Omega)$ be a distribution satisfying $\bar{\partial} u = 0$. Then there exists a holomorphic function f on Ω such that $u = T_f$.*

Proof. Let E be the fundamental solution to $\bar{\partial}$ given in (4.15). We have $\bar{\partial} E = \delta_0$. Now let $\Omega' \subset\subset \Omega$ be a relatively compact subset and choose $g \in \mathcal{D}(\Omega)$ such that $g = 1$ in some neighborhood of $\overline{\Omega'}$. Let $S := \bar{\partial}(gu)$. Then $S \in \mathcal{E}'(\Omega)$ with $\operatorname{supp}(S) \subset \mathbb{C} \setminus \overline{\Omega'}$, and hence

$$gu = \delta_0 * (gu) = (\bar{\partial} E) * (gu) = E * \bar{\partial}(gu) = E * S. \tag{4.16}$$

Now pick $z_0 \in \Omega'$ and $\epsilon > 0$ such that $V_\epsilon = \{z : \operatorname{dist}(z, \mathbb{C} \setminus \Omega') > \epsilon\}$ is a neighborhood of z_0. Let $\psi_\epsilon \in \mathcal{D}(\mathbb{C})$ be such that $\psi_\epsilon(z) = 1$ for $|z| \le \epsilon/2$ and $\phi_\epsilon(z) = 0$ for $|z| \ge \epsilon$. We can write (4.16) in the form

$$gu = (\psi_\epsilon E) * S + ((1-\psi_\epsilon)E) * S. \tag{4.17}$$

Since $\operatorname{supp}((\psi_\epsilon E) * S) \subseteq \operatorname{supp}(\psi_\epsilon E) + \operatorname{supp}(S) \subseteq \{z : \operatorname{dist}(z, \operatorname{supp}(S)) \leq \epsilon\}$, we have that $(\psi_\epsilon E) * S$ has to vanish in V_ϵ. This implies

$$u|_{V_\epsilon} = (gu)|_{V_\epsilon} = (((1 - \psi_\epsilon)E) * S)|_{V_\epsilon}. \tag{4.18}$$

The function $(1 - \psi_\epsilon)E$ is in $C^\infty(\mathbb{C})$, and $\operatorname{supp}(S)$ is compact; therefore $((1 - \psi_\epsilon)E) * S$ is smooth. Hence, using (4.18), we obtain that u is smooth in a neighborhood of z_0, and z_0 being arbitrary, u is smooth in Ω' and therefore also in Ω. This implies that u is holomorphic, since it also solves the Cauchy–Riemann equation. □

For an appropriate description of the appearing phenomena, we will further use Hilbert spaces of differentiable functions, the Sobolev spaces.

Definition 4.42. If Ω is a bounded open set in $\mathbb{R}^n$, and k is a nonnegative integer, then we define the Sobolev space

$$W^k(\Omega) = \{f \in L^2(\Omega) : \partial^\alpha f \in L^2(\Omega),\ |\alpha| \leq k\},$$

where the derivatives are taken in the sense of distributions, and endow the space with the norm

$$\|f\|_{k,\Omega} = \left[\sum_{|\alpha|\leq k} \int_\Omega |\partial^\alpha f|^2\, d\lambda\right]^{1/2}, \tag{4.19}$$

where $\alpha = (\alpha_1, \ldots, \alpha_n)$ is a multi-index, $|\alpha| = \sum_{j=1}^n \alpha_j$, and

$$\partial^\alpha f = \frac{\partial^{|\alpha|} f}{\partial x_1^{\alpha_1} \ldots \partial x_n^{\alpha_n}}.$$

By $W_0^k(\Omega)$ we denote the completion of $C_0^\infty(\Omega)$ under $W^k(\Omega)$-norm. Since $C_0^\infty(\Omega)$ is dense in $L^2(\Omega)$, it follows that $W_0^0(\Omega) = W^0(\Omega) = L^2(\Omega)$. Using the Fourier transform, it is also possible to introduce Sobolev spaces of noninteger exponent. (See [1, 31].)

In general, a function can belong to a Sobolev space and yet be discontinuous and unbounded.

Example 4.43. Take $\Omega = \mathbb{B}$ the open unit ball in $\mathbb{R}^n$ and

$$u(x) = |x|^{-\alpha}, \quad x \in \mathbb{B}, x \neq 0.$$

We claim that $u \in W^1(\mathbb{B})$ if and only if $\alpha < \frac{n-2}{2}$.

First note that u is smooth away from 0 and that

$$u_{x_j}(x) = \frac{-\alpha x_j}{|x|^{\alpha+2}}, \quad x \neq 0.$$

Hence

$$|\nabla u(x)| = \frac{|\alpha|}{|x|^{\alpha+1}}, \quad x \neq 0.$$

Now recall the Gauß–Green theorem: for a smoothly bounded $\omega \subseteq \mathbb{R}^n$, we have

$$\int_{\omega} \nabla . F(x)\, d\lambda(x) = \int_{b\omega} (F(x), \nu(x))\, d\sigma(x),$$

where $\nu(x) = \nabla r(x)$ is the normal to $b\omega$ at x, F is a C^1 vector field on $\overline{\omega}$, and

$$\nabla . F(x) = \sum_{j=1}^{n} \frac{\partial F_j}{\partial x_j}$$

(see [29]).

Let $\phi \in C_0^\infty(\mathbb{B})$, and let $\mathbb{B}_\epsilon$ be the open ball around 0 with radius $\epsilon > 0$. Take $\omega = \mathbb{B} \setminus \mathbb{B}_\epsilon$ and

$$F(x) = (0, \dots, 0, u\phi, 0, \dots, 0),$$

where $u\phi$ appears at the jth component. Then

$$\int_{\mathbb{B}\setminus\mathbb{B}_\epsilon} u(x)\, \phi_{x_j}(x)\, d\lambda(x) = -\int_{\mathbb{B}\setminus\mathbb{B}_\epsilon} u_{x_j}(x)\, \phi(x)\, d\lambda(x) + \int_{b\mathbb{B}_\epsilon} u(x)\phi(x)\nu_j(x)\, d\sigma(x),$$

where $\nu(x) = (\nu_1(x), \dots, \nu_n(x))$ denotes the inward pointing normal on $b\mathbb{B}_\epsilon$. If $\alpha < n-1$, then $|\nabla u(x)| \in L^1(\mathbb{B})$, and we obtain

$$\begin{aligned}\left|\int_{b\mathbb{B}_\epsilon} u(x)\phi(x)\nu_j(x)\, d\sigma(x)\right| &\le \|\phi\|_\infty \int_{b\mathbb{B}_\epsilon} \epsilon^{-\alpha}\, d\sigma(x) \\ &\le C\, \epsilon^{n-1-\alpha} \to 0\end{aligned}$$

as $\epsilon \to 0$. Thus

$$\int_{\mathbb{B}} u(x)\, \phi_{x_j}(x)\, d\lambda(x) = -\int_{\mathbb{B}} u_{x_j}(x)\, \phi(x)\, d\lambda(x)$$

for all $\phi \in C_0^\infty(\mathbb{B})$. As

$$|\nabla u(x)| = \frac{\alpha}{|x|^{\alpha+1}} \in L^2(\mathbb{B})$$

if and only if $2(\alpha + 1) < n$, we get that $u \in W^1(\mathbb{B})$ if and only if $\alpha < \frac{n-2}{2}$.

Before we proceed, we verify properties of weak (distributional) derivatives, which are obviously true for smooth functions. As functions in Sobolev spaces are not necessarily smooth, we must always rely upon the definition of weak derivatives.

Proposition 4.44. *Let $u \in W^k(\Omega)$ and $|\alpha| \le k$. Then*
(i) *$D^\alpha u \in W^{k-|\alpha|}(\Omega)$, and for multi-indices α, β with $|\alpha| + |\beta| \le k$, we have*

$$D^\beta(D^\alpha u) = D^\alpha(D^\beta u) = D^{\alpha+\beta} u.$$

(ii) *If ω is an open subset of Ω, then $u \in W^k(\omega)$.*
(iii) *If $\phi \in C_0^\infty(\Omega)$, then $\phi\, u \in W^k(\Omega)$, and*

$$D^\alpha(\phi\, u) = \sum_{\beta \le \alpha} \binom{\alpha}{\beta} D^\beta \phi\, D^{\alpha-\beta} u,$$

where $\binom{\alpha}{\beta} = \frac{\alpha!}{\beta!(\alpha-\beta)!}$.

Proof. To prove (i), fix $\phi \in C_0^\infty(\Omega)$. Then $D^\beta \phi \in C_0^\infty(\Omega)$, and

$$\begin{aligned}\int_\Omega D^\alpha u(x)\, D^\beta \phi(x)\, d\lambda(x) &= (-1)^{|\alpha|} \int_\Omega u(x)\, D^{\alpha+\beta}\phi(x)\, d\lambda(x) \\ &= (-1)^{|\alpha|}(-1)^{|\alpha+\beta|} \int_\Omega D^{\alpha+\beta} u(x)\, \phi(x)\, d\lambda(x) \\ &= (-1)^{|\beta|} \int_\Omega D^{\alpha+\beta} u(x)\, \phi(x)\, d\lambda(x).\end{aligned}$$

Hence $D^\beta(D^\alpha u) = D^{\alpha+\beta} u$ in the weak sense.

We omit the easy proof of (ii).

For (iii), we use induction on $|\alpha|$. Suppose first that $|\alpha| = 1$. Take any $\psi \in C_0^\infty(\Omega)$. Then

$$\begin{aligned}\int_\Omega \phi(x) u(x) D^\alpha \psi(x)\, d\lambda(x) &= \int_\Omega \big(u(x) D^\alpha(\phi(x)\psi(x)) - u(x)(D^\alpha \phi(x))\psi(x)\big)\, d\lambda(x) \\ &= -\int_\Omega \big(\phi(x) D^\alpha u(x) + u(x) D^\alpha \phi(x)\big)\psi(x)\, d\lambda(x).\end{aligned}$$

Therefore $D^\alpha(\phi u) = \phi D^\alpha u + u D^\alpha \phi$, as required. The induction step is carried out in a similar way. □

Proposition 4.45. *Let $k \in \mathbb{N}$. Then $W^k(\Omega)$ is a Hilbert space.*

Proof. It is clear that the norm of $W^k(\Omega)$ stems from an inner product. To prove the completeness, let $(u_m)_m$ be a Cauchy sequence in $W^k(\Omega)$. Then for each multi-index α with

$|\alpha| \leq k$, the sequence $(D^\alpha u_m)_m$ is a Cauchy sequence in $L^2(\Omega)$. Since $L^2(\Omega)$ is complete, there exist functions $u_\alpha \in L^2(\Omega)$ such that

$$D^\alpha u_m \to u_\alpha \quad \text{in } L^2(\Omega).$$

In particular, $u_m \to u_{(0,\dots,0)} := u$ in $L^2(\Omega)$.

Now we claim that $u \in W^k(\Omega)$ and $D^\alpha u = u_\alpha$ for $|\alpha| \leq k$. Fix $\phi \in C_0^\infty(\Omega)$. Then, by the Cauchy–Schwarz inequality,

$$\left| \int_\Omega (u(x) - u_m(x)) D^\alpha \phi(x)\, d\lambda(x) \right| \leq \|u - u_m\|_2 \, \|D^\alpha \phi\|_2,$$

where $\|\cdot\|_2$ denotes the norm in $L^2(\Omega)$. Hence

$$\begin{aligned} \int_\Omega u(x) D^\alpha \phi(x)\, d\lambda(x) &= \lim_{m\to\infty} \int_\Omega u_m(x) D^\alpha \phi(x)\, d\lambda(x) \\ &= \lim_{m\to\infty} (-1)^{|\alpha|} \int_\Omega D^\alpha u_m(x) \phi(x)\, d\lambda(x) \\ &= (-1)^{|\alpha|} \int_\Omega u_\alpha(x)\, \phi(x)\, d\lambda(x), \end{aligned}$$

which proves the claim. Since $D^\alpha u_m \to D^\alpha u$ in $L^2(\Omega)$ for all $|\alpha| \leq k$, we see that $u_m \to u$ in $W^k(\Omega)$. □

4.3 A finite-dimensional analog

We demonstrate the method for the $\bar{\partial}$-Neumann problem first in its finite-dimensional analog: let E, F, G be finite-dimensional vector spaces over $\mathbb{C}$ with inner product. We consider an exact sequence of linear maps

$$E \xrightarrow{S} F \xrightarrow{T} G,$$

which means that $\operatorname{im} S = \ker T$, and hence $TS = 0$.

Given $f \in \operatorname{im} S = \ker T$, we want to solve $Su = f$ with $u \perp \ker S$; then u will be called the canonical solution.

For this purpose, we investigate

$$E \underset{S^*}{\overset{S}{\rightleftarrows}} F \underset{T^*}{\overset{T}{\rightleftarrows}} G$$

and observe that $\ker T = (\operatorname{im} T^*)^\perp$ and $\ker T^* = (\operatorname{im} T)^\perp$. We claim that the operator

$$SS^* + T^*T : F \longrightarrow F$$

is bijective. Let $(SS^* + T^*T)g = 0$. Then $SS^*g = -T^*Tg$, which implies

$$SS^*g \in \operatorname{im} T^* \cap \operatorname{im} S = \operatorname{im} T^* \cap \ker T = \operatorname{im} T^* \cap (\operatorname{im} T^*)^\perp = \{0\},$$

and hence $SS^*g = T^*Tg = 0$, but this gives $S^*g \in \ker S \cap \operatorname{im} S^* = \ker S \cap (\ker S)^\perp = \{0\}$ and $g \in \ker S^* = (\operatorname{im} S)^\perp$; from $T^*Tg = 0$ we get $Tg \in \ker T^* \cap \operatorname{im} T = (\operatorname{im} T)^\perp \cap \operatorname{im} T = \{0\}$ and $g \in \ker T = \operatorname{im} S$, and therefore we obtain $g \in \operatorname{im} S \cap (\operatorname{im} S)^\perp = \{0\}$. So $SS^* + T^*T$ is injective, and as F is finite-dimensional, $SS^* + T^*T$ is bijective.

Let $N = (SS^* + T^*T)^{-1}$. We claim that

$$u = S^*Nf$$

is the canonical solution to $Su = f$. So we have to show that $SS^*Nf = f$ and $S^*Nf \perp \ker S$. The latter easily follows from the fact that $S^*Nf \in \operatorname{im} S^* = (\ker S)^\perp$.

We have

$$f = SS^*Nf + T^*TNf,$$

and therefore the assumption $Tf = 0$ implies

$$0 = Tf = TSS^*Nf + TT^*TNf = TT^*TNf,$$

since $TS = 0$. From here we obtain

$$0 = (TT^*TNf, TNf) = (T^*TNf, T^*TNf)$$

and $T^*TNf = 0$, hence $SS^*Nf = f$, and we are done.

In the following, we will use this method for the $\overline{\partial}$-operator.

4.4 The $\overline{\partial}$-Neumann operator

Let $\Omega = \{z \in \mathbb{C}^n : r(z) < 0\}$, where r is a real-valued C^1-function with

$$\nabla_z r := \left(\frac{\partial r}{\partial z_1}, \dots, \frac{\partial r}{\partial z_n}\right) \neq 0$$

on $b\Omega = \{z : r(z) = 0\}$. Then r is called a defining function for Ω. Without loss of generality, we can suppose that $|\nabla_z r| = |\nabla r| = 1$ on $b\Omega$. For $u, v \in C^\infty(\overline{\Omega})$ and

$$(u, v) = \int_\Omega u(z)\overline{v(z)}\, d\lambda(z),$$

we have

$$(u_{x_k}, v) = -(u, v_{x_k}) + \int_{b\Omega} u(z)\overline{v(z)}\, r_{x_k}(z)\, d\sigma(z),$$

where $d\sigma$ is the surface measure on $b\Omega$.

This follows from the Gauß–Green theorem: for $\omega \subseteq \mathbb{R}^n$, we have

$$\int_{\omega} \nabla . F(x)\, d\lambda(x) = \int_{b\omega} (F(x), \nu(x))\, d\sigma(x),$$

where $\nu(x) = \nabla r(x)$ is the normal to $b\omega$ at x, F is a C^1 vector field on $\overline{\omega}$, and

$$\nabla . F(x) = \sum_{j=1}^{n} \frac{\partial F_j}{\partial x_j}.$$

For $k = 1$ and $F = (u\bar{v}, 0, \dots, 0)$, we get

$$(u_{x_1}, v) = -(u, v_{x_1}) + \int_{b\Omega} u(z)\overline{v(z)}\, r_{x_1}(z)\, d\sigma(z).$$

Similarly, we obtain

$$\left(\frac{\partial u}{\partial z_k}, v\right) = -\left(u, \frac{\partial v}{\partial \overline{z}_k}\right) + \int_{b\Omega} u(z)\,\overline{v(z)}\,\frac{\partial r}{\partial z_k}(z)\, d\sigma(z). \tag{4.20}$$

We clarify the relationship between two defining functions, which will be important for the definition of the following geometric concepts.

Lemma 4.46. *If r and u are two local defining functions of $b\Omega$ of class C^2 in a neighborhood U of $b\Omega$, then there exists a positive C^1-function h such that $u = rh$ on U and $dr(z) = h(z)du(z)$ for $z \in U \cap d\Omega$.*

Proof. We can assume that $0 \in b\Omega$ and that $U \cap d\Omega = \{z \in U : \Im z_n = 0\}$. Then the defining function r can be written as $r(z) = \Im z_n$, and setting $z' = (z_1, \dots, z_{n-1})$, we get

$$\begin{aligned} u(z', \Re z_n, \Im z_n) &= u(z', \Re z_n, \Im z_n) - u(z', \Re z_n, 0) \\ &= \Im z_n \int_0^1 \frac{\partial u}{\partial \Im z_n}(z', \Re z_n, t\Im z_n)\, dt, \end{aligned}$$

and the last integral defines the desired positive function h such that $u = rh$. □

In the following, we introduce the concepts of pseudoconvex domains and plurisubharmonic functions, which are crucial for complex analysis of several variables.

Definition 4.47. Let Ω be a bounded domain in $\mathbb{C}^n$ with $n \geq 2$, and let r be a C^2 defining function for Ω. The Hermitian form

$$i\partial\bar\partial r(t,t)(p) = \sum_{j,k=1}^{n} \frac{\partial^2 r}{\partial z_j \partial \bar z_k}(p)\, t_j \bar t_k, \quad p \in b\Omega, \tag{4.21}$$

defined for all $t = (t_1, \dots, t_n) \in \mathbb{C}^n$ such that $\sum_{j=1}^n t_j(\partial r/\partial z_j)(p) = 0$ is called the Levi form of the function r at the point p.

By Lemma 4.46 the Levi form associated with Ω is independent of the defining function up to a positive factor.

For $p \in b\Omega$, let

$$T_p^{1,0}(b\Omega) = \left\{ t = (t_1, \dots, t_n) \in \mathbb{C}^n : \sum_{j=1}^n t_j(\partial r/\partial z_j)(p) = 0 \right\}.$$

Then $T_p^{1,0}(b\Omega)$ is the space of type $(1,0)$ vector fields that are tangent to the boundary at the point p.

Analogously,

$$T_p^{0,1}(b\Omega) = \left\{ t = (t_1, \dots, t_n) \in \mathbb{C}^n : \sum_{j=1}^n t_j(\partial r/\partial \bar z_j)(p) = 0 \right\}.$$

The tangent spaces $T_p^{1,0}(b\Omega)$ and $T_p^{0,1}(b\Omega)$ do not depend on the choice of the defining function r; see Lemma 4.46.

Smooth sections in $T_p^{0,1}(b\Omega)$ are the tangential Cauchy–Riemann operators, for instance,

$$\bar X = \frac{\partial r}{\partial \bar z_k}\frac{\partial}{\partial \bar z_j} - \frac{\partial r}{\partial \bar z_j}\frac{\partial}{\partial \bar z_k},$$

where $j \neq k$. So, if u is a C^2 defining function, then

$$\bar X(u) = \frac{\partial r}{\partial \bar z_k}\frac{\partial u}{\partial \bar z_j} - \frac{\partial r}{\partial \bar z_j}\frac{\partial u}{\partial \bar z_k} = 0. \tag{4.22}$$

For the rest of this chapter, we will suppose that $n \geq 2$.

Definition 4.48. Let Ω be a bounded domain in $\mathbb{C}^n$ with $n \geq 2$, and let r be a C^2 defining function for Ω. The domain Ω is called (Levi) pseudoconvex at $p \in b\Omega$ if the Levi form

$$i\partial\bar\partial r(t,t)(p) = \sum_{j,k=1}^{n} \frac{\partial^2 r}{\partial z_j \partial \bar z_k}(p)\, t_j \bar t_k \geq 0$$

for all $t \in T_p^{1,0}(b\Omega)$. The domain Ω is said to be strictly pseudoconvex at p if the Levi form is strictly positive for all such $t \neq 0$; it is called a (Levi) pseudoconvex domain if it is (Levi) pseudoconvex at every boundary point of Ω.

A C^2 real-valued function φ on Ω is plurisubharmonic if

$$\sum_{j,k=1}^{n} \frac{\partial^2 \varphi}{\partial z_j \partial \overline{z}_k}(z)\, t_j \overline{t}_k \geq 0$$

for all $t = (t_1, \dots, t_n) \in \mathbb{C}^n$ and all $z \in \Omega$.

A bounded domain Ω in $\mathbb{C}^n$ with $n \geq 2$ and C^2 boundary is pseudoconvex if and only if Ω has a smooth strictly plurisubharmonic exhaustion function φ, i. e., the sets $\{z \in \Omega : \varphi(z) < c\}$ are relatively compact in Ω for every $c \in \mathbb{R}$.

Definition 4.49. Let

$$L^2_{(0,1)}(\Omega) := \left\{ u = \sum_{j=1}^{n} u_j \, d\overline{z}_j : u_j \in L^2(\Omega), j = 1, \dots, n \right\}$$

be the space of $(0,1)$-forms with coefficients in $L^2(\Omega)$. For $u, v \in L^2_{(0,1)}(\Omega)$, we define the inner product by

$$(u, v) = \sum_{j=1}^{n} (u_j, v_j).$$

In this way, $L^2_{(0,1)}(\Omega)$ becomes a Hilbert space. $(0,1)$-forms with compactly supported C^∞ coefficients are dense in $L^2_{(0,1)}(\Omega)$.

Definition 4.50. Let $f \in C_0^\infty(\Omega)$ and set

$$\overline{\partial} f := \sum_{j=1}^{n} \frac{\partial f}{\partial \overline{z}_j} \, d\overline{z}_j.$$

Then

$$\overline{\partial} : C_0^\infty(\Omega) \longrightarrow L^2_{(0,1)}(\Omega);$$

$\overline{\partial}$ is a densely defined unbounded operator on $L^2(\Omega)$. It does not have a closed graph.

Definition 4.51. The domain $\mathrm{dom}(\overline{\partial})$ of $\overline{\partial}$ consists of all functions $f \in L^2(\Omega)$ such that $\overline{\partial} f$ in the sense of distributions belongs to $L^2_{(0,1)}(\Omega)$, i. e., $\overline{\partial} f = g = \sum_{j=1}^n g_j \, d\overline{z}_j$, and for each $\phi \in C_0^\infty(\Omega)$, we have

$$\int_\Omega f \overline{\left(\frac{\partial \phi}{\partial z_j}\right)} \, d\lambda = - \int_\Omega g_j \overline{\phi} \, d\lambda, \quad j = 1, \dots, n. \tag{4.23}$$

It is clear that $C_0^\infty(\Omega) \subseteq \mathrm{dom}(\overline{\partial})$, and hence $\mathrm{dom}(\overline{\partial})$ is dense in $L^2(\Omega)$. Since differentiation is a continuous operation in distribution theory, we have the following:

Lemma 4.52. *$\overline{\partial} : \mathrm{dom}(\overline{\partial}) \longrightarrow L^2_{(0,1)}(\Omega)$ has a closed graph, and $\mathrm{Ker}\,\overline{\partial}$ is a closed subspace of $L^2(\Omega)$.*

Proof. We use the arguments of the proof of Proposition 4.45. Let $(f_k)_k$ be a sequence in $\mathrm{dom}(\overline{\partial})$ such that $f_k \to f$ in $L^2(\Omega)$ and $\overline{\partial} f_k \to g$ in $L^2_{(0,1)}(\Omega)$. We have to show that $\overline{\partial} f = g$. From the proof of Proposition 4.45 we know that $\overline{\partial} f_k \to \overline{\partial} f$ as distributions. As $\overline{\partial} f_k \to g$ in $L^2_{(0,1)}(\Omega)$, it follows that $f \in \mathrm{dom}(\overline{\partial})$ and $\overline{\partial} f = g$.

Now we can apply Lemma 4.9 and get that $\mathrm{Ker}\,\overline{\partial}$ is a closed subspace of $L^2(\Omega)$. □

$\mathrm{Ker}\,\overline{\partial}$ coincides with the Bergman space $A^2(\Omega)$ of all holomorphic functions on Ω belonging to $L^2(\Omega)$. This is due to the fact that $\frac{\partial f}{\partial \overline{z}_k} = 0$ in the sense of distributions implies that f is already a holomorphic function (regularity of the Cauchy–Riemann operator; see Proposition 4.41).

More generally, for $q \geq 1$, $\overline{\partial} : L^2_{(0,q)}(\Omega) \longrightarrow L^2_{(0,q+1)}(\Omega)$ with domain as before is again a densely defined closed operator (see below). In this case, $\mathrm{Ker}\,\overline{\partial}$ is a closed subspace of $L^2_{(0,q)}(\Omega)$, which does not mean that all coefficients are holomorphic functions. The $(0,1)$-form $f(z_1, z_2) = \overline{z}_2\, d\overline{z}_1 + \overline{z}_1\, d\overline{z}_2$ satisfies $\overline{\partial} f = 0$ but has nonholomorphic coefficients.

Proposition 4.53. *Let Ω be a smoothly bounded domain in $\mathbb{C}^n$ with defining function r such that $|\nabla r(z)| = 1$ on $b\Omega$. We set $C^\infty(\overline{\Omega})$ for the restriction of $C^\infty(\mathbb{C}^n)$ to $\overline{\Omega}$ and $\mathcal{D}^{0,1} = C^\infty_{(0,1)}(\overline{\Omega}) \cap \mathrm{dom}(\overline{\partial}^*)$. A $(0,1)$-form $u = \sum_{j=1}^n u_j\, d\overline{z}_j$ with coefficients in $C^\infty(\overline{\Omega})$ belongs to $\mathcal{D}^{0,1}$ if and only if $\sum_{j=1}^n \frac{\partial r}{\partial z_j} u_j = 0$ on $b\Omega$.*

Proof. For $\psi \in C^\infty(\overline{\Omega}) \subset \mathrm{dom}(\overline{\partial})$, we have

$$\begin{aligned}\sum_{j=1}^n \left(-\frac{\partial u_j}{\partial z_j}, \psi \right) &= \sum_{j=1}^n \left(u_j, \frac{\partial \psi}{\partial \overline{z}_j} \right) - \sum_{j=1}^n \int_{b\Omega} u_j \overline{\psi} \frac{\partial r}{\partial z_j}\, d\sigma \\ &= (u, \overline{\partial} \psi) - \sum_{j=1}^n \int_{b\Omega} u_j \overline{\psi} \frac{\partial r}{\partial z_j}\, d\sigma.\end{aligned}$$

If ψ has in addition a compact support in Ω, then we have

$$(\overline{\partial}^* u, \psi) = (u, \overline{\partial} \psi).$$

Since the compactly supported smooth functions are dense in $L^2(\Omega)$, we must have

$$\sum_{j=1}^n \int_{b\Omega} u_j \overline{\psi} \frac{\partial r}{\partial z_j}\, d\sigma = 0$$

for all $\psi \in C^\infty(\overline{\Omega})$. This implies

$$\sum_{j=1}^{n} u_j \frac{\partial r}{\partial z_j} = 0$$

on $b\Omega$. □

Now we consider the $\overline{\partial}$-complex

$$L^2(\Omega) \xrightarrow{\overline{\partial}} L^2_{(0,1)}(\Omega) \xrightarrow{\overline{\partial}} \cdots \xrightarrow{\overline{\partial}} L^2_{(0,n)}(\Omega) \xrightarrow{\overline{\partial}} 0\,, \tag{4.24}$$

where $L^2_{(0,q)}(\Omega)$ denotes the space of $(0,q)$-forms on Ω with coefficients in $L^2(\Omega)$. The $\overline{\partial}$-operator on $(0,q)$-forms is given by

$$\overline{\partial}\Big(\sum_J{}' a_J\, d\overline{z}_J\Big) = \sum_{j=1}^{n} \sum_J{}' \frac{\partial a_J}{\partial \overline{z}_j} d\overline{z}_j \wedge d\overline{z}_J, \tag{4.25}$$

where $\sum'$ means that the sum is only taken over strictly increasing multi-indices $J = (j_1,\dots,j_q)$.

The derivatives are taken in the sense of distributions, and the domain of $\overline{\partial}$ consists of those $(0,q)$-forms for which the right-hand side belongs to $L^2_{(0,q+1)}(\Omega)$. So $\overline{\partial}$ is a densely defined closed operatorindexclosed operator, and therefore has an adjoint operator from $L^2_{(0,q+1)}(\Omega)$ into $L^2_{(0,q)}(\Omega)$ denoted by $\overline{\partial}^*$.

We consider the $\overline{\partial}$-complex

$$L^2_{(0,q-1)}(\Omega) \underset{\overline{\partial}^*}{\overset{\overline{\partial}}{\rightleftarrows}} L^2_{(0,q)}(\Omega) \underset{\overline{\partial}^*}{\overset{\overline{\partial}}{\rightleftarrows}} L^2_{(0,q+1)}(\Omega) \tag{4.26}$$

for $1 \le q \le n-1$.

We remark that a $(0,q+1)$-form $u = \sum_J' u_J\, d\overline{z}_J$ belongs to $C^\infty_{(0,q+1)}(\overline{\Omega}) \cap \mathrm{dom}(\overline{\partial}^*)$ if and only if

$$\sum_{k=1}^{n} u_{kK} \frac{\partial r}{\partial z_k} = 0 \tag{4.27}$$

on $b\Omega$ for all K with $|K| = q$. To see this, let $\alpha \in C^\infty_{(0,q)}(\overline{\Omega})$. Then

$$\begin{aligned}
(u, \overline{\partial}\alpha) &= \Bigg(\sum_{|J|=q+1}{}' u_J\, d\overline{z}_J, \sum_{j=1}^{n}\sum_{|K|=q}{}' \frac{\partial \alpha_K}{\partial \overline{z}_j} d\overline{z}_j \wedge d\overline{z}_K\Bigg) \\
&= \sum_{j=1}^{n}\sum_{|K|=q}{}' \int_\Omega u_{jK} \frac{\partial \overline{\alpha}_K}{\partial z_j}\, d\lambda \\
&= -\sum_{j=1}^{n}\sum_{|K|=q}{}' \int_\Omega \frac{\partial u_{jK}}{\partial z_j} \overline{\alpha}_K\, d\lambda + \sum_{j=1}^{n}\sum_{|K|=q}{}' \int_{b\Omega} u_{jK}\, \overline{\alpha}_K \frac{\partial r}{\partial z_j}\, d\sigma
\end{aligned}$$

$$
\begin{aligned}
&= \left(\sum_{|K|=q}' \left(-\sum_{j=1}^{n} \frac{\partial u_{jK}}{\partial z_j} \right) d\bar{z}_K, \quad \sum_{|K|=q}' \alpha_K \, d\bar{z}_K \right) \\
&\quad + \sum_{|K|=q}' \int_{b\Omega} \left(\sum_{j=1}^{n} u_{jK} \frac{\partial r}{\partial z_j} \right) \bar{\alpha}_K \, d\sigma \\
&= (\vartheta u, \alpha) - \int_{b\Omega} \langle \theta(\vartheta, dr) u, \alpha \rangle \, d\sigma,
\end{aligned}
$$

where

$$\vartheta u = \sum_{|K|=q}' \left(-\sum_{j=1}^{n} \frac{\partial u_{jK}}{\partial z_j} \right) d\bar{z}_K \tag{4.28}$$

and

$$\sum_{|K|=q}' \int_{b\Omega} \left(\sum_{j=1}^{n} u_{jK} \frac{\partial r}{\partial z_j} \right) \bar{\alpha}_K \, d\sigma = - \int_{b\Omega} \langle \theta(\vartheta, \partial r) u, \alpha \rangle \, d\sigma. \tag{4.29}$$

Hence we have

$$(\vartheta u, \alpha) = (u, \bar{\partial} \alpha) + \int_{b\Omega} \langle \theta(\vartheta, \partial r) u, \alpha \rangle \, d\sigma, \tag{4.30}$$

where $\theta(\vartheta, dr)u$ denotes the symbol of ϑ in the ∂r direction. Note that for $u \in \operatorname{dom}(\bar{\partial}^*)$, we have $\bar{\partial}^* u = \vartheta u$.

Similarly, for $u \in C^\infty_{(0,q)}(\bar{\Omega})$ and $\alpha \in C^\infty_{(0,q+1)}(\bar{\Omega})$, we have

$$(\bar{\partial} u, \alpha) = (u, \vartheta \alpha) + \int_{b\Omega} \langle \bar{\partial} r \wedge u, \alpha \rangle \, d\sigma, \tag{4.31}$$

where

$$\int_{b\Omega} \langle \bar{\partial} r \wedge u, \alpha \rangle \, d\sigma = \sum_{|K|=q}' \sum_{k=1}^{n} \int_{b\Omega} u_K \frac{\partial r}{\partial \bar{z}_k} \bar{\alpha}_{kK} \, d\sigma. \tag{4.32}$$

Proposition 4.54. *The complex Laplacian* $\square = \bar{\partial}\bar{\partial}^* + \bar{\partial}^*\bar{\partial}$, *defined on the domain* $\operatorname{dom}(\square) = \{u \in L^2_{(0,q)}(\Omega) : u \in \operatorname{dom}(\bar{\partial}) \cap \operatorname{dom}(\bar{\partial}^*), \bar{\partial} u \in \operatorname{dom}(\bar{\partial}^*), \bar{\partial}^* u \in \operatorname{dom}(\bar{\partial})\}$ *acts as an unbounded, densely defined, closed, and self-adjoint operator on* $L^2_{(0,q)}(\Omega)$ *for* $1 \le q \le n$, *which means that* $\square = \square^*$ *and* $\operatorname{dom}(\square) = \operatorname{dom}(\square^*)$.

Proof. $\operatorname{dom}(\square)$ contains all smooth forms with compact support, and hence $\square$ is densely defined. To show that $\square$ is closed depends on the fact that both $\bar{\partial}$ and $\bar{\partial}^*$ are closed: note that

$$(\square u, u) = (\bar\partial\bar\partial^* u + \bar\partial^*\bar\partial u, u) = \|\bar\partial u\|^2 + \|\bar\partial^* u\|^2 \tag{4.33}$$

for $u \in \mathrm{dom}(\square)$. We have to prove that for every sequence $u_k \in \mathrm{dom}(\square)$ such that $u_k \to u$ in $L^2_{(0,q)}(\Omega)$ and $\square u_k$ converges, we have $u \in \mathrm{dom}(\square)$ and $\square u_k \to \square u$. It follows from (4.33) that

$$(\square(u_k - u_\ell), u_k - u_\ell) = \|\bar\partial(u_k - u_\ell)\|^2 + \|\bar\partial^*(u_k - u_\ell)\|^2,$$

which implies that $\bar\partial u_k$ converges in $L^2_{(0,q+1)}(\Omega)$ and that $\bar\partial^* u_k$ converges in $L^2_{(0,q-1)}(\Omega)$. Since $\bar\partial$ and $\bar\partial^*$ are closed operators, we get $u \in \mathrm{dom}(\bar\partial) \cap \mathrm{dom}(\bar\partial^*)$ and $\bar\partial u_k \to \bar\partial u$ in $L^2_{(0,q+1)}(\Omega)$ and $\bar\partial^* u_k \to \bar\partial^* u$ in $L^2_{(0,q-1)}(\Omega)$.

To show that $\bar\partial u \in \mathrm{dom}(\bar\partial^*)$ and $\bar\partial^* u \in \mathrm{dom}(\bar\partial)$, we first notice that $\bar\partial\bar\partial^* u_k$ and $\bar\partial^*\bar\partial u_k$ are orthogonal, which follows from

$$(\bar\partial\bar\partial^* u_k, \bar\partial^*\bar\partial u_k) = (\bar\partial^2\bar\partial^* u_k, \bar\partial u_k) = 0.$$

Therefore the convergence of $\square u_k = \bar\partial\bar\partial^* u_k + \bar\partial^*\bar\partial u_k$ implies that both $\bar\partial\bar\partial^* u_k$ and $\bar\partial^*\bar\partial u_k$ converge. Now use again that $\bar\partial$ and $\bar\partial^*$ are closed operators to obtain that $\bar\partial\bar\partial^* u_k \to \bar\partial\bar\partial^* u$ and $\bar\partial^*\bar\partial u_k \to \bar\partial^*\bar\partial u$. This implies that $\square u_k \to \square u$. Hence $\square$ is closed.

To show that $\square$ is self-adjoint, we use Lemma 4.29. Define

$$R = \bar\partial\bar\partial^* + \bar\partial^*\bar\partial + I$$

on $\mathrm{dom}(\square)$. By Lemma 4.29 both $(I + \bar\partial\bar\partial^*)^{-1}$ and $(I + \bar\partial^*\bar\partial)^{-1}$ are bounded self-adjoint operators. Consider

$$L = (I + \bar\partial\bar\partial^*)^{-1} + (I + \bar\partial^*\bar\partial)^{-1} - I.$$

Then L is bounded and self-adjoint. We claim that $L = R^{-1}$. Since

$$(I + \bar\partial\bar\partial^*)^{-1} - I = (I - (I + \bar\partial\bar\partial^*))(I + \bar\partial\bar\partial^*)^{-1} = -\bar\partial\bar\partial^*(I + \bar\partial\bar\partial^*)^{-1},$$

we have that the range of $(I + \bar\partial\bar\partial^*)^{-1}$ is contained in $\mathrm{dom}(\bar\partial\bar\partial^*)$. Similarly, we have that the range of $(I + \bar\partial^*\bar\partial)^{-1}$ is contained in $\mathrm{dom}(\bar\partial^*\bar\partial)$, and we get

$$L = (I + \bar\partial^*\bar\partial)^{-1} - \bar\partial\bar\partial^*(I + \bar\partial\bar\partial^*)^{-1}.$$

Since $\bar\partial^2 = 0$, we have that the range of L is contained in $\mathrm{dom}(\bar\partial^*\bar\partial)$ and

$$\bar\partial^*\bar\partial L = \bar\partial^*\bar\partial(I + \bar\partial^*\bar\partial)^{-1}.$$

Similarly, we have that the range of L is contained in $\mathrm{dom}(\bar\partial\bar\partial^*)$ and

$$\bar\partial\bar\partial^* L = \bar\partial\bar\partial^*(I + \bar\partial\bar\partial^*)^{-1}.$$

This implies that the range of L is contained in dom($\square$). In addition, we have

$$RL = \bar{\partial}\bar{\partial}^*(I + \bar{\partial}\bar{\partial}^*)^{-1} + \bar{\partial}^*\bar{\partial}(I + \bar{\partial}^*\bar{\partial})^{-1} + L = I.$$

If $Ru = 0$, then we get $\square u = -u$ and $0 \leq (\square u, u) = -(u, u)$, which implies that $u = 0$. Hence R is injective, and we have that $L = R^{-1}$. By Lemma 4.29 we know that L is self-adjoint. Apply Lemma 4.28 to get that R is self-adjoint. Therefore $\square = R - I$ is self-adjoint. □

The two boundary conditions $u \in \text{dom}(\bar{\partial}^*)$ and $\bar{\partial}u \in \text{dom}(\bar{\partial}^*)$, which appear in the definition of dom($\square$)m are called the $\bar{\partial}$-Neumann boundary conditions. They amount to a Dirichlet boundary condition on the normal component of u and to the normal component of $\bar{\partial}u$, respectively; see [91] for more detail.

Example 4.55. Let Ω be a smoothly bounded domain in $\mathbb{C}^n$ with $0 \in b\Omega$. Assume that for some neighborhood U of 0,

$$\Omega \cap U = \{z \in \mathbb{C}^n : r(z) = \Im z_n = y_n < 0\} \cap U.$$

Let $u = \sum_{j=1}^n u_j \, d\bar{z}_j \in C^2_{(0,1)}(\overline{\Omega})$ and suppose that the support of u lies in $U \cap \overline{\Omega}$. Then $u \in \text{dom}(\square)$ if and only if

$$u_n = 0 \quad \text{on } b\Omega \cap U, \tag{4.34}$$

$$\frac{\partial u_j}{\partial \bar{z}_n} = 0 \quad \text{on } b\Omega \cap U, \quad j = 1, \ldots, n-1. \tag{4.35}$$

Equality (4.34) follows from (4.27), which means that $u \in \text{dom}(\bar{\partial}^*)$, and $\bar{\partial}u \in \text{dom}(\bar{\partial}^*)$ is equivalent to

$$\frac{\partial u_j}{\partial \bar{z}_n} - \frac{\partial u_n}{\partial \bar{z}_j} = 0 \quad \text{on } b\Omega \cap U, \quad j = 1, \ldots, n-1,$$

again by (4.27). Using (4.22) for u_n, we get

$$\bar{X} = \frac{\partial r}{\partial \bar{z}_j}\frac{\partial u_n}{\partial \bar{z}_n} - \frac{\partial r}{\partial \bar{z}_n}\frac{\partial u_n}{\partial \bar{z}_j} = 0,$$

which implies that $\frac{\partial u_n}{\partial \bar{z}_j} = 0$ on $b\Omega \cap U, j = 1, \ldots, n-1$, and we get (4.35).

It is the second boundary condition that makes the system noncoercive; see [66].

We continue investigating the boundary conditions.

Proposition 4.56. *Let Ω be a smoothly bounded domain in $\mathbb{C}^n$ with defining function r such that $|\nabla r(z)| = 1$ on $b\Omega$. Then for $u \in \mathcal{D}^{0,1}$, we have*

$$\|\bar{\partial}u\|^2 + \|\bar{\partial}^* u\|^2 = \sum_{j,k=1}^n \left\|\frac{\partial u_j}{\partial \bar{z}_k}\right\|^2 + \int_{b\Omega} \sum_{j,k=1}^n \frac{\partial^2 r}{\partial z_j \partial \bar{z}_k} u_j \bar{u}_k \, d\sigma. \tag{4.36}$$

Proof. For $u \in \mathcal{D}^{0,1}$, we have

$$\bar\partial u = \sum_{j<k} \left(\frac{\partial u_k}{\partial \bar z_j} - \frac{\partial u_j}{\partial \bar z_k} \right) d\bar z_j \wedge d\bar z_k$$

and

$$\bar\partial^* u = -\sum_{j=1}^{n} \frac{\partial u_j}{\partial z_j}.$$

For the norms, we get

$$\|\bar\partial u\|^2 = \sum_{j<k} \left\| \frac{\partial u_k}{\partial \bar z_j} - \frac{\partial u_j}{\partial \bar z_k} \right\|^2 = \sum_{j,k=1}^{n} \left\| \frac{\partial u_k}{\partial \bar z_j} \right\|^2 - \sum_{j,k=1}^{n} \left(\frac{\partial u_k}{\partial \bar z_j}, \frac{\partial u_j}{\partial \bar z_k} \right)$$

and

$$\|\bar\partial^* u\|^2 = \left\| \sum_{j=1}^{n} \frac{\partial u_j}{\partial z_j} \right\|^2 = \left(\sum_{j=1}^{n} \frac{\partial u_j}{\partial z_j}, \sum_{k=1}^{n} \frac{\partial u_k}{\partial z_k} \right).$$

Noting that the commutator $[\frac{\partial}{\partial z_k}, \frac{\partial}{\partial \bar z_j}] = 0$, integrate by parts

$$\begin{aligned}
-\sum_{j,k=1}^{n} \left(\frac{\partial u_k}{\partial \bar z_j}, \frac{\partial u_j}{\partial \bar z_k} \right) &= \sum_{j,k=1}^{n} \left\{ \left(\frac{\partial}{\partial z_k} \frac{\partial u_k}{\partial \bar z_j}, u_j \right) - \int_{b\Omega} \frac{\partial r}{\partial z_k} \frac{\partial u_k}{\partial \bar z_j} \bar u_j \, d\sigma \right\} \\
&= \sum_{j,k=1}^{n} \left\{ \left(\frac{\partial}{\partial \bar z_j} \frac{\partial u_k}{\partial z_k}, u_j \right) - \int_{b\Omega} \frac{\partial r}{\partial z_k} \frac{\partial u_k}{\partial \bar z_j} \bar u_j \, d\sigma \right\} \\
&= \sum_{j,k=1}^{n} \left\{ -\left(\frac{\partial u_k}{\partial z_k}, \frac{\partial u_j}{\partial z_j} \right) + \int_{b\Omega} \frac{\partial r}{\partial \bar z_j} \frac{\partial u_k}{\partial z_k} \bar u_j \, d\sigma \right. \\
&\qquad \left. - \int_{b\Omega} \frac{\partial r}{\partial z_k} \frac{\partial u_k}{\partial \bar z_j} \bar u_j \, d\sigma \right\} \\
&= -\|\bar\partial^* u\|^2 + \sum_{j,k=1}^{n} \int_{b\Omega} \frac{\partial r}{\partial \bar z_j} \frac{\partial u_k}{\partial z_k} \bar u_j \, d\sigma \\
&\qquad - \sum_{j,k=1}^{n} \int_{b\Omega} \frac{\partial r}{\partial z_k} \frac{\partial u_k}{\partial \bar z_j} \bar u_j \, d\sigma.
\end{aligned}$$

Since $u \in \mathcal{D}^{0,1}$, which means that $\sum_{j=1}^{n} \frac{\partial r}{\partial z_j} u_j = 0$ on $b\Omega$, the second term of the last line is 0.

Also, since $u \in \mathcal{D}^{0,1}$, the vector field $X := \sum_{k=1}^{n} u_k \frac{\partial}{\partial z_k}$ is tangent to $b\Omega$. Thus, if g is any function vanishing on $b\Omega$, then $X(g) = 0$ on $b\Omega$. The function $g_0 = \sum_{j=1}^{n} \frac{\partial r}{\partial \overline{z}_j} \overline{u}_j = 0$ on $b\Omega$. Hence we get

$$X(g_0) = \sum_{k=1}^{n} u_k \frac{\partial}{\partial z_k}\left(\sum_{j=1}^{n} \frac{\partial r}{\partial \overline{z}_j} \overline{u}_j\right) = \sum_{j,k=1}^{n} u_k \frac{\partial^2 r}{\partial \overline{z}_j \partial z_k} \overline{u}_j + \sum_{j,k=1}^{n} u_k \frac{\partial r}{\partial \overline{z}_j} \frac{\partial \overline{u}_j}{\partial z_k} = 0.$$

Taking the complex conjugates, we see that

$$\sum_{j,k=1}^{n} \frac{\partial r}{\partial z_k} \frac{\partial u_k}{\partial \overline{z}_j} \overline{u}_j = -\sum_{j,k=1}^{n} \frac{\partial^2 r}{\partial z_j \partial \overline{z}_k} u_j \overline{u}_k,$$

and since the right side is real, this implies that

$$\|\overline{\partial} u\|^2 = \sum_{j,k=1}^{n} \left\|\frac{\partial u_k}{\partial \overline{z}_j}\right\|^2 - \|\overline{\partial}^* u\|^2 + \int_{b\Omega} \sum_{j,k=1}^{n} \frac{\partial^2 r}{\partial z_j \partial \overline{z}_k} u_j \overline{u}_k \, d\sigma. \qquad \square$$

Corollary 4.57. *Let Ω be a smoothly bounded pseudoconvex domain in $\mathbb{C}^n$ with defining function r such that $|\nabla r(z)| = 1$ on $b\Omega$. Then for $u \in \mathcal{D}^{0,1}$, we have*

$$\|\overline{\partial} u\|^2 + \|\overline{\partial}^* u\|^2 \geq \sum_{j,k=1}^{n} \left\|\frac{\partial u_j}{\partial \overline{z}_k}\right\|^2. \tag{4.37}$$

Proof. By Definition 4.48 the Levi form

$$\sum_{j,k=1}^{n} \frac{\partial^2 r}{\partial z_j \partial \overline{z}_k}(p)\, u_j(p)\overline{u_k(p)} \geq 0$$

for all $p \in b\Omega$ and all $u \in \mathcal{D}^{0,1}$. Hence the result follows immediately from Proposition 4.56. $\square$

We return to the $\square$-operator on $(0,q)$-forms. For the rest of this chapter, we will now suppose that Ω is a smoothly bounded pseudoconvex domain in $\mathbb{C}^n$. In Theorem 8.1, we will show that there is $c > 0$ such that

$$\|\overline{\partial} u\|^2 + \|\overline{\partial}^* u\|^2 \geq c\,\|u\|^2 \tag{4.38}$$

for each $u \in \operatorname{dom}(\overline{\partial}) \cap \operatorname{dom}(\overline{\partial}^*)$.

First, we will show that (4.38) implies that $\overline{\partial}$ and $\overline{\partial}^*$ have a closed range.

Proposition 4.58. *Let $\Omega \subset \mathbb{C}^n$ be a smoothly bounded pseudoconvex domain. Then $\overline{\partial}$ and $\overline{\partial}^*$ have closed ranges.*

Proof. We notice that $\ker\overline{\partial} = (\operatorname{im}\overline{\partial}^*)^\perp$, which implies that

$$(\ker\overline{\partial})^\perp = \overline{\operatorname{im}\overline{\partial}^*} \subseteq \ker\overline{\partial}^*.$$

If $u \in \ker\overline{\partial} \cap \ker\overline{\partial}^*$, then by (4.38) we have that $u = 0$. Hence

$$(\ker\overline{\partial})^\perp = \ker\overline{\partial}^*. \tag{4.39}$$

If $u \in \operatorname{dom}(\overline{\partial}) \cap (\ker\overline{\partial})^\perp$, then $u \in \ker\overline{\partial}^*$, and (4.38) implies

$$\|u\| \le \frac{1}{c}\,\|\overline{\partial}u\|.$$

Now we can apply Lemma 4.20 to conclude that $\operatorname{im}\overline{\partial}$ is closed. Finally, Proposition 4.22 gives that $\operatorname{im}\overline{\partial}^*$ is also closed. □

The next result describes the implication of the basic estimates (4.38) for the □-operator.

Proposition 4.59. *Let $\Omega \subset \mathbb{C}^n$ be a smoothly bounded pseudoconvex domain. Then $\Box : \operatorname{dom}(\Box) \longrightarrow L^2_{(0,q)}(\Omega)$ is bijective and has a bounded inverse*

$$N : L^2_{(0,q)}(\Omega) \longrightarrow \operatorname{dom}(\Box),$$

which is called the $\overline{\partial}$-Neumann operator. In addition,

$$\|Nu\| \le \frac{1}{c}\,\|u\|. \tag{4.40}$$

Proof. Since $(\Box u, u) = \|\overline{\partial}u\|^2 + \|\overline{\partial}^* u\|^2$, it follows that for a convergent sequence $(\Box u_n)_n$,

$$\|\Box u_n - \Box u_m\|\,\|u_n - u_m\| \ge (\Box(u_n - u_m), u_n - u_m) \ge c\|u_n - u_m\|^2,$$

which implies that $(u_n)_n$ is convergent, and since □ is a closed operator, we obtain that □ has a closed range. If $\Box u = 0$, then $\overline{\partial}u = 0$ and $\overline{\partial}^* u = 0$, and by (4.38) $u = 0$, and hence □ is injective. By Lemma 4.28(ii) the range of □ is dense, and therefore □ is surjective.

We showed that

$$\Box : \operatorname{dom}(\Box) \longrightarrow L^2_{(0,q)}(\Omega)$$

is bijective and therefore, by Lemma 4.28(iv), has a bounded inverse

$$N : L^2_{(0,q)}(\Omega) \longrightarrow \operatorname{dom}(\Box).$$

For $u \in L^2_{(0,q)}(\Omega)$, we use (4.38) for Nu to obtain

$$c\|Nu\|^2 \le \|\bar{\partial}Nu\|^2 + \|\bar{\partial}^* Nu\|^2 = (\bar{\partial}^*\bar{\partial}Nu, Nu) + (\bar{\partial}\bar{\partial}^* Nu, Nu) = (u, Nu) \le \|u\| \, \|Nu\|,$$

which implies (4.40). □

For $u \in L^2_{(0,q)}(\Omega)$ and $v \in \operatorname{dom}(\bar{\partial}) \cap \operatorname{dom}(\bar{\partial}^*)$, we get

$$(u, v) = (\Box Nu, v) = ((\bar{\partial}\bar{\partial}^* + \bar{\partial}^*\bar{\partial})Nu, v) = (\bar{\partial}^* Nu, \bar{\partial}^* v) + (\bar{\partial}Nu, \bar{\partial}v). \tag{4.41}$$

Now we discuss a different approach to the $\bar{\partial}$-Neumann operator, which is related to the quadratic form

$$Q(u, v) = (\bar{\partial}u, \bar{\partial}v) + (\bar{\partial}^* u, \bar{\partial}^* v).$$

For this purpose, we consider the embedding

$$j : \operatorname{dom}(\bar{\partial}) \cap \operatorname{dom}(\bar{\partial}^*) \longrightarrow L^2_{(0,q)}(\Omega),$$

where $\operatorname{dom}(\bar{\partial}) \cap \operatorname{dom}(\bar{\partial}^*)$ is endowed with the graph norm

$$u \mapsto (\|\bar{\partial}u\|^2 + \|\bar{\partial}^* u\|^2)^{1/2}.$$

The graph norm stems from the inner product

$$Q(u, v) = (u, v)_Q = (\Box u, v) = (\bar{\partial}u, \bar{\partial}v) + (\bar{\partial}^* u, \bar{\partial}^* v).$$

The basic estimates (4.38) imply that j is a bounded operator with operator norm

$$\|j\| \le \frac{1}{\sqrt{c}}.$$

By (4.38) it follows in addition that $\operatorname{dom}(\bar{\partial}) \cap \operatorname{dom}(\bar{\partial}^*)$ endowed with the graph norm $u \mapsto (\|\bar{\partial}u\|^2 + \|\bar{\partial}^* u\|^2)^{1/2}$ is a Hilbert space.

Since $(u, v) = (u, jv)$, we have that $(u, v) = (j^* u, v)_Q$. Equation (4.41) suggests that as an operator to $\operatorname{dom}(\bar{\partial}) \cap \operatorname{dom}(\bar{\partial}^*)$, N coincides with j^*, and as an operator to $L^2_{(0,q)}(\Omega)$, N should be equal to $j \circ j^*$. For this purpose, set

$$N' = j \circ j^*, \tag{4.42}$$

and note that $N'^* = (j \circ j^*)^* = j \circ j^* = N'$, i. e., N' is self-adjoint (and of course also bounded). We claim that the range of N' is contained in $\operatorname{dom}(\Box)$. To show this, we use an approach due to Berger (see [5]): since $\Box$ is self-adjoint, it suffices to show that $N'u \in \operatorname{dom}(\Box^*)$ for all $u \in L^2_{(0,q)}(\Omega)$, that is, to show that the functional $v \mapsto (\Box v, N'u)$ is bounded on $\operatorname{dom}(\Box)$:

$$|(\Box v, N'u)| = |((\bar{\partial}\bar{\partial}^* + \bar{\partial}^*\bar{\partial})v, N'u)| = |(\bar{\partial}v, \bar{\partial}N'u) + (\bar{\partial}^* v, \bar{\partial}^* N'u)|$$

$$= |Q(v, j^*u)| = |(jv, u)| = |(v, u)| \leq \|v\| \, \|u\|.$$

For $v \in \mathrm{dom}(\overline{\partial}) \cap \mathrm{dom}(\overline{\partial}^*)$, we have

$$(\Box N'u, v) = (N'u, v)_Q = (j^*u, v)_Q = (u, jv) = (u, v),$$

and hence $\Box N'u = u$; in a similar way, for $u \in \mathrm{dom}(\Box)$,

$$(N'\Box u, v) = (\Box u, N'v) = (u, N'v)_Q = (u, j^*v)_Q = (ju, v) = (u, v),$$

which implies that $N'\Box u = u$. Altogether we obtain that $N = N'$.

Remark 4.60. In a similar way, we can use Proposition 4.36 to prove the existence and the main properties of the $\overline{\partial}$-Neumann operator. For this purpose, let $H = L^2_{(0,q)}(\Omega)$, and let $\mathcal{V} = \mathrm{dom}(\overline{\partial}) \cap \mathrm{dom}(\overline{\partial}^*)$ endowed with the graph norm $u \mapsto (\|\overline{\partial}u\|^2 + \|\overline{\partial}^* u\|^2)^{1/2}$. The corresponding sesquilinear form $a : \mathcal{V} \times \mathcal{V} \longrightarrow \mathbb{C}$ is given by

$$a(u, v) = Q(u, v) = (\overline{\partial}u, \overline{\partial}v) + (\overline{\partial}^* u, \overline{\partial}^* v)$$

for $u, v \in \mathrm{dom}(\overline{\partial}) \cap \mathrm{dom}(\overline{\partial}^*)$.

The basic estimates (4.38) show that $\mathcal{V}$ is continuously embedded into H, which gives (4.5), and by definition the sesquilinear form $a(u, v)$ is $\mathcal{V}$-elliptic (4.7). The domain of the operator S in Proposition 4.36 coincides with the domain of the $\Box$-operator, because the mapping $v \mapsto a(u, v)$ is continuous in the L^2-norm exactly if we can write

$$|a(u, v)| = |(\Box u, v)| \leq \|\Box u\| \, \|v\|.$$

We have $S = \Box$, and by Proposition 4.36 we immediately get that

$$\Box : \mathrm{dom}(\Box) \longrightarrow L^2_{(0,q)}(\Omega)$$

is bijective and self-adjoint and has a bounded inverse.

Proposition 4.61. *The operators*

$$\begin{aligned} \overline{\partial}N : L^2_{(0,q)}(\Omega) &\longrightarrow L^2_{(0,q+1)}(\Omega), \quad 1 \leq q \leq n-1, \quad \textit{and} \\ \overline{\partial}^* N : L^2_{(0,q)}(\Omega) &\longrightarrow L^2_{(0,q-1)}(\Omega), \quad 2 \leq q \leq n, \end{aligned}$$

are both bounded.

Proof. From the above considerations on N we get

$$\|\overline{\partial}Nu\|^2 + \|\overline{\partial}^* Nu\|^2 = (j^*u, j^*u)_Q \leq \|j^*\|^2 \, \|u\|^2$$

for $u \in L^2_{(0,q)}(\Omega)$, which implies the result. ☐

Proposition 4.62. *Let N_q denote the $\bar\partial$-Neumann operator on $L^2_{(0,q)}(\Omega)$. Then*

$$N_{q+1}\bar\partial = \bar\partial N_q, \quad 1 \le q \le n-1, \tag{4.43}$$

on $\text{dom}(\bar\partial)$*, and*

$$N_{q-1}\bar\partial^* = \bar\partial^* N_q, \quad 2 \le q \le n, \tag{4.44}$$

on $\text{dom}(\bar\partial^*)$.

In addition, we have that $\bar\partial^* N_q$ is zero on $(\ker\bar\partial)^\perp$.

Proof. For $u \in \text{dom}(\bar\partial)$, we have $\bar\partial u = \bar\partial\bar\partial^*\bar\partial N_q u$ and

$$N_{q+1}\bar\partial u = N_{q+1}\bar\partial\bar\partial^*\bar\partial N_q u = N_{q+1}(\bar\partial\bar\partial^* + \bar\partial^*\bar\partial)\bar\partial N_q u = \bar\partial N_q u,$$

which proves (4.43). In a similar way, we get (4.44).

Now let $k \in (\ker\bar\partial)^\perp$ and $u \in \text{dom}(\bar\partial)$. Then

$$(\bar\partial^* N_q k, u) = (N_q k, \bar\partial u) = (k, N_q\bar\partial u) = (k, \bar\partial N_{q-1}u) = 0,$$

since $\bar\partial N_{q-1}u \in \ker(\bar\partial)$, which gives $\bar\partial^* N_q k = 0$. □

Since we already know that both operators $\bar\partial N_q$ and $\bar\partial^* N_q$ are bounded, we can extend both operators $N_{q+1}\bar\partial$ and $N_{q-1}\bar\partial^*$ to bounded operators on $L^2_{(0,q)}(\Omega)$.

Proposition 4.63. *Let $\alpha \in L^2_{(0,q)}(\Omega)$ with $\bar\partial\alpha = 0$. Then $u_0 = \bar\partial^* N_q\alpha$ is the canonical solution of $\bar\partial u = \alpha$, which means that $\bar\partial u_0 = \alpha$, $u_0 \perp \ker\bar\partial$, and*

$$\|\bar\partial^* N_q\alpha\| \le c^{-1/2}\|\alpha\|. \tag{4.45}$$

Proof. For $\alpha \in L^2_{(0,q)}(\Omega)$ with $\bar\partial\alpha = 0$, we get

$$\alpha = \bar\partial\bar\partial^* N_q\alpha + \bar\partial^*\bar\partial N_q\alpha. \tag{4.46}$$

Applying $\bar\partial$ to the last equality, we obtain

$$0 = \bar\partial\alpha = \bar\partial\bar\partial^*\bar\partial N_q\alpha,$$

and since $\bar\partial N_q\alpha \in \text{dom}(\bar\partial^*)$, we have

$$0 = (\bar\partial\bar\partial^*\bar\partial N_q\alpha, \bar\partial N_q\alpha) = (\bar\partial^*\bar\partial N_q\alpha, \bar\partial^*\bar\partial N_q\alpha) = \|\bar\partial^*\bar\partial N_q\alpha\|^2. \tag{4.47}$$

Finally, we set $u_0 = \bar\partial^* N_q\alpha$ and derive from (4.46) and (4.47) that for $\bar\partial\alpha = 0$,

$$\alpha = \overline{\partial} u_0,$$

and we see that $u_0 \perp \ker \overline{\partial}$, since for $h \in \ker \overline{\partial}$, we get

$$(u_0, h) = (\overline{\partial}^* N_q \alpha, h) = (N_q \alpha, \overline{\partial} h) = 0.$$

It follows that

$$\begin{aligned}
\|\overline{\partial}^* N_q \alpha\|^2 &= (\overline{\partial}\,\overline{\partial}^* N_q \alpha, N_q \alpha) \\
&= (\overline{\partial}\,\overline{\partial}^* N_q \alpha, N_q \alpha) + (\overline{\partial}^* \overline{\partial} N_q \alpha, N_q \alpha) \\
&= (\alpha, N_q \alpha) \le \|\alpha\| \, \|N_q \alpha\|,
\end{aligned}$$

and using (4.40), we obtain

$$\|\overline{\partial}^* N_q \alpha\| \le c^{-1/2} \, \|\alpha\|. \qquad \square$$

We showed that the canonical solution operator S_q for $\overline{\partial}$ coincides with $\overline{\partial}^* N_q$ as an operator on

$$L^2_{(0,q)}(\Omega) \cap \ker \overline{\partial}$$

and is a bounded operator.

The $\overline{\partial}$-Neumann operator N can be expressed in terms of the canonical solution operators:

Proposition 4.64. *For* $1 \le q \le n-1$, *we have*

$$N_q = S_q^* \, S_q + S_{q+1} \, S_{q+1}^*. \tag{4.48}$$

Proof. We use (4.43) and (4.44) to show that

$$\overline{\partial}^* N_q = \overline{\partial}^* N_q^* = (N_q \overline{\partial})^* \quad \text{and} \quad (\overline{\partial}^* N_q)^* = N_q \overline{\partial}$$

and

$$\overline{\partial} N_q = \overline{\partial}^{**} N_q^* = (N_q \overline{\partial}^*)^* = (\overline{\partial}^* N_{q+1})^* \quad \text{and} \quad \overline{\partial}^* N_{q+1} = (\overline{\partial} N_q)^* = N_q \overline{\partial}^*,$$

where we applied Lemma 4.3. Hence it follows that for $u \in L^2_{(0,q)}(\Omega)$, we have

$$\begin{aligned}
N_q u &= N_q (\overline{\partial}\,\overline{\partial}^* + \overline{\partial}^* \overline{\partial}) N_q u \\
&= (N_q \overline{\partial})(\overline{\partial}^* N_q) u + (N_q \overline{\partial}^*)(\overline{\partial} N_q) u \\
&= (\overline{\partial}^* N_q)^* (\overline{\partial}^* N_q) u + (\overline{\partial}^* N_{q+1})(\overline{\partial}^* N_{q+1})^* u
\end{aligned}$$

$$= S_q^* S_q u + S_{q+1} S_{q+1}^* u. \qquad \square$$

Proposition 4.65. *Let $P_q : L^2_{(0,q)}(\Omega) \longrightarrow \ker\bar\partial$ denote the orthogonal projection, $0 \le q \le n-1$, which is the Bergman projection for $q = 0$. Then*

$$P_q = I - \bar\partial^* N_{q+1}\bar\partial \tag{4.49}$$

on $\mathrm{dom}(\bar\partial)$.

Proof. First, we show that the range of the right-hand side of (4.49), which we denote by $\tilde P$, coincides with $\ker\bar\partial$: for $u \in \mathrm{dom}(\bar\partial)$, we have

$$\bar\partial u - \bar\partial\bar\partial^* N_{q+1}\bar\partial u = \bar\partial u - \Box N_{q+1}\bar\partial u + \bar\partial^*\bar\partial N_{q+1}\bar\partial u = \bar\partial u - \bar\partial u = 0,$$

where we used (4.43) to show that $\bar\partial N_{q+1}\bar\partial u = N_{q+2}\bar\partial\bar\partial u = 0$, and since

$$u - \bar\partial^* N_{q+1}\bar\partial u = u$$

for $u \in \ker\bar\partial$, we have shown the first claim. Now we obtain

$$\tilde P^* = (I - \bar\partial^* N_{q+1}\bar\partial)^* = I - \bar\partial^* N_{q+1}\bar\partial^{**} = \tilde P$$

and

$$\begin{aligned}\tilde P^2 u &= \tilde P u - \bar\partial^* N_{q+1}\bar\partial\tilde P u\\ &= \tilde P u - \bar\partial^* N_{q+1}\bar\partial u + \bar\partial^* N_{q+1}\bar\partial\bar\partial^* N_{q+1}\bar\partial u\\ &= \tilde P u - \bar\partial^* N_{q+1}\bar\partial u + \bar\partial^* N_{q+1}(\Box - \bar\partial^*\bar\partial)N_{q+1}\bar\partial u\\ &= \tilde P u.\end{aligned}$$

This means that $\tilde P$ coincides with P_q on $\mathrm{dom}(\bar\partial)$. $\square$

Finally, we remark that $\tilde P$ can be extended to a unique bounded operator on $L^2_{(0,q)}(\Omega)$, which coincides with P_q: for $u \in \mathrm{dom}(\bar\partial)$, we have by (4.43) that $\bar\partial^* N_{q+1}\bar\partial u = \bar\partial^*\bar\partial N_q u$ and $u = \Box N_q u = \bar\partial\bar\partial^* N_q u + \bar\partial^*\bar\partial N_q u$ is an orthogonal decomposition, which follows from

$$(\bar\partial\bar\partial^* N_q u, \bar\partial^*\bar\partial N_q u) = (\bar\partial\bar\partial\bar\partial^* N_q u, \bar\partial N_q u) = 0.$$

Hence

$$\|\bar\partial^* N_{q+1}\bar\partial u\| = \|\bar\partial^*\bar\partial N_q u\| \le \|u\|, \quad u \in \mathrm{dom}(\bar\partial),$$

which proves the claim since $\mathrm{dom}(\bar\partial)$ is dense in $L^2_{(0,q)}(\Omega)$.

4.5 Notes

The basics on functional analysis, such as the open-mapping, the closed-graph theorem, and the uniform boundedness principle were taken from [29]. A thorough treatment of unbounded operators on Hilbert spaces can be found in [94]. The construction of self-adjoint operators by means of sesquilinear forms acting on different Hilbert spaces is mainly due to the assumption of the $\mathcal{V}$-ellipticity, sometimes also known as the Lax–Milgram situation; see [12, 51]. Basic facts about pseudoconvex domains are, for instance, in [57, 69, 60]. We point out that Proposition 4.59 and the following results on the $\bar{\partial}$-Neumann operator also hold for general bounded pseudoconvex domains without regularity conditions for the boundary of the domain; see [16, 91]. Propositions 4.63 and 4.65 are due to Kohn [64, 65]. The approach to the $\bar{\partial}$-Neumann operator N using the quadratic form Q can be found in [91]. Later on, the formula $N = j \circ j^*$ will be very important to handle the problem of compactness of N. This approach is also inspired by the general study of quadratic forms associated with nonnegative self-adjoint operators and their quadratic roots; see [24]. We will explain these ideas, which are related to the spectral theorem, in more detail in the chapters about compactness of N and on weighted L^2-spaces.

5 Density of smooth forms

To obtain the basic estimates

$$\|\overline{\partial}u\|^2 + \|\overline{\partial}^* u\|^2 \geq c\,\|u\|^2$$

on $\mathrm{dom}(\overline{\partial}) \cap \mathrm{dom}(\overline{\partial}^*)$, it is crucial to show that $C^k_{(0,q)}(\overline{\Omega}) \cap \mathrm{dom}(\overline{\partial}^*)$ is dense in $\mathrm{dom}(\overline{\partial}) \cap \mathrm{dom}(\overline{\partial}^*)$ in the graph norm $u \mapsto (\|u\|^2 + \|\overline{\partial}u\|^2 + \|\overline{\partial}^* u\|^2)^{1/2}$.

For this purpose we will need some general results on differential operators.

5.1 Friedrichs' lemma and Sobolev spaces

First, we approximate solutions of a first-order differential operator by regularization using convolutions. It turns out that the following generalization of Minkowski's inequality is useful.

Lemma 5.1. *Let $F : \mathbb{R}^n \times \mathbb{R}^n \to \mathbb{R}$ be a nonnegative measurable function. Then*

$$\left[\int_{\mathbb{R}^n}\left(\int_{\mathbb{R}^n} F(x,y)\,d\lambda(y)\right)^2 d\lambda(x)\right]^{1/2} \leq \int_{\mathbb{R}^n}\left(\int_{\mathbb{R}^n} F(x,y)^2\,d\lambda(x)\right)^{1/2} d\lambda(y), \tag{5.1}$$

where we suppose that the right side is finite.

Proof. We use the duality for L^2-spaces:

$$\|f\|_2 = \sup\left\{\left|\int_{\mathbb{R}^n} f(x)\,\overline{g(x)}\,d\lambda(x)\right| : \|g\|_2 = 1\right\}, \tag{5.2}$$

where $f \in L^2(\mathbb{R}^n)$; see Corollary 1.7. Let

$$f(x) = \int_{\mathbb{R}^n} F(x,y)\,d\lambda(y).$$

Then

$$\begin{aligned}
\|f\|_2 &= \left[\int_{\mathbb{R}^n}\left(\int_{\mathbb{R}^n} F(x,y)\,d\lambda(y)\right)^2 d\lambda(x)\right]^{1/2} \\
&= \sup\left\{\left|\int_{\mathbb{R}^n}\int_{\mathbb{R}^n} F(x,y)\overline{g(x)}\,d\lambda(y)\,d\lambda(x)\right| : \|g\|_2 = 1\right\} \\
&= \sup\left\{\left|\int_{\mathbb{R}^n}\int_{\mathbb{R}^n} F(x,y)\overline{g(x)}\,d\lambda(x)\,d\lambda(y)\right| : \|g\|_2 = 1\right\}
\end{aligned}$$

https://doi.org/10.1515/9783111182926-005

$$\leq \int_{\mathbb{R}^n} \Big(\int_{\mathbb{R}^n} F(x,y)^2 \, d\lambda(x) \Big)^{1/2} \, d\lambda(y),$$

where we used the Cauchy–Schwarz inequality in the last step. □

To begin with, for a function f on $\mathbb{R}^n$ and $x \in \mathbb{R}^n$, we define the function f_x as $f_x(y) = f(x+y)$.

Lemma 5.2. *If* $f \in L^2(\mathbb{R}^n)$, *then* $\lim_{x\to 0} \|f_x - f\|_2 = 0$.

Proof. If g is continuous with compact support, then g is uniformly continuous, so $g_x \to g$ uniformly as $x \to 0$. Since g_x and g are supported in a common compact set for $|x| \leq 1$, it follows that $\|g_x - g\|_2 \to 0$. Given $f \in L^2(\mathbb{R}^n)$ and $\epsilon > 0$, choose a continuous function g with compact support such that $\|f - g\|_2 < \epsilon/3$. Then also $\|f_x - g_x\|_2 < \epsilon/3$, so

$$\|f_x - f\|_2 \leq \|f_x - g_x\|_2 + \|g_x - g\|_2 + \|g - f\|_2 < \|g_x - g\|_2 + 2\epsilon/3.$$

For $|x|$ sufficiently small, $\|g_x - g\|_2 < \epsilon/3$, and hence $\|f_x - f\|_2 < \epsilon$. □

Let $\chi \in C_0^\infty(\mathbb{R}^n)$ be a function with support in the unit ball such that $\chi \geq 0$ and

$$\int_{\mathbb{R}^n} \chi(x) \, d\lambda(x) = 1.$$

We define $\chi_\epsilon(x) = \epsilon^{-n}\chi(x/\epsilon)$ for $\epsilon > 0$. Let $f \in L^2(\mathbb{R}^n)$, and for $x \in \mathbb{R}^n$, define

$$\begin{aligned} f_\epsilon(x) = (f * \chi_\epsilon)(x) &= \int_{\mathbb{R}^n} f(x')\chi_\epsilon(x - x') \, d\lambda(x') \\ &= \int_{\mathbb{R}^n} f(x - x')\chi_\epsilon(x') \, d\lambda(x') \\ &= \int_{\mathbb{R}^n} f(x - \epsilon x')\chi(x') \, d\lambda(x'). \end{aligned}$$

In the first integral, we can differentiate under the integral sign to show that $f_\epsilon \in C^\infty(\mathbb{R}^n)$.

The family of functions $(\chi_\epsilon)_\epsilon$ is called an approximation to the identity.

Lemma 5.3. $\|f_\epsilon - f\|_2 \to 0$ *as* $\epsilon \to 0$.

Proof.

$$f_\epsilon(x) - f(x) = \int_{\mathbb{R}^n} [f(x - \epsilon x') - f(x)]\chi(x') \, d\lambda(x').$$

We use Minkowski's inequality (5.1) to get

$$\|f_\epsilon - f\|_2 \le \int_{\mathbb{R}^n} \|f_{-\epsilon x'} - f\|_2 \, |\chi(x')| \, d\lambda(x').$$

However, $\|f_{-\epsilon x'} - f\|_2$ is bounded by $2\|f\|_2$ and tends to 0 as $\epsilon \to 0$ by Lemma 5.3. Now set

$$F_\epsilon(x') = \|f_{-\epsilon x'} - f\|_2 \chi(x').$$

Then $F_\epsilon(x') \to 0$ as $\epsilon \to 0$, and

$$|F_\epsilon(x')| \le 2\|f\|_2 \chi(x'),$$

and we can apply the dominated convergence theorem to get the desired result. □

If $u \in C_0^\infty(\mathbb{R}^n)$, then we have

$$D_j(u * \chi_\epsilon) = (D_j u) * \chi_\epsilon,$$

where $D_j = \partial/\partial x_j$. This is also true if $u \in L^2(\mathbb{R}^n)$ and $D_j u$ is defined in the sense of distributions. We will show even more using these methods for approximating a function in a Sobolev space by smooth functions.

Let $\Omega \subseteq \mathbb{R}^n$ be an open subset, and let

$$\Omega_\epsilon = \{x \in \Omega : \operatorname{dist}(x, b\Omega) > \epsilon\}.$$

Lemma 5.4. *Let $u \in W^k(\Omega)$ and set $u_\epsilon = u * \chi_\epsilon$ in Ω_ϵ. Then*
(i) *$u_\epsilon \in C^\infty(\Omega_\epsilon)$ for each $\epsilon > 0$,*
(ii) *$D^\alpha u_\epsilon = D^\alpha u * \chi_\epsilon$ in Ω_ϵ for $|\alpha| \le k$.*

Proof. (i) has already been shown.

(ii) means that the ordinary αth partial derivative of the smooth functions u_ϵ is the ϵ-mollification of the αth weak partial derivative of u. To see this, we take $x \in \Omega_\epsilon$ and compute

$$\begin{aligned} D^\alpha u_\epsilon(x) &= D^\alpha \int_\Omega u(y) \chi_\epsilon(x-y) \, d\lambda(y) \\ &= \int_\Omega D_x^\alpha \chi_\epsilon(x-y) u(y) \, d\lambda(y) \\ &= (-1)^{|\alpha|} \int_\Omega D_y^\alpha \chi_\epsilon(x-y) u(y) \, d\lambda(y). \end{aligned}$$

For a fixed $x \in \Omega_\epsilon$ the function $\phi(y) := \chi_\epsilon(x-y)$ belongs to $C^\infty(\Omega)$. The definition of the αth weak partial derivative implies

$$\int_\Omega D_y^\alpha \chi_\epsilon(x-y) u(y) \, d\lambda(y) = (-1)^{|\alpha|} \int_\Omega \chi_\epsilon(x-y) \, D^\alpha u(y) \, d\lambda(y).$$

Thus

$$\begin{aligned} D^\alpha u_\epsilon(x) &= (-1)^{|\alpha|+|\alpha|} \int_\Omega \chi_\epsilon(x-y)\, D^\alpha u(y)\, d\lambda(y) \\ &= (D^\alpha u * \chi_\epsilon)(x), \end{aligned}$$

which proves (ii). □

We are now ready to prove the following:

Lemma 5.5 (Friedrichs' lemma). *If $v \in L^2(\mathbb{R}^n)$ is a function with compact support and a is a C^1-function in a neighborhood of the support of v, then*

$$\|aD_j(v * \chi_\epsilon) - (aD_jv) * \chi_\epsilon\|_2 \to 0 \quad \text{as } \epsilon \to 0,$$

where $D_j = \partial/\partial x_j$ and aD_jv is defined in the sense of distributions.

Proof. If $v \in C_0^\infty(\mathbb{R}^n)$, then we have

$$D_j(v * \chi_\epsilon) = (D_jv) * \chi_\epsilon \to D_jv, \quad (aD_jv) * \chi_\epsilon \to aD_jv,$$

with uniform convergence. We claim that

$$\|aD_j(v * \chi_\epsilon) - (aD_jv) * \chi_\epsilon\|_2 \le C\|v\|_2, \tag{5.3}$$

where $v \in L^2(\mathbb{R}^n)$, and C is some positive constant independent of ϵ and v. Since $C_0^\infty(\mathbb{R}^n)$ is dense in $L^2(\mathbb{R}^n)$, the lemma will follow as in the proof of Lemma 5.3 from (5.3) and the dominated convergence theorem.

To show (5.3), we may assume that $a \in C_0^1(\mathbb{R}^n)$, since v has a compact support. For $v \in C_0^\infty(\mathbb{R}^n)$, we have

$$\begin{aligned} &a(x)D_j(v * \chi_\epsilon)(x) - ((aD_jv) * \chi_\epsilon)(x) \\ &\quad = a(x)\, D_j \int v(x-y)\chi_\epsilon(y)\, d\lambda(y) - \int a(x-y)\frac{\partial v}{\partial x_j}(x-y)\chi_\epsilon(y)\, d\lambda(y) \\ &\quad = \int (a(x) - a(x-y))\, \frac{\partial v}{\partial x_j}(x-y)\chi_\epsilon(y)\, d\lambda(y) \\ &\quad = -\int (a(x) - a(x-y))\, \frac{\partial v}{\partial y_j}(x-y)\chi_\epsilon(y)\, d\lambda(y) \\ &\quad = \int (a(x) - a(x-y))v(x-y)\, \frac{\partial}{\partial y_j}\chi_\epsilon(y)\, d\lambda(y) \\ &\quad\quad - \int \left(\frac{\partial}{\partial y_j} a(x-y)\right) v(x-y)\chi_\epsilon(y)\, d\lambda(y). \end{aligned}$$

Let M be the Lipschitz constant for a, that is, $|a(x) - a(x-y)| \le M|y|$ for all $x, y \in \mathbb{R}^n$. Then

$$|a(x)D_j(v*\chi_\epsilon)(x)-((aD_jv)*\chi_\epsilon)(x)|\le M\int|v(x-y)|\left(\chi_\epsilon(y)+\left|y\frac{\partial}{\partial y_j}\chi_\epsilon(y)\right|\right)d\lambda(y).$$

By Minkowski's inequality (5.1) we obtain

$$\begin{aligned}\|aD_j(v*\chi_\epsilon)-(aD_jv)*\chi_\epsilon\|_2&\le M\,\|v\|_2\int\left(\chi_\epsilon(y)+\left|y\frac{\partial}{\partial y_j}\chi_\epsilon(y)\right|\right)d\lambda(y)\\&=M(1+m_j)\|v\|_2,\end{aligned}$$

where

$$m_j=\int\left|y\frac{\partial}{\partial y_j}\chi_\epsilon(y)\right|d\lambda(y)=\int\left|y\frac{\partial}{\partial y_j}\chi(y)\right|d\lambda(y).$$

This shows (5.3) when $v\in C_0^\infty(\mathbb{R}^n)$. Since $C_0^\infty(\mathbb{R}^n)$ is dense in $L^2(\mathbb{R}^n)$, we have proved (5.3) and the lemma. □

Lemma 5.6. *Let*

$$L=\sum_{j=1}^n a_jD_j+a_0$$

be a first-order differential operator with variable coefficients $a_j\in C^1(\mathbb{R}^n)$ and $a_0\in C(\mathbb{R}^n)$. If $v\in L^2(\mathbb{R}^n)$ is a function with compact support and $Lv=f\in L^2(\mathbb{R}^n)$, where Lv is defined in the distribution sense, then the convolution $v_\epsilon=v*\chi_\epsilon$ is in $C_0^\infty(\mathbb{R}^n)$, $v_\epsilon\to v$, and $Lv_\epsilon\to f$ in $L^2(\mathbb{R}^n)$ as $\epsilon\to 0$.

Proof. Since $a_0v\in L^2(\mathbb{R}^n)$, we have

$$\lim_{\epsilon\to 0}a_0(v*\chi_\epsilon)=\lim_{\epsilon\to 0}(a_0v*\chi_\epsilon)=a_0v$$

in $L^2(\mathbb{R}^n)$. Using Friedrichs' Lemma 5.5, we have

$$Lv_\epsilon-Lv*\chi_\epsilon=Lv_\epsilon-f*\chi_\epsilon\to 0$$

in $L^2(\mathbb{R}^n)$ as $\epsilon\to 0$. The lemma follows easily since $f*\chi_\epsilon\to f$ in $L^2(\mathbb{R}^n)$. □

Before we proceed with results about Sobolev spaces, we prove an important inequality for the sgn-function.

Let $z\in\mathbb{C}$. Define

$$\operatorname{sgn} z=\begin{cases}\bar{z}/|z|, & z\ne 0,\\ 0, & z=0.\end{cases}$$

Proposition 5.7. *Suppose that $f\in L^1_{\mathrm{loc}}(\mathbb{R}^n)$ with $\nabla f\in L^1_{\mathrm{loc}}(\mathbb{R}^n)$. Then*

$$\nabla|f| \in L^1_{\mathrm{loc}}(\mathbb{R}^n),$$

and

$$\nabla|f|(x) = \Re[\operatorname{sgn}(f(x))\,\nabla f(x)] \tag{5.4}$$

almost everywhere. In particular, we have

$$\big|\nabla|f|\big| \le |\nabla f| \tag{5.5}$$

almost everywhere.

Proof. Let $z \in \mathbb{C}$ and $\epsilon > 0$. We define

$$|z|_\epsilon := \sqrt{|z|^2 + \epsilon^2} - \epsilon$$

and observe that

$$0 \le |z|_\epsilon \le |z| \quad \text{and} \quad \lim_{\epsilon\to 0} |z|_\epsilon = |z|.$$

If $u \in C^\infty(\mathbb{R}^n)$, then $|u|_\epsilon \in C^\infty(\mathbb{R}^n)$, and as $|u|^2 = u\,\overline{u}$, we get

$$\nabla|u|_\epsilon = \frac{\Re(\overline{u}\,\nabla u)}{\sqrt{|u|^2 + \epsilon^2}}. \tag{5.6}$$

Now let f be as assumed, take an approximation to the identity $(\chi_\delta)_\delta$, and define

$$f_\delta = f * \chi_\delta.$$

By Lemmas 5.2, 5.3, and 5.4 we obtain that $f_\delta \to f$, $|f_\delta| \to |f|$, and $\nabla f_\delta \to \nabla f$ in $L^1_{\mathrm{loc}}(\mathbb{R}^n)$ as $\delta \to 0$.

Let $\phi \in C_0^\infty(\mathbb{R}^n)$ be a test function. There exists a subsequence $\delta_k \to 0$ such that $f_{\delta_k}(x) \to f(x)$ for almost every $x \in \operatorname{supp}(\phi)$. For simplicity, we now omit the index k. Using the dominated convergence theorem and (5.6), we get

$$\begin{aligned}\int (\nabla\phi)\,|f|\,d\lambda &= \lim_{\epsilon\to 0}\int (\nabla\phi)\,|f|_\epsilon\,d\lambda\\ &= \lim_{\epsilon\to 0}\lim_{\delta\to 0}\int (\nabla\phi)\,|f_\delta|_\epsilon\,d\lambda\\ &= -\lim_{\epsilon\to 0}\lim_{\delta\to 0}\int \phi\,\frac{\Re(\overline{f}_\delta\nabla f_\delta)}{\sqrt{|f_\delta|^2 + \epsilon^2}}\,d\lambda.\end{aligned}$$

Since $\nabla f_\delta \to \nabla f$ in $L^1_{\mathrm{loc}}(\mathbb{R}^n)$, taking the limit as $\delta \to 0$, we get

$$\int (\nabla\phi)\,|f|\,d\lambda = -\lim_{\epsilon\to 0}\int \phi\,\frac{\Re(\bar{f}\nabla f)}{\sqrt{|f|^2+\epsilon^2}}\,d\lambda,$$

and since $\phi\nabla f \in L^1(\mathbb{R}^n)$ and $\bar{f}/\sqrt{|f|^2+\epsilon^2} \to \operatorname{sgn} f$ as $\epsilon \to 0$, we get the desired result by applying once more dominated convergence. □

In the sequel, we still use the methods from above for approximating a function in a Sobolev space by smooth functions. In a similar way, as in the last lemma, we get the following:

Lemma 5.8. *If $u \in W^k(\Omega)$, then $u_\epsilon \to u$ in $W^k_{\mathrm{loc}}(\Omega)$ as $\epsilon \to 0$, which means that this happens in each space $W^k(\omega)$, where ω is an open subset such that $\omega \Subset \Omega$.*

Using a smooth partition of unity, we still show that we can find smooth functions that approximate in the $W^k(\Omega)$-norm, and not just in $W^k_{\mathrm{loc}}(\Omega)$.

Lemma 5.9. *Let $\Omega \subset \mathbb{R}^n$ be a bounded open set, and let $u \in W^k(\Omega)$. Then there exist functions $u_m \in C^\infty(\Omega) \cap W^k(\Omega)$ such that $u_m \to u$ in $W^k(\Omega)$.*

Note that we do not assume that $u_m \in C^\infty(\overline{\Omega})$.

Proof. We write $\Omega = \bigcup_{j=1}^\infty \omega_j$, where

$$\omega_j := \{x \in \Omega : \operatorname{dist}(x, b\Omega) > 1/j\}, \quad j = 1, 2, \ldots.$$

Set $U_j := \omega_{j+3} \setminus \overline{\omega}_{j+1}$ and choose any open set $U_0 \Subset \Omega$ so that $\Omega = \bigcup_{j=0}^\infty U_j$. Let $(\phi_j)_j$ be a smooth partition of unity subordinate to the open sets $(U_j)_j$, that is, $0 \le \phi_j \le 1$, $\phi_j \in C_0^\infty(U_j)$, and $\sum_{j=0}^\infty \phi_j = 1$ on Ω.

According to Proposition 4.44, $\phi_j u \in W^k(\Omega)$, and the support of $\phi_j u$ is contained in U_j.

Now we use Lemma 5.8: fix $\epsilon > 0$ and choose $\epsilon_j > 0$ so small that $u_j := (\phi_j u) * \chi_{\epsilon_j}$ satisfies

$$\|u_j - \phi_j u\|_{W^k(\Omega)} \le \frac{\epsilon}{2^{j+1}}, \quad j = 0, 1, \ldots,$$

and u_j has support in $V_j := \omega_{j+4} \setminus \overline{\omega}_j \supset U_j$ for $j = 1, 2, \ldots$.

Now define $v := \sum_{j=0}^\infty u_j$. This function belongs to $C^\infty(\Omega)$, since for each open set $\omega \Subset \Omega$, there are at most finitely many nonzero terms in the sum. Since $u = \sum_{j=0}^\infty \phi_j u$, for each $\omega \Subset \Omega$, we have

$$\|v - u\|_{W^k(\omega)} \le \sum_{j=0}^\infty \|u_j - \phi_j u\|_{W^k(\Omega)} \le \epsilon \sum_{j=0}^\infty \frac{1}{2^{j+1}} = \epsilon.$$

Finally, take the supremum over all sets $\omega \Subset \Omega$ to conclude that

$$\|v - u\|_{W^k(\Omega)} \le \epsilon.$$

□

Before we proceed to prove the density result for the $\bar\partial$-setting, we show that a function $u \in W^k(\Omega)$ can be approximated by functions in $C^\infty(\overline{\Omega})$, where all derivatives extend continuously to $\overline{\Omega}$. This of course requires some conditions on the boundary $b\Omega$.

Proposition 5.10. *Let Ω be a bounded open set in $\mathbb{R}^n$ and assume that $b\Omega$ is C^1. Let $u \in W^k(\Omega)$. Then there exist functions $u_m \in C^\infty(\overline{\Omega})$ such that $u_m \to u$ in $W^k(\Omega)$.*

Proof. Let $x_0 \in b\Omega$. As $b\Omega$ is C^1, there exist a radius $r > 0$ and a C^1-function $\gamma : \mathbb{R}^{n-1} \longrightarrow \mathbb{R}$ such that

$$\Omega \cap B(x_0, r) = \{x \in B(x_0, r) : x_n > \gamma(x_1, \dots, x_{n-1})\}.$$

We set $V := \Omega \cap B(x_0, r/2)$ and define the shifted point $x_\epsilon := x + \mu\epsilon e_n$ for $x \in V$ and $\epsilon > 0$. We see that for some fixed, sufficiently large number $\mu > 0$, the ball $B(x_\epsilon, \epsilon)$ lies in $\Omega \cap B(x_0, r)$ for all $x \in V$ and all small $\epsilon > 0$.

Now we define $u_\epsilon(x) := u(x_\epsilon)$ for $x \in V$; this is the function u translated a distance $\mu\epsilon$ in the e_n-direction. Next, we write $v_\epsilon = u_\epsilon * \chi_\epsilon$. The idea is that we have moved up enough so that there is room to mollify within Ω. We have $v_\epsilon \in C^\infty(\overline{V})$.

We now claim that $v_\epsilon \to u$ in $W^k(V)$ as $\epsilon \to 0$. Let α be a multi-index with $|\alpha| \le k$. Then

$$\|D^\alpha v_\epsilon - D^\alpha u\|_{L^2(V)} \le \|D^\alpha v_\epsilon - D^\alpha u_\epsilon\|_{L^2(V)} + \|D^\alpha u_\epsilon - D^\alpha u\|_{L^2(V)}.$$

The second term on the right-hand side goes to zero with ϵ, since by Lemma 5.2 translation is continuous in the L^2-norm. The first term also vanishes as $\epsilon \to 0$ by a similar reasoning as in Lemma 5.6.

Let $\delta > 0$. Since $b\Omega$ is compact, we can find finitely many points $x_j \in b\Omega$, radii $r_j > 0$, corresponding sets $V_j = \Omega \cap B(x_j, r_j/2)$, and functions $v_j \in C^\infty(\overline{V}_j)$, $j = 1, \dots, N$, such that

$$b\Omega \subset \bigcup_{j=1}^{N} B(x_j, r_j/2) \quad \text{and} \quad \|v_j - u\|_{W^k(V_j)} \le \delta. \tag{5.7}$$

Now we take an open set $V_0 \Subset \Omega$ such that

$$\Omega \subset \bigcup_{j=0}^{N} V_j$$

and select, using Lemma 5.8, a function $v_0 \in C^\infty(\overline{V}_0)$ satisfying

$$\|v_0 - u\|_{W^k(V_0)} \le \delta. \tag{5.8}$$

Finally, we take a smooth partition $(\phi_j)_j$ of unity subordinate to the open sets $(V_j)_j$ in Ω for $j = 0, \dots, N$. Define $v := \sum_{j=0}^N \phi_j v_j$. Then $v \in C^\infty(\overline{\Omega})$. Since $u = \sum_{j=0}^N \phi_j u$, we see that for each $|\alpha| \le k$,

$$\begin{aligned}\|D^\alpha v - D^\alpha u\|_{L^2(\Omega)} &\le \sum_{j=0}^{N} \|D^\alpha(\phi_j v_j) - D^\alpha(\phi_j u)\|_{L^2(V_j)}\\ &\le \sum_{j=0}^{N} \|v_j - u\|_{W^k(V_j)}\\ &\le (N+1)\delta,\end{aligned}$$

where we used (5.7) and (5.8). □

In the next step, we show that functions in the Sobolev space $W^1(\Omega)$ can be extended to functions in $W^1(\mathbb{R}^n)$, provided that Ω is a bounded domain with C^1-boundary.

Proposition 5.11. *Assume that Ω is a bounded domain with C^1-boundary. Select a bounded open set V such that $\Omega \Subset V$. Then there exists a bounded linear operator*

$$E : W^1(\Omega) \longrightarrow W^1(\mathbb{R}^n)$$

such that for each $u \in W^1(\Omega)$:
(i) *$Eu = u$ almost everywhere in Ω,*
(ii) *Eu has a support within V,*
(iii) *there exists a constant C, depending only on Ω and V, such that*

$$\|Eu\|_{W^1(\mathbb{R}^n)} \le C\|u\|_{W^1(\Omega)}.$$

Proof. We use the method of a higher-order reflection for the extension. Let $x_0 \in b\Omega$ and suppose first that $b\Omega$ is flat near x_0, lying in the plane $\{x_n = 0\}$. Then we may assume that there exists an open ball B centered in x_0 with radius r such that

$$B^+ := B \cap \{x_n \ge 0\} \subset \overline{\Omega}, \quad B^- := B \cap \{x_n \le 0\} \subset \mathbb{R}^n \setminus \Omega.$$

Temporarily, we suppose that $u \in C^\infty(\overline{\Omega})$. We define then

$$\tilde{u}(x) = \begin{cases} u(x) & \text{if } x \in B^+,\\ -3u(x_1, \dots, x_{n-1}, -x_n) + 4u(x_1, \dots, x_{n-1}, -\frac{x_n}{2}) & \text{if } x \in B^-.\end{cases}$$

This is called a higher-order reflection of u from B^+ to B^-.

First, we show that $\tilde{u} \in C^1(B)$. To check this, we write

$$u^- := \tilde{u}|_{B^-} \quad \text{and} \quad u^+ := \tilde{u}|_{B^+}.$$

By definition we have

$$\frac{\partial u^-}{\partial x_n}(x) = 3\frac{\partial u}{\partial x_n}(x_1, \dots, x_{n-1}, -x_n) - 2\frac{\partial u}{\partial x_n}\left(x_1, \dots, x_{n-1}, -\frac{x_n}{2}\right),$$

and so

$$u^-_{x_n}|_{\{x_n=0\}} = u^+_{x_n}|_{\{x_n=0\}}.$$

Now since $u^+ = u^-$ on $\{x_n = 0\}$, we see that also

$$u^-_{x_j}|_{\{x_n=0\}} = u^+_{x_j}|_{\{x_n=0\}}$$

for $j = 1, \dots, n-1$. Hence we have

$$D^\alpha u^-|_{\{x_n=0\}} = D^\alpha u^+|_{\{x_n=0\}}$$

for each $|\alpha| \leq 1$, which implies $\tilde{u} \in C^1(B)$.

Using these computations, we readily see that

$$\|\tilde{u}\|_{W^1(B)} \leq C\|u\|_{W^1(B^+)} \tag{5.9}$$

for some constant $C > 0$ that does not depend on u.

If $b\Omega$ is not flat near x_0, then we can find a C^1-mapping Φ, with inverse Ψ, which straightens out $b\Omega$ near x_0. We write $y = \Phi(x)$ and $x = \Psi(y)$ and define $u^*(y) := u(\Psi(y))$. We choose a small ball B and use the same reasoning as before to extend u^* from B^+ to a function $\tilde{u}^*$ defined on all of B and such that $\tilde{u}^* \in C^1(B)$, and as in (5.9), we get

$$\|\tilde{u}^*\|_{W^1(B)} \leq C\|u^*\|_{W^1(B^+)}. \tag{5.10}$$

Let $W := \Psi(B)$. Then converting back to the x-variables, we obtain an extension $\tilde{u}$ of u to W such that

$$\|\tilde{u}\|_{W^1(W)} \leq C\|u\|_{W^1(\Omega)}. \tag{5.11}$$

Since $b\Omega$ is compact, there exist finitely many points $x_0^j \in b\Omega$, open sets W_j, and extensions $\tilde{u}_j$ of u to W_j for $j = 1, \dots, N$ such that $b\Omega \subset \bigcup_{j=1}^N W_j$. Take $W_0 \Subset \Omega$ such that $\Omega \subset \bigcup_{j=0}^N W_j$, and let $(\phi_j)_j$ be an associated partition of unity.

Write

$$\tilde{u} = \sum_{j=0}^N \phi_j \tilde{u}_j,$$

where $\tilde{u}_0 = u$. Then by (5.11) we obtain the estimate

$$\|\tilde{u}\|_{W^1(\mathbb{R}^n)} \leq C\|u\|_{W^1(\Omega)} \tag{5.12}$$

for some constant $C > 0$ independent of u. In addition, we arrange the support of $\tilde{u}$ to lie within $V \Supset \Omega$.

We define $Eu := \tilde{u}$ and observe that the mapping $u \mapsto Eu$ is linear. So far we have assumed that $u \in C^\infty(\overline{\Omega})$. Now take $u \in W^1(\Omega)$, and choose a sequence $u_m \in C^\infty(\overline{\Omega})$ converging to u in $W^1(\Omega)$ (see Proposition 5.10). Estimate (5.12) implies

$$\|Eu_m - Eu_\ell\|_{W^1(\mathbb{R}^n)} \leq C\|u_m - u_\ell\|_{W^1(\Omega)}.$$

Hence $(Eu_m)_m$ is a Cauchy sequence and so converges to $\tilde{u} =: Eu$. This extension does not depend on the particular choice of the approximating sequence $(u_m)_m$. □

In a similar way, we treat the problem how to assign boundary values along $b\Omega$ to a function $u \in W^1(\Omega)$ assuming that $b\Omega$ is C^1.

Proposition 5.12. *Let Ω be a bounded domain with C^1-boundary. Then there exists a bounded linear operator*

$$T : W^1(\Omega) \longrightarrow L^2(b\Omega)$$

such that

(i) $Tu = u|_{b\Omega}$ *if* $u \in W^1(\Omega) \cap C(\overline{\Omega})$

and

(ii) $\|Tu\|_{L^2(b\Omega)} \leq C\|u\|_{W^1(\Omega)}$

for each $u \in W^1(\Omega)$, with the constant C depending only on Ω.

We call Tu the trace of u on $b\Omega$.

Proof. First, we assume that $u \in C^1(\overline{\Omega})$ and proceed as in the proof of Proposition 5.11. We suppose that $x_0 \in b\Omega$ and that $b\Omega$ is flat near x_0, lying in the plane $\{x_n = 0\}$. We choose an open ball B as in the previous proof and let $\tilde{B}$ denote the concentric ball of radius $r/2$. Select a function $\chi \in C_0^\infty(B)$ such that $\chi \geq 0$ in B and $\chi = 1$ on $\tilde{B}$. Denote by Γ the portion of $b\Omega$ within $\tilde{B}$. Set $x' = (x_1, \dots, x_{n-1}) \in \mathbb{R}^{n-1} = \{x_n = 0\}$. Then we have

$$\begin{aligned}\int_\Gamma |u|^2 \, d\lambda(x') &\leq \int_{\{x_n=0\}} \chi|u|^2 \, d\lambda(x') = -\int_{B^+} (\chi|u|^2)_{x_n} \, d\lambda(x) \\ &= -\int_{B^+} (|u|^2\chi_{x_n} + |u|(\operatorname{sgn} u)\, u_{x_n}\chi) \, d\lambda(x) \\ &\leq C \int_{B^+} (|u|^2 + |\nabla u|^2) \, d\lambda(x),\end{aligned}$$

where we used Proposition 5.7 and the inequality $ab \leq a^2/2 + b^2/2$ for $a, b \geq 0$.

After this, we straighten out the boundary near x_0 to get the estimate

$$\int_\Gamma |u|^2 \, d\lambda(x') \leq C \int_\Omega (|u|^2 + |\nabla u|^2) \, d\lambda(x),$$

where Γ is some open subset of $b\Omega$ containing x_0.

Since $b\Omega$ is compact, there exist finitely many points $x_{0,k} \in b\Omega$ and open subsets $\Gamma_k \subset b\Omega$ ($k = 1, \dots, N$) such that

$$b\Omega = \bigcup_{k=1}^{N} \Gamma_k$$

and

$$\|u\|_{L^2(\Gamma_k)} \leq C\|u\|_{W^1(\Omega)}$$

for $k = 1, \dots, N$. Hence, writing

$$Tu := u|_{b\Omega},$$

we get

$$\|Tu\|_{L^2(b\Omega)} \leq C\|u\|_{W^1(\Omega)} \tag{5.13}$$

for some constant C, which does not depend on u.

Inequality (5.13) holds for $u \in C^1(\overline{\Omega})$. Assume now that $u \in W^1(\Omega)$. Then by Proposition 5.10 there exist functions $u_m \in C^\infty(\overline{\Omega})$ converging to u in $W^1(\Omega)$. By (5.13) we have

$$\|Tu_m - Tu_\ell\|_{L^2(b\Omega)} \leq C\|u_m - u_\ell\|_{W^1(\Omega)}, \tag{5.14}$$

and hence $(Tu_m)_m$ is a Cauchy sequence in $L^2(b\Omega)$. Set

$$Tu := \lim_{m\to\infty} Tu_m,$$

where the limit is taken in $L^2(b\Omega)$. By (5.14) this definition does not depend on the particular choice of the smooth functions approximating u.

If $u \in W^1(\Omega) \cap C(\overline{\Omega})$, then we can use the fact that the functions $u_m \in C^\infty(\overline{\Omega})$ constructed in the proof of Proposition 5.10 converge uniformly to u on $\overline{\Omega}$. This implies $Tu = u|_{b\Omega}$. □

Proposition 5.13. *Let Ω be a bounded domain with C^1-boundary. Then $u \in W_0^1(\Omega)$ if and only if $Tu = 0$ on $b\Omega$.*

Proof. If $u \in W_0^1(\Omega)$, then there exist functions $u_m \in C_0^\infty(\Omega)$ such that $u_m \to u$ in $W^1(\Omega)$. As $Tu_m = 0$ on $b\Omega$ and

$$T : W^1(\Omega) \longrightarrow L^2(b\Omega)$$

is a continuous linear operator (Proposition 5.12), we obtain $Tu = 0$ on $b\Omega$.

The converse statement is more difficult, but it will not be important in the following, so we omit the proof; see, for instance, [27]. □

5.2 Density in the graph norm

We are now able to prove the density result for the basic estimates.

Proposition 5.14. *If $b\Omega$ is $\mathcal{C}^{k+1}$, then $\mathcal{C}^k_{(0,q)}(\overline{\Omega}) \cap \mathrm{dom}(\overline{\partial}^*)$ is dense in $\mathrm{dom}(\overline{\partial}) \cap \mathrm{dom}(\overline{\partial}^*)$ in the graph norm $u \mapsto (\|u\|^2 + \|\overline{\partial}u\|^2 + \|\overline{\partial}^* u\|^2)^{1/2}$. The statement also holds with $k+1$ and k replaced by ∞.*

Before we begin with the proof of this important approximation result, we gather some observations about $\mathrm{dom}(\overline{\partial})$ and $\mathrm{dom}(\overline{\partial}^*)$.

Remark 5.15.

(a) It is useful to know that $\mathrm{dom}(\overline{\partial}^*)$ is preserved under multiplication by a function in $\mathcal{C}^1(\overline{\Omega})$. Indeed, let $u \in \mathrm{dom}(\overline{\partial}^*)$, $v \in \mathrm{dom}(\overline{\partial})$, and $\psi \in \mathcal{C}^1(\overline{\Omega})$. Then

$$(\overline{\partial}v, \psi u) = (\overline{\psi}\,\overline{\partial}v, u) = (\overline{\partial}(\overline{\psi}v), u) - (\overline{\partial}\,\overline{\psi} \wedge v, u) = (\overline{\psi}v, \overline{\partial}^* u) - (\overline{\partial}\,\overline{\psi} \wedge v, u).$$

The right-hand side is bounded by $\|v\|$, and hence $\psi u \in \mathrm{dom}(\overline{\partial}^*)$ (see [91]).

(b) Compactly supported forms are not dense in $\mathrm{dom}(\overline{\partial}) \cap \mathrm{dom}(\overline{\partial}^*)$ in the graph norm: for compactly supported forms, Proposition 4.56 gives

$$\|\overline{\partial}u\|^2 + \|\overline{\partial}^* u\|^2 = \sum_{j,k=1}^{n} \left\| \frac{\partial u_j}{\partial \overline{z}_k} \right\|^2,$$

and integration by parts also shows that in this case

$$\left\| \frac{\partial u_j}{\partial z_k} \right\|^2 = \left\| \frac{\partial u_j}{\partial \overline{z}_k} \right\|^2.$$

Hence

$$\|u\|_1^2 \le 2\,(\|\overline{\partial}u\|^2 + \|\overline{\partial}^* u\|^2),$$

where $\|u\|_1^2$ denotes the standard Sobolev-1 norm of u on Ω. Therefore the closure of the compactly supported forms in $\mathrm{dom}(\overline{\partial}) \cap \mathrm{dom}(\overline{\partial}^*)$ in the graph norm is contained in the Sobolev space $W^1_0(\Omega)$ for forms that are $\mathcal{C}^\infty$ on $\overline{\Omega}$, which means that they are zero on the boundary (Proposition 5.13), which is stronger than the condition

$$\sum_{j=1}^{n} \frac{\partial r}{\partial z_j} u_j = 0$$

on $b\Omega$ from Proposition 4.53.

(c) If Ω is a smoothly bounded pseudoconvex domain, then $\operatorname{dom}(\bar\partial) \cap \operatorname{dom}(\bar\partial^*)$ is a Hilbert space in the graph norm $u \mapsto (\|\bar\partial u\|^2 + \|\bar\partial^* u\|^2)^{1/2}$. This follows from (4.38).

We carry out the proof of Proposition 5.14 in several steps.

Lemma 5.16. *Let Ω be as in Proposition 5.14. Then $\mathcal{C}^\infty_{(0,q)}(\overline{\Omega})$ is dense in $\operatorname{dom}(\bar\partial) \cap \operatorname{dom}(\bar\partial^*)$ in the graph norm $u \mapsto (\|u\|^2 + \|\bar\partial u\|^2 + \|\bar\partial^* u\|^2)^{1/2}$.*

Proof. By this we mean that if $u \in \operatorname{dom}(\bar\partial) \cap \operatorname{dom}(\bar\partial^*)$, then we can construct a sequence $u_m \in \mathcal{C}^\infty_{(0,q)}(\overline{\Omega})$ such that $u_m \to u$, $\bar\partial u_m \to \bar\partial u$, and $\vartheta u_m \to \vartheta u$ in $L^2(\Omega)$; recall (4.28).

We use a method closely related to Friedrichs' Lemma 5.5 and use the notation from there.

Let $(\chi_\epsilon)_\epsilon$ be an approximation of the identity, let $(\delta_\nu)_\nu$ be a sequence of small positive numbers such that $\delta_\nu \to 0$, and let

$$\Omega_{\delta_\nu} = \{z \in \Omega : r(z) < -\delta_\nu\}.$$

Then $(\Omega_{\delta_\nu})_\nu$ is a sequence of relatively compact open subsets of Ω with union equal to Ω. The forms $u_\epsilon = u * \chi_\epsilon$ belong to $\mathcal{C}^\infty_{(0,q)}(\overline{\Omega}_{\delta_\nu})$, and $u_\epsilon \to u$, $\bar\partial u_\epsilon \to \bar\partial u$, and $\vartheta u_\epsilon \to \vartheta u$ in $L^2(\Omega_{\delta_\nu})$; see Lemmas 5.3 and 5.5.

To see that this can be done up to the boundary, we first assume that the domain Ω is star-shaped and $0 \in \Omega$ is a center. Let $\Omega^\epsilon = \{(1+\epsilon)z : z \in \Omega\}$ and

$$u^\epsilon(z) = u\left(\frac{z}{1+\epsilon}\right),$$

where the dilation is performed for each coefficient of u. Then $\Omega \Subset \Omega^\epsilon$ and $u^\epsilon \in L^2(\Omega^\epsilon)$. Also, by the dominated convergence theorem, $u^\epsilon \to u$, $\bar\partial u^\epsilon \to \bar\partial u$, and $\vartheta u^\epsilon \to \vartheta u$ in $L^2(\Omega)$. Now we regularize u^ϵ defining

$$u_{(\epsilon)} = u^\epsilon * \chi_{\delta_\epsilon}, \tag{5.15}$$

where $\delta_\epsilon \to 0$ as $\epsilon \to 0$, and δ_ϵ are chosen sufficiently small. Then $u_{(\epsilon)} \in \mathcal{C}^\infty_{(0,q)}(\overline{\Omega})$, and $u_{(\epsilon)} \to u$, $\bar\partial u_{(\epsilon)} \to \bar\partial u$, and $\vartheta u_{(\epsilon)} \to \vartheta u$ in $L^2(\Omega)$. Thus $\mathcal{C}^\infty_{(0,q)}(\overline{\Omega})$ is dense in the graph norm when Ω is star-shaped. The general case follows by using a partition of unity since we assume that our domain has at least $\mathcal{C}^2$ boundary. □

Lemma 5.17. *Let Ω be as in Proposition 5.14. Then compactly supported smooth forms are dense in $\operatorname{dom}(\bar\partial^*)$ in the graph norm $u \mapsto (\|u\|^2 + \|\bar\partial^* u\|^2)^{1/2}$.*

Proof. We remark that if $u \in \operatorname{dom}(\bar\partial^*)$ and if we extend u to $\tilde u$ on the whole space $\mathbb{C}^n$ by setting $\tilde u$ to be zero outside of Ω, then $\vartheta \tilde u \in L^2(\mathbb{C}^n)$ in the distribution sense; in fact, for $u \in \operatorname{dom}(\bar\partial^*)$, we have

$$\vartheta \tilde u = \widetilde{\vartheta u},$$

where $\widetilde{\vartheta u} = \vartheta u$ in Ω and $\widetilde{\vartheta u} = 0$ outside of Ω. This can be checked from the definition of $\bar{\partial}^*$, since for any $v \in C^\infty_{(0,q-1)}(\mathbb{C}^n)$,

$$(\tilde{u}, \bar{\partial} v)_{L^2(\mathbb{C}^n)} = (u, \bar{\partial} v)_{L^2(\Omega)} = (\vartheta u, v)_{L^2(\Omega)} = (\widetilde{\vartheta u}, v)_{L^2(\mathbb{C}^n)}.$$

We assume again without loss of generality that Ω is star-shaped with 0 as a center. We first approximate $\tilde{u}$ by

$$\tilde{u}^{-\epsilon}(z) = \tilde{u}\left(\frac{z}{1-\epsilon}\right).$$

Now we have forms $\tilde{u}^{-\epsilon}$ with compact support in Ω and $\vartheta \tilde{u}^{-\epsilon} \to \vartheta \tilde{u}$ in $L^2(\mathbb{C}^n)$. Regularizing $\tilde{u}^{-\epsilon}$ as before, we define

$$u_{(-\epsilon)} = \tilde{u}^{-\epsilon} * \chi_{\delta_\epsilon}. \tag{5.16}$$

Then the $u_{(-\epsilon)}$ are $(0,q)$-forms with coefficients in $C_0^\infty(\Omega)$ such that $u_{(-\epsilon)} \to u$ and $\vartheta u_{(-\epsilon)} \to \vartheta u$ in $L^2(\Omega)$. □

However, compactly supported smooth forms are not dense in $\mathrm{dom}(\bar{\partial})$ in the graph norm $u \mapsto (\|u\|^2 + \|\bar{\partial} u\|^2)^{1/2}$. Nevertheless, we have the following:

Lemma 5.18. *Let Ω be as in Proposition* 5.14. *Then $C^k_{(0,q)}(\overline{\Omega}) \cap \mathrm{dom}(\bar{\partial}^*)$ is dense in $\mathrm{dom}(\bar{\partial})$ in the graph norm $u \mapsto (\|u\|^2 + \|\bar{\partial} u\|^2)^{1/2}$.*

Proof. By Lemma 5.16 it suffices to show that for any $u \in C^\infty_{(0,q)}(\overline{\Omega})$, we can find a sequence $u_m \in C^k_{(0,q)}(\overline{\Omega}) \cap \mathrm{dom}(\bar{\partial}^*)$ such that $u_m \to u$ and $\bar{\partial} u_m \to \bar{\partial} u$ in $L^2(\Omega)$.

Let r be a C^{k+1} defining function such that $|dr| = 1$ on $b\Omega$. We now introduce some special vector fields and $(1,0)$-forms associated with $b\Omega$. Near a point $p \in b\Omega$, we choose fields $L_1, L_2, \dots, L_{n-1}$ of type $(1,0)$ that are orthonormal and span $T^{1,0}_p(b\Omega)$. This can be done by choosing a basis and then using the Gram–Schmidt process. To this collection add L_n, the complex normal, normalized to have length 1. So L_n is a smooth multiple of

$$\sum_{j=1}^n \frac{\partial r}{\partial \bar{z}_j} \frac{\partial}{\partial z_j}.$$

Now denote by $w_1, w_2, \dots, w_n$ the $(1,0)$-forms such that $w_j(L_k) = \delta_{jk}$. Note that L_n is defined globally, in contrast to $L_1, \dots, L_{n-1}$. The w^j then form an orthonormal basis for the $(1,0)$-forms near p. The $(1,0)$-form w_n is a smooth multiple of $\sum_{j=1}^n \frac{\partial r}{\partial z_j} dz_j$ and is again globally defined. Taking the wedge products of the w^j yields (local) orthonormal bases for the $(1,0)$-forms.

We will regularize near a boundary point $p \in b\Omega$. Let U be a small neighborhood of p. By a partition of unity we may assume that $\Omega \cap U$ is star-shaped and u is supported in $U \cap \overline{\Omega}$. Shrinking U if necessary, we can choose a special boundary chart

$(t_1, t_2, \ldots, t_{2n-1}, r)$, where $(t_1, t_2, \ldots, t_{2n-1}, 0)$ are coordinates on $b\Omega$ near p. Let $\overline{w}_1, \ldots, \overline{w}_n$ be an orthonormal basis for the $(0,1)$-forms on U such that $\overline{\partial} r = \overline{w}_n$.

Let $L_j = \sum_{s=1}^n a_{js} \frac{\partial}{\partial z_s}$, $w_j = \sum_{s=1}^n b_{js} dz_s$, $1 \le j \le n$. Then

$$\delta_{jk} = w_j(L_k) = \sum_{s=1}^n b_{js}\, dz_s \left(\sum_{\ell=1}^n a_{k\ell} \frac{\partial}{\partial z_\ell} \right) = \sum_{s=1}^n b_{js} a_{ks}.$$

Consequently, if f is a function, then

$$\overline{\partial} f = \sum_{s=1}^n \frac{\partial f}{\partial \overline{z}_s}\, d\overline{z}_s = \sum_{j,k,s=1}^n \overline{a^{sk}} (\overline{L}_k f)\, \overline{b^{sj}}\, \overline{w}_j = \sum_{j=1}^n (\overline{L}_j f)\, \overline{w}_j, \tag{5.17}$$

where the superscripts denote the entries of the inverses of the corresponding matrices with subscripts. Since multiplication by functions in $C^1(\overline{\Omega})$ preserves $\operatorname{dom}(\overline{\partial}^*)$, we may assume that the form u is supported in a special boundary chart. So $u = \sum'_{|J|=q} u_J \overline{w}_J$, where $\overline{w}_J = \overline{w}_{j_1} \wedge \cdots \wedge \overline{w}_{j_q}$, and each u_J is a function in $C^k(\overline{\Omega})$. Then, in view of (5.17),

$$\overline{\partial} u = \overline{\partial}\Big(\sum_{|J|=q}' u_J \overline{w}_J \Big) = \sum_{|J|=q}' (\overline{\partial} u_J \wedge \overline{w}_J + u_J \overline{\partial}\, \overline{w}_J) \tag{5.18}$$

$$= \sum_{|J|=q}' \sum_{j=1}^n (\overline{L}_j u_J)\, \overline{w}_j \wedge \overline{w}_J + \sum_{|J|=q}' u_J \overline{\partial}\, \overline{w}_J. \tag{5.19}$$

Using the special boundary chart, from (4.27) we get that

$$u \in \operatorname{dom}(\overline{\partial}^*) \iff u_J = 0 \quad \text{on } b\Omega \text{ when } n \in J. \tag{5.20}$$

Indeed, the only boundary terms that arise when proving (4.27) come from integrating

$$\int_\Omega \overline{L}_n a_K\, \overline{u}_{nK}\, d\lambda$$

by parts, and they equal

$$\int_{b\Omega} a_K\, \overline{u}_{nK} (\overline{L}_n r)\, d\sigma.$$

Since a_K can be arbitrary on $b\Omega$ and $\overline{L}_n r \neq 0$ on $b\Omega$, we conclude that $u_{nK} = 0$ on $b\Omega$ for all K. To see that the condition is sufficient, note that the computation to prove (4.30) shows that $(u, \overline{\partial}\alpha) = (\vartheta u, \alpha)$ when (4.27) holds and $\alpha \in C^\infty_{(0,q)}(\overline{\Omega})$. In view of Lemma 5.16, $C^\infty_{(0,q)}(\overline{\Omega})$ is dense in $\operatorname{dom}(\overline{\partial})$ in the graph norm $u \mapsto (\|u\|^2 + \|\overline{\partial} u\|^2)^{1/2}$. Hence $(u, \overline{\partial}\alpha) = (\vartheta u, \alpha)$ for all $\alpha \in \operatorname{dom}(\overline{\partial})$, which implies $u \in \operatorname{dom}(\overline{\partial}^*)$ and $\overline{\partial}^* u = \vartheta u$.

These arguments also give a formula for ϑ and $\overline{\partial}^*$ in special boundary frames:

$$\vartheta u = \vartheta\Big(\sum_{|J|=q}{}' u_J \overline{w}_J\Big) = -\sum_{|K|=q-1}{}' \Big(\sum_{j=1}^{n} L_j u_{jK}\Big)\overline{w}_K + 0\text{th order}(u); \tag{5.21}$$

0th order(u) indicates the terms that contain no derivatives of the u_J.

We note that both $\overline{\partial}$ and ϑ are first-order differential operators with variable coefficients in $\mathcal{C}^k(\overline{\Omega})$ when computed in the special frame $\overline{w}_1, \dots, \overline{w}_n$. We write

$$u = u^\tau + u^\nu,$$

where

$$u^\tau = \sum_{|J|=q, n\notin J}{}' u_J \overline{w}_J, \quad u^\nu = \sum_{|J|=q, n\in J}{}' u_J \overline{w}_J;$$

u^τ is the complex tangential part of u, and u^ν is the complex normal part of u.

Our arguments from above imply that

$$u \in \mathcal{C}^k_{(0,q)}(\overline{\Omega}) \cap \operatorname{dom}(\overline{\partial}^*) \iff u^\nu = 0 \quad \text{on } b\Omega.$$

For $u \in \mathcal{C}^\infty_{(0,q)}(\overline{\Omega})$ and $\alpha \in \mathcal{C}^\infty_{(0,q+1)}(\overline{\Omega})$, by (4.31) we have

$$(\overline{\partial} u, \alpha) = (u, \vartheta\alpha) + \int_{b\Omega} \langle \overline{\partial} r \wedge u, \alpha\rangle \, d\sigma$$

and $\overline{\partial} r \wedge u = \overline{\partial} r \wedge u^\tau$ on $b\Omega$, which follows from the representation in special boundary charts:

$$\overline{\partial} r \wedge u^\nu = c\overline{w}_n \wedge \sum_{|J|=q, n\in J}{}' u_J \overline{w}_J = 0. \tag{5.22}$$

To approximate a form $u \in \mathcal{C}^\infty_{(0,q)}(\overline{\Omega})$ by forms in $\mathcal{C}^k_{(0,q)}(\overline{\Omega}) \cap \operatorname{dom}(\overline{\partial}^*)$, we only change the complex normal part u^ν and leave the complex tangential part u^τ unchanged: for $u \in \mathcal{C}^\infty_{(0,q)}(\overline{\Omega})$, it follows that $u^\tau \in \mathcal{C}^k_{(0,q)}(\overline{\Omega}) \cap \operatorname{dom}(\overline{\partial}^*)$, and we denote by $\tilde{u}^\nu$ the extension of u^ν to $\mathbb{C}^n$ by setting $\tilde{u}^\nu$ equal to zero outside of Ω. We approximate $\tilde{u}^\nu$ as in Lemma 5.17 by

$$u^\nu_{(-\epsilon)} = (\tilde{u}^\nu)^{-\epsilon} * \chi_{\delta_\epsilon}.$$

Then $u^\nu_{(-\epsilon)}$ is smooth and supported in a compact subset of $\Omega \cap U$. By this we approximate u^ν by $u^\nu_{(-\epsilon)} \in \mathcal{C}^\infty_0(\Omega \cap U)$ in the L^2-norm. Furthermore, by extending $\overline{\partial} u^\nu$ to be zero outside $\Omega \cap U$ and denoting the extension by $\widetilde{\overline{\partial} u^\nu}$ we have

$$\overline{\partial}\tilde{u}^\nu = \widetilde{\overline{\partial}u^\nu}$$

in $L^2(\mathbb{C}^n)$ in the sense of distributions. This follows from (4.31) and (5.22), since $u^\nu \in C^k_{(0,q)}(\overline{\Omega})$, and for $\alpha \in C^\infty_{(0,q+1)}(\mathbb{C}^n)$, we have

$$(\tilde{u}^\nu, \vartheta\alpha)_{L^2(\mathbb{C}^n)} = (\overline{\partial}u^\nu, \alpha)_{L^2(\Omega)} - \int_{b\Omega} \langle \overline{\partial}r \wedge u^\nu, \alpha \rangle \, d\sigma = (\widetilde{\overline{\partial}u^\nu}, \alpha)_{L^2(\mathbb{C}^n)}.$$

Since $\overline{\partial}$ is a first-order differential operator with variable coefficients, by Friedrichs' Lemma 5.6 we get

$$\overline{\partial}u^\nu_{(-\epsilon)} \to \overline{\partial}\tilde{u}^\nu \quad \text{in } L^2(\mathbb{C}^n). \tag{5.23}$$

We set

$$u_{(-\epsilon)} = u^\tau + u^\nu_{(-\epsilon)}.$$

It follows that $u_{(-\epsilon)} \in C^k_{(0,q)}(\overline{\Omega}) \cap \operatorname{dom}(\overline{\partial}^*)$, since each coefficient of u^τ, $u^\nu_{(-\epsilon)}$, and w_j is in $C^k(\overline{\Omega \cap U})$. Therefore we get $u_{(-\epsilon)} \in C^k_{(0,q)}(\overline{\Omega}) \cap \operatorname{dom}(\overline{\partial}^*)$ and $u_{(-\epsilon)} \to u$ in $L^2(\Omega)$.

To see that $\overline{\partial}u_{(-\epsilon)} \to \overline{\partial}u$ in the $L^2(\Omega)$-norm, using (5.23), we find that

$$\overline{\partial}u_{(-\epsilon)} = \overline{\partial}u^\tau + \overline{\partial}u^\nu_{(-\epsilon)} \to \overline{\partial}u$$

in $L^2(\Omega)$ as $\epsilon \to 0$. □

To finish the proof of Proposition 5.14, we consider an arbitrary $u \in \operatorname{dom}(\overline{\partial}) \cap \operatorname{dom}(\overline{\partial}^*)$ and use a partition of unity and the same notation as before to regularize u in each small star-shaped neighborhood near the boundary. We regularize the complex tangential and normal part separately by setting

$$u_{((\epsilon))} = u^\tau_{(\epsilon)} + u^\nu_{(-\epsilon)},$$

which means that we first consider $u_{(\epsilon)}$ as it was defined in (5.15), then take the tangential components $u^\tau_{(\epsilon)}$, then we consider $u_{(-\epsilon)}$ as it is defined in (5.16), and finally take the normal components $u^\nu_{(-\epsilon)}$. It follows that for sufficiently small $\epsilon > 0$, $u^\nu_{(-\epsilon)}$ has coefficients in $C^\infty_0(\Omega)$ and $u^\tau_{(\epsilon)}$ has coefficients in $C^\infty(\overline{\Omega})$.

Thus we see that $u_{((\epsilon))} \in C^k_{(0,q)}(\overline{\Omega}) \cap \operatorname{dom}(\overline{\partial}^*)$. We get from Lemma 5.16 that $u_{(\epsilon)} \to u$ in the graph norm $u \mapsto (\|u\|^2 + \|\overline{\partial}u\|^2 + \|\overline{\partial}^* u\|^2)^{1/2}$, and hence $u^\tau_{(\epsilon)} \to u^\tau$ in the graph norm $u \mapsto (\|u\|^2 + \|\overline{\partial}u\|^2 + \|\overline{\partial}^* u\|^2)^{1/2}$. By Lemma 5.17 we obtain $u_{(-\epsilon)} \to u$ in the graph norm $u \mapsto (\|u\|^2 + \|\overline{\partial}^* u\|^2)^{1/2}$, and hence $u^\nu_{(-\epsilon)} \to u^\nu$ in the graph norm $u \mapsto (\|u\|^2 + \|\overline{\partial}^* u\|^2)^{1/2}$. Finally, we use Lemma 5.18, in particular, formula (5.23), and see that $\overline{\partial}u^\nu_{(-\epsilon)} \to \overline{\partial}\tilde{u}^\nu$ in $L^2(\mathbb{C}^n)$, and hence $u_{((\epsilon))} \to u$ in the graph norm $u \mapsto (\|u\|^2 + \|\overline{\partial}u\|^2 + \|\overline{\partial}^* u\|^2)^{1/2}$.

This shows that $C^k_{(0,q)}(\overline{\Omega}) \cap \operatorname{dom}(\overline{\partial}^*)$ is dense in $\operatorname{dom}(\overline{\partial}) \cap \operatorname{dom}(\overline{\partial}^*)$ in the graph norm $u \mapsto (\|u\|^2 + \|\overline{\partial} u\|^2 + \|\overline{\partial}^* u\|^2)^{1/2}$.

5.3 Notes

Friedrichs' lemma can be found in [16]. It will also be important in the following chapters. In the proof of the density result, we mainly follow the presentation in [16], Straube [91] uses a different approach for the same result, whereas Hörmander [57] proves a corresponding density result in weighted L^2-spaces.

6 The weighted $\overline{\partial}$-complex

In this chapter, we investigate the basic properties of the $\overline{\partial}$-operator on weighted L^2-spaces on the whole of $\mathbb{C}^n$. It turns out that the behavior of the $\overline{\partial}$-operator depends very much on properties of the Levi matrix of the weight function. For $n = 1$, there is an interesting connection between $\overline{\partial}$ and the theory of Schrödinger operators with magnetic fields, and for $n \geq 2$, the corresponding notion is the Witten Laplacian. The corresponding density result from the last chapter is much easier to handle in the weighted case, therefore also the basic estimate and existence of the $\overline{\partial}$-Neumann operator.

6.1 The $\overline{\partial}$-Neumann operator on $(0,1)$-forms

Let $\varphi : \mathbb{C}^n \longrightarrow \mathbb{R}^+$ be a plurisubharmonic $\mathcal{C}^2$-weight function and define the space

$$L^2(\mathbb{C}^n, e^{-\varphi}) = \left\{ f : \mathbb{C}^n \longrightarrow \mathbb{C} : \int_{\mathbb{C}^n} |f|^2 e^{-\varphi}\, d\lambda < \infty \right\},$$

where λ denotes the Lebesgue measure, the space $L^2_{(0,1)}(\mathbb{C}^n, e^{-\varphi})$ of $(0,1)$-forms with coefficients in $L^2(\mathbb{C}^n, e^{-\varphi})$, and the space $L^2_{(0,2)}(\mathbb{C}^n, e^{-\varphi})$ of $(0,2)$-forms with coefficients in $L^2(\mathbb{C}^n, e^{-\varphi})$. Let

$$(f, g)_\varphi = \int_{\mathbb{C}^n} f \overline{g} e^{-\varphi}\, d\lambda$$

denote the inner product and

$$\|f\|_\varphi^2 = \int_{\mathbb{C}^n} |f|^2 e^{-\varphi}\, d\lambda$$

the norm in $L^2(\mathbb{C}^n, e^{-\varphi})$.

We define $\operatorname{dom}(\overline{\partial})$ to be the space of all functions $f \in L^2(\mathbb{C}^n, e^{-\varphi})$ such that $\overline{\partial} f$, in the sense of distributions, belongs to $L^2_{(0,1)}(\mathbb{C}^n, e^{-\varphi})$, and consider the weighted $\overline{\partial}$-complex

$$L^2(\mathbb{C}^n, e^{-\varphi}) \underset{\overline{\partial}_\varphi^*}{\overset{\overline{\partial}}{\rightleftarrows}} L^2_{(0,1)}(\mathbb{C}^n, e^{-\varphi}) \underset{\overline{\partial}_\varphi^*}{\overset{\overline{\partial}}{\rightleftarrows}} L^2_{(0,2)}(\mathbb{C}^n, e^{-\varphi}), \tag{6.1}$$

where $\overline{\partial}_\varphi^*$ is the adjoint operator to $\overline{\partial}$ with respect to the weighted inner product. For a smooth $(0,1)$-form $u = \sum_{j=1}^n u_j d\overline{z}_j \in \operatorname{dom}(\overline{\partial}_\varphi^*)$, we have

$$\overline{\partial}_\varphi^* u = -\sum_{j=1}^n \left(\frac{\partial}{\partial z_j} - \frac{\partial \varphi}{\partial z_j} \right) u_j. \tag{6.2}$$

https://doi.org/10.1515/9783111182926-006

The complex Laplacian on (0, 1)-forms is defined as

$$\Box_\varphi := \bar\partial\bar\partial_\varphi^* + \bar\partial_\varphi^*\bar\partial,$$

and $\operatorname{dom}(\Box_\varphi)$ is the space of all $f \in L^2_{(0,1)}(\mathbb{C}^n, e^{-\varphi})$ such that

$$f \in \operatorname{dom}(\bar\partial) \cap \operatorname{dom}(\bar\partial_\varphi^*),$$

$\bar\partial f \in \operatorname{dom}(\bar\partial_\varphi^*)$, and $\bar\partial_\varphi^* f \in \operatorname{dom}(\bar\partial)$.

The operator $\Box_\varphi$ is self-adjoint and positive (see Chapter 4), which means that

$$(\Box_\varphi f, f)_\varphi \geq 0 \quad \text{for } f \in \operatorname{dom}(\Box_\varphi).$$

The associated Dirichlet form is denoted by

$$Q_\varphi(f, g) = (\bar\partial f, \bar\partial g)_\varphi + (\bar\partial_\varphi^* f, \bar\partial_\varphi^* g)_\varphi \tag{6.3}$$

for $f, g \in \operatorname{dom}(\bar\partial) \cap \operatorname{dom}(\bar\partial_\varphi^*)$. The weighted $\bar\partial$-Neumann operator N_φ, if it exists, is the bounded inverse of $\Box_\varphi$.

In the weighted space $L^2_{(0,1)}(\mathbb{C}^n, e^{-\varphi})$, we can give a simple characterization of $\operatorname{dom}(\bar\partial_\varphi^*)$.

Proposition 6.1. *Let $f = \sum f_j d\bar z_j \in L^2_{(0,1)}(\mathbb{C}^n, e^{-\varphi})$. Then $f \in \operatorname{dom}(\bar\partial_\varphi^*)$ if and only if*

$$e^\varphi \sum_{j=1}^n \frac{\partial}{\partial z_j}(f_j e^{-\varphi}) \in L^2(\mathbb{C}^n, e^{-\varphi}), \tag{6.4}$$

where the derivative is to be taken in the sense of distributions.

Proof. Suppose first that $e^\varphi \sum_{j=1}^n \frac{\partial}{\partial z_j}(f_j e^{-\varphi}) \in L^2(\mathbb{C}^n, e^{-\varphi})$. We have to show that there exists a constant C such that $|(\bar\partial g, f)_\varphi| \leq C\|g\|_\varphi$ for all $g \in \operatorname{dom}(\bar\partial)$. To this end, let $(\chi_R)_{R\in\mathbb{N}}$ be a family of radially symmetric smooth cut-off functions, which are identically one on $\mathbb{B}_R$, the ball with radius R, such that the support of χ_R is contained in $\mathbb{B}_{R+1}$, $\operatorname{supp}(\chi_R) \subset \mathbb{B}_{R+1}$, and such that all first-order derivatives of all functions in this family are uniformly bounded by a constant M. Then for all $g \in C_0^\infty(\mathbb{C}^n)$,

$$(\bar\partial g, \chi_R f)_\varphi = \sum_{j=1}^n \left(\frac{\partial g}{\partial \bar z_j}, \chi_R f_j\right)_\varphi = -\int_{\mathbb{C}^n} \sum_{j=1}^n g \frac{\partial}{\partial \bar z_j}(\chi_R \bar f_j e^{-\varphi})\, d\lambda$$

by integration by parts, which in particular means that

$$|(\bar\partial g, f)_\varphi| = \lim_{R\to\infty} |(\bar\partial g, \chi_R f)_\varphi| = \lim_{R\to\infty} \left| \int_{\mathbb{C}^n} \sum_{j=1}^n g \frac{\partial}{\partial \bar z_j}(\chi_R \bar f_j e^{-\varphi})\, d\lambda \right|.$$

Now we use the triangle inequality and then the Cauchy–Schwarz inequality to get

$$\begin{aligned}\left|\int_{\mathbb{C}^n}\sum_{j=1}^n g\frac{\partial}{\partial\overline{z}_j}(\chi_R\overline{f}_je^{-\varphi})\,d\lambda\right| &\leq \left|\int_{\mathbb{C}^n}\chi_R g\sum_{j=1}^n\frac{\partial}{\partial\overline{z}_j}(\overline{f}_je^{-\varphi})\,d\lambda\right| + \left|\int_{\mathbb{C}^n}\sum_{j=1}^n\overline{f}_jg\frac{\partial\chi_R}{\partial\overline{z}_j}e^{-\varphi}\,d\lambda\right|\\ &\leq \|\chi_R g\|_\varphi\left\|e^\varphi\sum_{j=1}^n\frac{\partial}{\partial z_j}(f_je^{-\varphi})\right\|_\varphi + M\|g\|_\varphi\|f\|_\varphi\\ &\leq \|g\|_\varphi\left\|e^\varphi\sum_{j=1}^n\frac{\partial}{\partial z_j}(f_je^{-\varphi})\right\|_\varphi + M\|g\|_\varphi\|f\|_\varphi.\end{aligned}$$

Hence by assumption,

$$|(\overline{\partial}g,f)_\varphi| \leq \|g\|_\varphi\left\|e^\varphi\sum_{j=1}^n\frac{\partial}{\partial z_j}(f_je^{-\varphi})\right\|_\varphi + M\|g\|_\varphi\|f\|_\varphi \leq C\|g\|_\varphi$$

for all $g \in C_0^\infty(\mathbb{C}^n)$, and by the denseness of $C_0^\infty(\mathbb{C}^n)$ this is true for all $g \in \operatorname{dom}(\overline{\partial})$.

Conversely, let $f \in \operatorname{dom}(\overline{\partial}_\varphi^*)$, which means that there exists a uniquely determined element $\overline{\partial}_\varphi^*f \in L^2(\mathbb{C}^n, e^{-\varphi})$ such that for each $g \in \operatorname{dom}(\overline{\partial})$, we have

$$(\overline{\partial}g,f)_\varphi = (g,\overline{\partial}_\varphi^*f)_\varphi.$$

Now take $g \in C_0^\infty(\mathbb{C}^n)$. Then $g \in \operatorname{dom}(\overline{\partial})$, and

$$\begin{aligned}(g,\overline{\partial}_\varphi^*f)_\varphi &= (\overline{\partial}g,f)_\varphi\\ &= \sum_{j=1}^n\left(\frac{\partial g}{\partial\overline{z}_j},f_j\right)_\varphi\\ &= -\left(g,\sum_{j=1}^n\frac{\partial}{\partial z_j}(f_je^{-\varphi})\right)_{L^2}\\ &= -\left(g,e^\varphi\sum_{j=1}^n\frac{\partial}{\partial z_j}(f_je^{-\varphi})\right)_\varphi.\end{aligned}$$

Since $C_0^\infty(\mathbb{C}^n)$ is dense in $L^2(\mathbb{C}^n, e^{-\varphi})$, we conclude that

$$\overline{\partial}_\varphi^*f = -e^\varphi\sum_{j=1}^n\frac{\partial}{\partial z_j}(f_je^{-\varphi}),$$

which in particular implies that $e^\varphi\sum_{j=1}^n\frac{\partial}{\partial z_j}(f_je^{-\varphi}) \in L^2(\mathbb{C}^n, e^{-\varphi})$. □

The following lemma will be important for our considerations.

Lemma 6.2. *The forms with coefficients in* $C_0^\infty(\mathbb{C}^n)$ *are dense in* $\mathrm{dom}(\bar{\partial}) \cap \mathrm{dom}(\bar{\partial}_\varphi^*)$ *in the graph norm* $f \mapsto (\|f\|_\varphi^2 + \|\bar{\partial} f\|_\varphi^2 + \|\bar{\partial}_\varphi^* f\|_\varphi^2)^{\frac{1}{2}}$.

Proof. First, we show that compactly supported L^2-forms are dense in the graph norm. So let $\{\chi_R\}_{R\in\mathbb{N}}$ be a family of smooth radially symmetric cut-offs identically one on $\mathbb{B}_R$ and supported in $\mathbb{B}_{R+1}$, such that all first-order derivatives of the functions in this family are uniformly bounded in R by a constant M.

Let $f \in \mathrm{dom}(\bar{\partial}) \cap \mathrm{dom}(\bar{\partial}_\varphi^*)$. Then, clearly, $\chi_R f \in \mathrm{dom}(\bar{\partial}) \cap \mathrm{dom}(\bar{\partial}_\varphi^*)$ and $\chi_R f \to f$ in $L^2_{(0,1)}(\mathbb{C}^n, e^{-\varphi})$ as $R \to \infty$. As observed in Proposition 6.1, we have

$$\bar{\partial}_\varphi^* f = -e^\varphi \sum_{j=1}^n \frac{\partial}{\partial z_j}(f_j e^{-\varphi}),$$

and hence

$$\bar{\partial}_\varphi^*(\chi_R f) = -e^\varphi \sum_{j=1}^n \frac{\partial}{\partial z_j}(\chi_R f_j e^{-\varphi}).$$

We need to estimate the difference of these expressions,

$$\bar{\partial}_\varphi^* f - \bar{\partial}_\varphi^*(\chi_R f) = \bar{\partial}_\varphi^* f - \chi_R \bar{\partial}_\varphi^* f + \sum_{j=1}^n \frac{\partial \chi_R}{\partial z_j} f_j.$$

By the triangle inequality,

$$\|\bar{\partial}_\varphi^* f - \bar{\partial}_\varphi^*(\chi_R f)\|_\varphi \leq \|\bar{\partial}_\varphi^* f - \chi_R \bar{\partial}_\varphi^* f\|_\varphi + M \sum_{j=1}^n \int_{\mathbb{C}^n \setminus \mathbb{B}_R} |f_j|^2 e^{-\varphi}\, d\lambda.$$

Now both terms tend to 0 as $R \to \infty$, and we can similarly see that also $\bar{\partial}(\chi_R f) \to \bar{\partial} f$ as $R \to \infty$.

So we have the denseness of compactly supported forms in the graph norm, and the denseness of forms with coefficients in $C_0^\infty(\mathbb{C}^n)$ will follow by applying Friedrichs' Lemma 5.6. □

As in the case of bounded domains, the canonical solution operator to $\bar{\partial}$, which we denote by S_φ, is given by $\bar{\partial}_\varphi^* N_\varphi$. The existence and compactness of N_φ and S_φ are closely related. For the rest of this chapter, we will suppose that $n \geq 2$. The case $n = 1$ will be discussed separately in Chapter 10.

Now we suppose that the lowest eigenvalue μ_φ of the Levi matrix

$$M_\varphi = \left(\frac{\partial^2 \varphi}{\partial z_j \partial \bar{z}_k}\right)_{jk}$$

of φ satisfies

$$\mu_\varphi(z) > \epsilon \tag{6.5}$$

for all $z \in \mathbb{C}^n$, where $\epsilon > 0$. Now we can prove the basic estimate.

Proposition 6.3. *For a plurisubharmonic weight function φ satisfying* (6.5), *there is* $C > 0$ *such that*

$$\|u\|_\varphi^2 \leq C(\|\bar\partial u\|_\varphi^2 + \|\bar\partial_\varphi^* u\|_\varphi^2) \tag{6.6}$$

for each $(0,1)$*-form* $u \in \operatorname{dom}(\bar\partial) \cap \operatorname{dom}(\bar\partial_\varphi^*)$.

Proof. By Lemma 6.2 and the assumption on φ it suffices to show that

$$\int_{\mathbb{C}^n} \sum_{j,k=1}^n \frac{\partial^2 \varphi}{\partial z_j \partial \bar z_k} u_j \bar u_k \, e^{-\varphi} \, d\lambda \leq \|\bar\partial u\|_\varphi^2 + \|\bar\partial_\varphi^* u\|_\varphi^2 \tag{6.7}$$

for each $(0,1)$-form $u = \sum_{k=1}^n u_k \, d\bar z_k$ with coefficients $u_k \in \mathcal{C}_0^\infty(\mathbb{C}^n)$, $k = 1, \dots, n$.
For this purpose, we set $\delta_k = \frac{\partial}{\partial z_k} - \frac{\partial\varphi}{\partial z_k}$, and since

$$\bar\partial u = \sum_{j<k} \left(\frac{\partial u_j}{\partial \bar z_k} - \frac{\partial u_k}{\partial \bar z_j} \right) d\bar z_j \wedge d\bar z_k,$$

we get that

$$\begin{aligned}
\|\bar\partial u\|_\varphi^2 + \|\bar\partial_\varphi^* u\|_\varphi^2 &= \int_{\mathbb{C}^n} \sum_{j<k} \left| \frac{\partial u_j}{\partial \bar z_k} - \frac{\partial u_k}{\partial \bar z_j} \right|^2 e^{-\varphi} \, d\lambda + \int_{\mathbb{C}^n} \sum_{j,k=1}^n \delta_j u_j \, \overline{\delta_k u_k} \, e^{-\varphi} \, d\lambda \\
&= \sum_{j,k=1}^n \int_{\mathbb{C}^n} \left| \frac{\partial u_j}{\partial \bar z_k} \right|^2 e^{-\varphi} \, d\lambda + \sum_{j,k=1}^n \int_{\mathbb{C}^n} \left(\delta_j u_j \, \overline{\delta_k u_k} - \frac{\partial u_j}{\partial \bar z_k} \overline{\frac{\partial u_k}{\partial \bar z_j}} \right) e^{-\varphi} \, d\lambda \\
&= \sum_{j,k=1}^n \int_{\mathbb{C}^n} \left| \frac{\partial u_j}{\partial \bar z_k} \right|^2 e^{-\varphi} \, d\lambda + \sum_{j,k=1}^n \int_{\mathbb{C}^n} \left[\delta_j, \frac{\partial}{\partial \bar z_k} \right] u_j \, \bar u_k \, e^{-\varphi} \, d\lambda,
\end{aligned}$$

where we used the fact that for $f, g \in \mathcal{C}_0^\infty(\mathbb{C}^n)$,

$$\left(\frac{\partial f}{\partial \bar z_k}, g \right)_\varphi = -(f, \delta_k g)_\varphi,$$

and hence

$$\left(\left[\delta_j, \frac{\partial}{\partial \bar z_k} \right] u_j, u_k \right)_\varphi = -\left(\frac{\partial u_j}{\partial \bar z_k}, \frac{\partial u_k}{\partial \bar z_j} \right)_\varphi + (\delta_j u_j, \delta_k u_k)_\varphi.$$

Since

$$\left[\delta_j, \frac{\partial}{\partial \overline{z}_k}\right] = \frac{\partial^2 \varphi}{\partial z_j \partial \overline{z}_k},$$

we have

$$\|\overline{\partial} u\|_\varphi^2 + \|\overline{\partial}_\varphi^* u\|_\varphi^2 = \sum_{j,k=1}^n \int_{\mathbb{C}^n} \left|\frac{\partial u_j}{\partial \overline{z}_k}\right|^2 e^{-\varphi}\, d\lambda + \sum_{j,k=1}^n \int_{\mathbb{C}^n} \frac{\partial^2 \varphi}{\partial z_j \partial \overline{z}_k} u_j \overline{u}_k\, e^{-\varphi}\, d\lambda, \tag{6.8}$$

and since φ satisfies (6.5), we are done. □

Remark 6.4. For $n = 1$, we have $\overline{\partial} u = 0$ if u is a $(0, 1)$-form, and (6.8) implies

$$\|\overline{\partial}_\varphi^* u\|_\varphi^2 = \int_{\mathbb{C}} \left|\frac{\partial u}{\partial \overline{z}}\right|^2 e^{-\varphi}\, d\lambda + \int_{\mathbb{C}} \frac{\partial^2 \varphi}{\partial z \partial \overline{z}} |u|^2\, e^{-\varphi}\, d\lambda \tag{6.9}$$

for $u \in \operatorname{dom}(\overline{\partial}_\varphi^*)$.

6.2 $(0, q)$-forms

At this stage, we first generalize formula (6.8) for $(0, q)$-forms $u = \sum_{|J|=q}' u_J\, d\overline{z}_J$ with coefficients in $\mathcal{C}_0^\infty(\mathbb{C}^n)$. We notice that

$$\overline{\partial} u = \sum_{|J|=q}' \sum_{j=1}^n \frac{\partial u_J}{\partial \overline{z}_j}\, d\overline{z}_j \wedge d\overline{z}_J$$

and

$$\overline{\partial}_\varphi^* u = -\sum_{|K|=q-1}' \sum_{k=1}^n \delta_k u_{kK}\, d\overline{z}_K.$$

We obtain

$$\begin{aligned}\|\overline{\partial} u\|_\varphi^2 + \|\overline{\partial}_\varphi^* u\|_\varphi^2 = &\sum_{|J|=|M|=q}' \sum_{j,k=1}^n \epsilon_{jJ}^{kM} \int_{\mathbb{C}^n} \frac{\partial u_J}{\partial \overline{z}_j} \overline{\frac{\partial u_M}{\partial \overline{z}_k}}\, e^{-\varphi}\, d\lambda \\ &+ \sum_{|K|=q-1}' \sum_{j,k=1}^n \int_{\mathbb{C}^n} \delta_j u_{jK} \overline{\delta_k u_{kK}}\, e^{-\varphi}\, d\lambda,\end{aligned}$$

where $\epsilon_{jJ}^{kM} = 0$ if $j \in J$ or $k \in M$ or if $k \cup M \neq j \cup J$, and equals the sign of the permutation $\binom{kM}{jJ}$ otherwise. The right-hand side of the last formula can be rewritten as

$$\sum_{|J|=q}' \sum_{j=1}^{n} \left\| \frac{\partial u_J}{\partial \overline{z}_j} \right\|_\varphi^2 + \sum_{|K|=q-1}' \sum_{j,k=1}^{n} \int_{\mathbb{C}^n} \left(\delta_j u_{jK} \overline{\delta_k u_{kK}} - \frac{\partial u_{jK}}{\partial \overline{z}_k} \overline{\frac{\partial u_{kK}}{\partial \overline{z}_j}} \right) e^{-\varphi} \, d\lambda. \tag{6.10}$$

To prove this, we first consider the (nonzero) terms where $j = k$ (and hence $M = J$). These terms result in the portion of the first sum in (6.10) where $j \notin J$. On the other hand, if $j \neq k$, then $j \in M$ and $k \in J$, and deletion of j from M and k from J results in the strictly increasing multi-index K of length $q-1$. Consequently, these terms can be collected into the second sum in (6.10) (in the part with the minus sign, we have interchanged the summation indices j and k). In this sum the terms where $j = k$ compensate for the terms in the first sum where $j \in J$.

Now we can use the same reasoning as in the last proof to get

$$\|\overline{\partial} u\|_\varphi^2 + \|\overline{\partial}_\varphi^* u\|_\varphi^2 = \sum_{|J|=q}' \sum_{j=1}^{n} \left\| \frac{\partial u_J}{\partial \overline{z}_j} \right\|_\varphi^2 + \sum_{|K|=q-1}' \sum_{j,k=1}^{n} \int_{\mathbb{C}^n} \frac{\partial^2 \varphi}{\partial z_j \partial \overline{z}_k} u_{jK} \overline{u}_{kK} \, e^{-\varphi} \, d\lambda. \tag{6.11}$$

Proposition 6.5. *For a plurisubharmonic weight function φ satisfying* (6.5), *there exists a uniquely determined bounded linear operator*

$$N_\varphi : L^2_{(0,1)}(\mathbb{C}^n, e^{-\varphi}) \longrightarrow L^2_{(0,1)}(\mathbb{C}^n, e^{-\varphi})$$

such that $\Box_\varphi \circ N_\varphi u = u$ for all $u \in L^2_{(0,1)}(\mathbb{C}^n, e^{-\varphi})$. If $u \in L^2_{(0,1)}(\mathbb{C}^n, e^{-\varphi})$ satisfies $\overline{\partial} u = 0$, then $\overline{\partial}_\varphi^* N_\varphi u$ is the canonical solution of $\overline{\partial} f = u$, which means that $\overline{\partial}_\varphi^* N_\varphi u \perp A^2(\mathbb{C}^n, \varphi)$, where

$$A^2(\mathbb{C}^n, e^{-\varphi}) = \{f : \mathbb{C}^n \longrightarrow \mathbb{C} \text{ entire} : f \in L^2(\mathbb{C}^n, e^{-\varphi})\}.$$

Proof. First, we mention that $\Box_\varphi$ is a self-adjoint operator, which is proved in a similar way as in the case without weights in Chapter 4.

For a given $v \in L^2_{(0,1)}(\mathbb{C}^n, \varphi)$, consider the linear functional L on $\mathrm{dom}(\overline{\partial}) \cap \mathrm{dom}(\overline{\partial}_\varphi^*)$ given by $L(u) = (u, v)_\varphi$. Notice that $\mathrm{dom}(\overline{\partial}) \cap \mathrm{dom}(\overline{\partial}_\varphi^*)$ is a Hilbert space with the inner product Q_φ. By Proposition 6.3 we have

$$|L(u)| = |(u, v)_\varphi| \leq \|u\|_\varphi \, \|v\|_\varphi \leq C Q_\varphi(u, u)^{1/2} \, \|v\|_\varphi.$$

Hence by the Riesz representation theorem there exists a uniquely determined $(0,1)$-form $N_\varphi v \in \mathrm{dom}(\overline{\partial}) \cap \mathrm{dom}(\overline{\partial}_\varphi^*)$ such that

$$(u, v)_\varphi = Q_\varphi(u, N_\varphi v) = (\overline{\partial} u, \overline{\partial} N_\varphi v)_\varphi + (\overline{\partial}_\varphi^* u, \overline{\partial}_\varphi^* N_\varphi v)_\varphi,$$

and we claim that $N_\varphi v \in \mathrm{dom}(\Box_\varphi) = \mathrm{dom}(\Box_\varphi^*)$, for which we have to show that $w \mapsto (\Box_\varphi w, N_\varphi v)_\varphi$ is bounded on $\mathrm{dom}(\Box_\varphi)$:

$$|(\Box_\varphi w, N_\varphi v)_\varphi| = |(\overline{\partial} w, \overline{\partial} N_\varphi v)_\varphi + (\overline{\partial}_\varphi^* w, \overline{\partial}_\varphi^* N_\varphi v)_\varphi|$$

$$= |Q_\varphi(w, N_\varphi v)| = |(w,v)_\varphi| \le \|w\|_\varphi \|v\|_\varphi.$$

Now we get

$$(u,v)_\varphi = Q_\varphi(u, N_\varphi v) = (u, \Box_\varphi N_\varphi v)_\varphi,$$

and hence $\Box_\varphi N_\varphi v = v$ for all $v \in L^2_{(0,1)}(\mathbb{C}^n, \varphi)$. If we set $u = N_\varphi v$, then again from Proposition 6.3 we get

$$\begin{aligned} \|\overline{\partial} N_\varphi v\|_\varphi^2 + \|\overline{\partial}_\varphi^* N_\varphi v\|_\varphi^2 &= Q_\varphi(N_\varphi v, N_\varphi v) = (N_\varphi v, v)_\varphi \le \|N_\varphi v\|_\varphi \, \|v\|_\varphi \\ &\le C_1 \big(\|\overline{\partial} N_\varphi v\|_\varphi^2 + \|\overline{\partial}_\varphi^* N_\varphi v\|_\varphi^2\big)^{1/2} \, \|v\|_\varphi, \end{aligned}$$

hence

$$\big(\|\overline{\partial} N_\varphi v\|_\varphi^2 + \|\overline{\partial}_\varphi^* N_\varphi v\|_\varphi^2\big)^{1/2} \le C_2 \|v\|_\varphi,$$

and finally, again by Proposition 6.3,

$$\|N_\varphi v\|_\varphi \le C_3 \big(\|\overline{\partial} N_\varphi v\|_\varphi^2 + \|\overline{\partial}_\varphi^* N_\varphi v\|_\varphi^2\big)^{1/2} \le C_4 \|v\|_\varphi,$$

where $C_1, C_2, C_3, C_4 > 0$ are constants. Hence we get that N_φ is a continuous linear operator from $L^2_{(0,1)}(\mathbb{C}^n, \varphi)$ into itself (see also [57] or [16]). The rest is clear from the remarks made for the $\overline{\partial}$-Neumann operator without weights. □

Remark 6.6. In this case, we can also show that $N_\varphi = j_\varphi \circ j_\varphi^*$, where

$$j_\varphi : \operatorname{dom}(\overline{\partial}) \cap \operatorname{dom}(\overline{\partial}_\varphi^*) \longrightarrow L^2_{(0,1)}(\mathbb{C}^n, e^{-\varphi})$$

is the embedding, and $\operatorname{dom}(\overline{\partial}) \cap \operatorname{dom}(\overline{\partial}_\varphi^*)$ is endowed with the graph norm

$$u \mapsto \big(\|\overline{\partial} u\|_\varphi^2 + \|\overline{\partial}_\varphi^* u\|_\varphi^2\big)^{1/2}.$$

Proposition 6.7. *Let $1 \le q \le n$ and suppose that the sum s_q of any q (equivalently, the smallest q) eigenvalues of M_φ satisfies*

$$s_q(z) > \epsilon \tag{6.12}$$

for all $z \in \mathbb{C}^n$, where $\epsilon > 0$. Then there exists a uniquely determined bounded linear operator

$$N_{\varphi,q} : L^2_{(0,q)}(\mathbb{C}^n, e^{-\varphi}) \longrightarrow L^2_{(0,q)}(\mathbb{C}^n, e^{-\varphi})$$

such that $\Box_\varphi \circ N_{\varphi,q} u = u$ for all $u \in L^2_{(0,q)}(\mathbb{C}^n, e^{-\varphi})$.

Proof. Let $\mu_{\varphi,1} \le \mu_{\varphi,2} \le \cdots \le \mu_{\varphi,n}$ denote the eigenvalues of M_φ and suppose that M_φ is diagonalized. Then, in a suitable basis,

$$\begin{aligned}\sum_{|K|=q-1}' \sum_{j,k=1}^{n} \frac{\partial^2\varphi}{\partial z_j \partial \overline{z}_k} u_{jK}\overline{u}_{kK} &= \sum_{|K|=q-1}' \sum_{j=1}^{n} \mu_{\varphi,j}|u_{jK}|^2 \\ &= \sum_{J=(j_1,\dots,j_q)}' (\mu_{\varphi,j_1} + \cdots + \mu_{\varphi,j_q})|u_J|^2 \\ &\ge s_q|u|^2.\end{aligned}$$

The last equality results as follows: for $J = (j_1, \dots, j_q)$ fixed, $|u_J|^2$ occurs precisely q times in the second sum, once as $|u_{j_1K_1}|^2$, once as $|u_{j_2K_2}|^2$, etc. At each occurrence, it is multiplied by μ_{φ,j_ℓ}. For the rest of the proof, we use (6.11) and proceed as in the proof of Proposition 6.5. □

Remark 6.8. For the $\overline{\partial}$-Neumann operator $N_{\varphi,q}$ on $(0,q)$-forms, we obtain in a similar way as in Proposition 6.3 that $N_{\varphi,q} = j_{\varphi,q} \circ j^*_{\varphi,q}$, where

$$j_{\varphi,q} : \operatorname{dom}(\overline{\partial}) \cap \operatorname{dom}(\overline{\partial}^*_\varphi) \longrightarrow L^2_{(0,q)}(\mathbb{C}^n, e^{-\varphi}),$$

and $\operatorname{dom}(\overline{\partial}) \cap \operatorname{dom}(\overline{\partial}^*_\varphi)$ is endowed with the graph norm $(\|\overline{\partial}u\|^2_\varphi + \|\overline{\partial}^*_\varphi u\|^2_\varphi)^{1/2}$.

It also follows in this case that both $\overline{\partial}$ and $\overline{\partial}^*_\varphi$ have closed ranges; see Proposition 4.58.

6.3 Notes

Most of the material in this chapter is taken from [35, 36, 46]. It will be fundamental for a closer analysis of the $\overline{\partial}$-Neumann operator and its connection to Schrödinger operators with magnetic fields and the Witten Laplacian.

Using spectral analysis and Persson's theorem on the bottom of the essential spectrum, we can show that assumption (6.5) in Proposition 6.3 can be replaced by the more general condition $\liminf_{|z|\to\infty} \mu_\varphi(z) > 0$; see [46]. We indicate that formula (6.11) has an interesting differential geometric interpretation: let (M,ω) be a Kähler manifold with fundamental form ω, and let (E, M, π) be a holomorphic vector bundle over M.

Let

$$\nabla : \Gamma(T^{\mathbb{C}}(M)) \times \Gamma(E) \longrightarrow \Gamma(E)$$

be the uniquely determined connection on E that is both holomorphic and compatible with the metric. The operator

$$\Theta := \nabla^2$$

is called the curvature of the connection ∇.

We consider the weighted $\overline{\partial}$-complex on $\mathbb{C}^n$ with fundamental form

$$\omega = i\sum_{k=1}^{n} dz_k \wedge d\overline{z}_k.$$

The weight factor $e^{-\varphi}$ can be interpreted as a metric on the trivial line bundle over $\mathbb{C}^n$ and

$$\Theta = \partial\overline{\partial}\varphi = \sum_{j,k=1}^{n} \frac{\partial^2\varphi}{\partial z_j \partial\overline{z}_k}\, dz_j \wedge d\overline{z}_k;$$

see [94] for the details. Let Λ denote the interior multiplication with the fundamental form ω,

$$(\Lambda\alpha, w) = (\alpha, \omega \wedge w)$$

for suitable differential forms α and w, and let $u = \sum'_{|J|=q} u_J\, d\overline{z}_J$ be a $(0,q)$-form with coefficients in $\mathcal{C}_0^\infty(\mathbb{C}^n)$. We want to interpret the term

$$\sum_{|K|=q-1}{}' \sum_{j,k=1}^{n} \int_{\mathbb{C}^n} \frac{\partial^2\varphi}{\partial z_j \partial\overline{z}_k}\, u_{jK}\overline{u}_{kK}\, e^{-\varphi}\, d\lambda$$

of (6.11) by the curvature Θ and operator Λ. For this purpose, we consider (n,q)-forms

$$\xi = \sum_{|I|=q}{}' \xi_I\, dz \wedge d\overline{z}_I,$$

instead of $(0,q)$-forms, where $dz = dz_1 \wedge \cdots \wedge dz_n$. We use the notation

$$d\hat{z}_j := dz_1 \wedge \cdots \wedge \widehat{dz_j} \wedge \cdots \wedge dz_n,$$

which means that dz_j is excluded. It follows that

$$\Lambda\xi = i\sum_{j=1}^{n} \sum_{|J|=q-1}{}' \xi_{jJ}\, d\hat{z}_j \wedge d\overline{z}_J,$$

and since $\Theta\xi = 0$, we obtain for the commutator $[\Theta, \Lambda]$ that

$$(i[\Theta,\Lambda]\xi, \xi)_\varphi = \sum_{|J|=q-1}{}' \sum_{j,k=1}^{n} \int_{\mathbb{C}^n} \frac{\partial^2\varphi}{\partial z_j \partial\overline{z}_k}\, \xi_{jJ}\overline{\xi}_{kJ}\, e^{-\varphi}\, d\lambda.$$

The commutator $[\Theta, \Lambda]$ appears in the Nakano vanishing theorem; see [94].

7 The twisted $\overline{\partial}$-complex

We consider the twisted $\overline{\partial}$-complex

$$L^2(\Omega) \xrightarrow{T} L^2_{(0,1)}(\Omega) \xrightarrow{S} L^2_{(0,2)}(\Omega) \tag{7.1}$$

for operators $T = \overline{\partial} \circ \sqrt{\tau}$ and $S = \sqrt{\tau} \circ \overline{\partial}$, where $\tau \in \mathcal{C}^2(\overline{\Omega})$ and $\tau > 0$ on Ω. It turns out that this approach leads to generalized basic estimates, which are seminal for the applications explained in the next chapter. The key method consists of a keen use of integration by parts.

7.1 An exact sequence of unbounded operators

First, we prove a general result about operators like T and S.

Proposition 7.1. *Let H_1, H_2, H_3 be Hilbert spaces, let $T : H_1 \longrightarrow H_2$ and $S : H_2 \longrightarrow H_3$ be densely defined linear operators such that $S(T(f)) = 0$ for each $f \in \operatorname{dom}(T)$, and let $P : H_2 \longrightarrow H_2$ be a positive invertible operator such that*

$$\|Pu\|_2^2 \leq \|T^*u\|_1^2 + \|Su\|_3^2 \tag{7.2}$$

for all $u \in \operatorname{dom}(S) \cap \operatorname{dom}(T^*)$, where

$$\operatorname{dom}(T^*) = \{u \in H_2 : |(u, Tf)_2| \leq C\,\|f\|_1 \text{ for all } f \in \operatorname{dom}(T)\}.$$

Suppose (7.2) *holds and let $\alpha \in H_2$ be such that $S\alpha = 0$. Then there exists $\sigma \in H_1$ such that*
(i) $T(\sigma) = \alpha$

and
(ii) $\|\sigma\|_1^2 \leq \|P^{-1}\alpha\|_2^2$.

Proof. Since P is positive, it follows that $P = P^*$. Now let $\alpha \in H_2$ be such that $S\alpha = 0$. We consider the linear functional $T^*u \mapsto (u, \alpha)_2$ for $u \in \operatorname{dom}(T^*)$. We show that this linear functional is well defined and continuous. If $u \in \ker S$, then

$$\begin{aligned}|(u, \alpha)_2| &= |(Pu, P^{-1}\alpha)_2| \leq \|Pu\|_2\,\|P^{-1}\alpha\|_2 \leq \left(\|T^*u\|_1^2 + \|Su\|_3^2\right)^{1/2}\|P^{-1}\alpha\|_2\\ &= \|T^*u\|_1\,\|P^{-1}\alpha\|_2,\end{aligned}$$

and if $u \perp_2 \ker S$, then $(u, \alpha)_2 = 0$. In addition, we have that $T^*w = 0$ for all $w \perp_2 \ker S$; this follows from the assumption that $Tf \in \ker S$, so $0 = (w, Tf)_2 \leq C\|f\|_1$, which means that

https://doi.org/10.1515/9783111182926-007

$w \in \operatorname{dom}(T^*)$ and $T^* w = 0$, since $(T^* w, f)_1 = (w, Tf)_2 = 0$ for all $f \in \operatorname{dom}(T)$. If $T^* u = 0$, then from the above estimate it follows that $(u, \alpha)_2 = 0$.

We apply the Hahn–Banach theorem, where we keep the constant for the estimate of the functional and the Riesz representation theorem to get $\sigma \in H_1$ such that $(T^* u, \sigma)_1 = (u, \alpha)_2$, which implies that $(u, T\sigma)_2 = (u, \alpha)_2$. Hence $T\sigma = \alpha$, and, again by the above estimate, $\|\sigma\|_1 \leq \|P^{-1}\alpha\|_2$. □

7.2 The twisted basic estimates

Let Ω be a smoothly bounded pseudoconvex domain in $\mathbb{C}^n$ with defining function r such that $|\nabla r(z)| = 1$ on $b\Omega$. Let $\tau \in \mathcal{C}^2(\overline{\Omega})$ and $\tau > 0$ on Ω. For $f \in \mathcal{C}^\infty(\overline{\Omega})$, we define

$$Tf = (\overline{\partial} \circ \sqrt{\tau})f = \sum_{k=1}^n \frac{\partial}{\partial \overline{z}_k}(\sqrt{\tau} f)\, d\overline{z}_k, \tag{7.3}$$

and for $u = \sum_{j=1}^n u_j\, d\overline{z}_j$ with coefficients u_j in $\mathcal{C}^\infty(\overline{\Omega})$ (we will write $u \in \Lambda^{0,1}(\overline{\Omega})$), we define

$$Su = \sum_{j<k} \sqrt{\tau}\left(\frac{\partial u_k}{\partial \overline{z}_j} - \frac{\partial u_j}{\partial \overline{z}_k}\right) d\overline{z}_j \wedge d\overline{z}_k. \tag{7.4}$$

We call τ a twist factor. We also introduce a weight factor φ: for $f \in \mathcal{C}^\infty(\overline{\Omega})$ and $u \in \mathcal{D}^{0,1}$ (see Proposition 4.53),

$$\begin{aligned}(Tf, u)_\varphi = (\overline{\partial}(\sqrt{\tau} f), u)_\varphi &= (\overline{\partial}(\sqrt{\tau} f), e^{-\varphi} u) \\ &= (\sqrt{\tau} f, \overline{\partial}^*(e^{-\varphi} u)) \\ &= (f, \sqrt{\tau}\, e^{\varphi}(\overline{\partial}^*(e^{-\varphi} u)))_\varphi,\end{aligned}$$

which implies that

$$T^* u = \sqrt{\tau}\, \overline{\partial}_\varphi^* u,$$

where (see Proposition 6.1)

$$\overline{\partial}_\varphi^* u = -\sum_{\ell=1}^n e^\varphi \frac{\partial}{\partial z_\ell}(e^{-\varphi} u_\ell) =: -\sum_{\ell=1}^n \delta_\ell u_\ell.$$

In the sequel, we will use the following equations: for $f, g \in \mathcal{C}^\infty(\overline{\Omega})$,

$$(\delta_\ell f, g)_\varphi = -\left(f, \frac{\partial g}{\partial \overline{z}_\ell}\right)_\varphi + \int_{b\Omega} f \overline{g} \frac{\partial r}{\partial z_\ell} e^{-\varphi}\, d\sigma, \tag{A}$$

$$\left(\frac{\partial f}{\partial \bar z_k}, g\right)_\varphi = -(f, \delta_k g)_\varphi + \int_{b\Omega} f \bar g \frac{\partial r}{\partial \bar z_k} e^{-\varphi}\, d\sigma, \tag{B}$$

$$\left[\delta_\ell, \frac{\partial}{\partial \bar z_k}\right] f = \frac{\partial^2 \varphi}{\partial \bar z_k \partial z_\ell} f, \tag{C}$$

$$\delta_\ell(f g) = f(\delta_\ell g) + \frac{\partial f}{\partial z_\ell} g. \tag{D}$$

We introduce the notations

$$i\partial\bar\partial g(\xi, \xi)(p) := \sum_{k,\ell=1}^{n} \frac{\partial^2 g}{\partial \bar z_k \partial z_\ell}(p) \bar\xi_k \xi_\ell \tag{7.5}$$

and

$$\langle \partial g, \xi\rangle(p) := \sum_{k=1}^{n} \frac{\partial g}{\partial z_k}(p)\, \xi_k \tag{7.6}$$

for $g \in \mathcal{C}^\infty(\Omega)$ and $\xi \in \mathbb{C}^n$.

If $u \in \mathcal{D}^{0,1}$, then we set

$$\|\sqrt{\tau} u\|^2_{\varphi,\bar z} := \sum_{j,k=1}^{n} \left\| \sqrt{\tau} \frac{\partial u_j}{\partial \bar z_k} \right\|^2_\varphi. \tag{7.7}$$

Now we prove the "twisted" basic estimates.

Theorem 7.2. *Let Ω be a smoothly bounded pseudoconvex domain in $\mathbb{C}^n$ with defining function r such that $|\nabla r(z)| = 1$ on $b\Omega$. Let $\tau, \varphi, A \in \mathcal{C}^2(\bar\Omega)$ and $\tau, A > 0$ on Ω, and let $u \in \mathcal{D}^{0,1}$. Then*

$$\|\sqrt{\tau + A}\, \bar\partial^*_\varphi u\|^2_\varphi + \|\sqrt{\tau}\,\bar\partial u\|^2_\varphi \geq \|\sqrt{\tau} u\|^2_{\varphi,\bar z} + \int_\Omega \Theta(u,u)\, e^{-\varphi}\, d\lambda + \int_{b\Omega} \tau i\partial\bar\partial r(u,u)\, e^{-\varphi}\, d\sigma,$$

where

$$\Theta(u,u) = \tau i\partial\bar\partial\varphi(u,u) - i\partial\bar\partial\tau(u,u) - \frac{|\langle \partial\tau, u\rangle|^2}{A}. \tag{7.8}$$

Proof. As in the case without twists, we get

$$\begin{aligned} \|\sqrt{\tau}\,\bar\partial u\|^2_\varphi &= \sum_{j<k} \left\| \sqrt{\tau}\left(\frac{\partial u_k}{\partial \bar z_j} - \frac{\partial u_j}{\partial \bar z_k}\right)\right\|^2_\varphi \\ &= \sum_{j,k=1}^{n} \left\| \sqrt{\tau} \frac{\partial u_k}{\partial \bar z_j}\right\|^2_\varphi - \sum_{j,k=1}^{n} \left(\sqrt{\tau}\frac{\partial u_k}{\partial \bar z_j}, \sqrt{\tau}\frac{\partial u_j}{\partial \bar z_k}\right)_\varphi \end{aligned}$$

and

$$\|\sqrt{\tau}\,\bar\partial_\varphi^* u\|_\varphi^2 = \sum_{j,k=1}^n (\sqrt{\tau}\,\delta_k u_k, \sqrt{\tau}\,\delta_j u_j)_\varphi,$$

and hence

$$\|\sqrt{\tau}\,\bar\partial u\|_\varphi^2 + \|\sqrt{\tau}\,\bar\partial_\varphi^* u\|_\varphi^2 = \|\sqrt{\tau} u\|_{\varphi,\bar z}^2 + \sum_{j,k=1}^n \left\{(\tau\,\delta_k u_k, \delta_j u_j)_\varphi - \left(\tau\,\frac{\partial u_k}{\partial \bar z_j}, \frac{\partial u_j}{\partial \bar z_k}\right)_\varphi\right\}.$$

Using (A) and (B) and integrating by parts, we have

$$= \|\sqrt{\tau} u\|_{\varphi,\bar z}^2 + \sum_{j,k=1}^n \left\{-\left(\frac{\partial}{\partial \bar z_j}(\tau\,\delta_k u_k), u_j\right)_\varphi + \left(\delta_k\left(\tau\,\frac{\partial u_k}{\partial \bar z_j}\right), u_j\right)_\varphi\right\} + T_1 + T_2,$$

where

$$T_1 = \sum_{j,k=1}^n \int_{b\Omega} \frac{\partial r}{\partial \bar z_j}\,\tau(\delta_k u_k)\,\bar u_j\, e^{-\varphi}\, d\sigma,$$

and

$$T_2 = -\sum_{j,k=1}^n \int_{b\Omega} \frac{\partial r}{\partial z_k}\,\tau\frac{\partial u_k}{\partial \bar z_j}\,\bar u_j\, e^{-\varphi}\, d\sigma.$$

Next, by (C) and (D) we obtain

$$\begin{aligned}
\|\sqrt{\tau}\,\bar\partial u\|_\varphi^2 + \|\sqrt{\tau}\,\bar\partial_\varphi^* u\|_\varphi^2 &= \|\sqrt{\tau} u\|_{\varphi,\bar z}^2 + \sum_{j,k=1}^n \left(\tau\delta_k\left(\frac{\partial u_k}{\partial \bar z_j}\right) + \frac{\partial \tau}{\partial z_k}\frac{\partial u_k}{\partial \bar z_j}, u_j\right)_\varphi \\
&\quad - \sum_{j,k=1}^n \left(\tau\,\frac{\partial}{\partial \bar z_j}(\delta_k u_k) + \frac{\partial \tau}{\partial \bar z_j}\,\delta_k u_k, u_j\right)_\varphi + T_1 + T_2 \\
&= \|\sqrt{\tau} u\|_{\varphi,\bar z}^2 + \sum_{j,k=1}^n \left(\tau\left[\delta_k, \frac{\partial}{\partial \bar z_j}\right] u_k, u_j\right)_\varphi \\
&\quad + \sum_{j,k=1}^n \left(\frac{\partial \tau}{\partial z_k}\frac{\partial u_k}{\partial \bar z_j} - \frac{\partial \tau}{\partial \bar z_j}\,\delta_k u_k, u_j\right)_\varphi + T_1 + T_2 \\
&= \underbrace{\|\sqrt{\tau} u\|_{\varphi,\bar z}^2}_{(A1)} + \underbrace{\int_\Omega \tau\, i\partial\bar\partial\varphi(u,u)\, e^{-\varphi}\, d\lambda}_{(A2)} \\
&\quad - \sum_{j,k=1}^n \left(\delta_k u_k, \frac{\partial \tau}{\partial z_j}\,u_j\right)_\varphi + \sum_{j,k=1}^n \left(\frac{\partial u_k}{\partial \bar z_j}, \frac{\partial \tau}{\partial \bar z_k}\,u_j\right)_\varphi + \underbrace{T_1 + T_2}_{(A3)} \\
&= (A1) + (A2) + (A3) \\
&\quad - \sum_{j,k=1}^n \left\{\left(\delta_k u_k, \frac{\partial \tau}{\partial z_j}\,u_j\right)_\varphi + \left(u_k, \delta_j\left(\frac{\partial \tau}{\partial \bar z_k}\,u_j\right)\right)_\varphi\right\} + T_3,
\end{aligned}$$

where

$$T_3 = \sum_{j,k=1}^{n} \int_{b\Omega} \frac{\partial r}{\partial \overline{z}_j} u_k \frac{\partial \tau}{\partial z_k} \overline{u}_j e^{-\varphi} d\sigma.$$

Now we get

$$\begin{aligned}\| \sqrt{\tau}\, \overline{\partial} u\|_\varphi^2 + \| \sqrt{\tau}\, \overline{\partial}_\varphi^* u\|_\varphi^2 &= (A1) + (A2) + (A3) + T_3 - \sum_{j,k=1}^{n} \left(\delta_k u_k, \frac{\partial \tau}{\partial z_j} u_j \right)_\varphi \\ &\quad - \sum_{j,k=1}^{n} \left(u_k, \frac{\partial \tau}{\partial \overline{z}_k} \delta_j u_j + \frac{\partial^2 \tau}{\partial \overline{z}_k \partial z_j} u_j \right)_\varphi \\ &= (A1) + (A2) + (A3) + T_3 \\ &\quad - \int_\Omega i\partial\overline{\partial}\tau(u,u)\, e^{-\varphi}\, d\lambda - 2\Re \sum_{j,k=1}^{n} \left(\delta_k u_k, \frac{\partial \tau}{\partial z_j} u_j \right)_\varphi .\end{aligned}$$

We now estimate the last term:

$$\begin{aligned}\left| -2\Re \sum_{j,k=1}^{n} \left(\delta_k u_k, \frac{\partial \tau}{\partial z_j} u_j \right)_\varphi \right| &= \left| -2\Re \int_\Omega \sqrt{A}\, \overline{\partial}_\varphi^* u \frac{\langle \partial\tau, u\rangle}{\sqrt{A}} e^{-\varphi}\, d\lambda \right| \\ &\le 2\| \sqrt{A}\, \overline{\partial}_\varphi^* u\|_\varphi \, \|\langle \partial\tau, u\rangle / \sqrt{A}\|_\varphi \le \| \sqrt{A}\, \overline{\partial}_\varphi^* u\|_\varphi^2 + \|\langle \partial\tau, u\rangle / \sqrt{A}\|_\varphi^2 ,\end{aligned}$$

which means that

$$-2\Re \sum_{j,k=1}^{n} \left(\delta_k u_k, \frac{\partial \tau}{\partial z_j} u_j \right)_\varphi \ge -\| \sqrt{A}\, \overline{\partial}_\varphi^* u\|_\varphi^2 - \|\langle \partial\tau, u\rangle / \sqrt{A}\|_\varphi^2 .$$

Now we move the first term in the last expression to the other side and get the desired result, since

$$\| \sqrt{\tau + A}\, \overline{\partial}_\varphi^* u\|_\varphi^2 = \int_\Omega (\tau + A) |\overline{\partial}_\varphi^* u|^2 e^{-\varphi}\, d\lambda,$$

$T_1 = T_3 = 0$ for $u \in \mathcal{D}^{0,1}$, and

$$T_2 = \sum_{j,k=1}^{n} \int_{b\Omega} \tau \frac{\partial^2 r}{\partial \overline{z}_j \partial z_k} \overline{u}_j u_k e^{-\varphi} d\sigma = \int_{b\Omega} \tau\, i\partial\overline{\partial} r(u,u)\, e^{-\varphi}\, d\sigma. \qquad \square$$

Remark 7.3. In a similar way, we can prove an analogous twisted estimate for $(0,q)$-forms, $1 \le q \le n$, again using integration by parts repeatedly; see [91].

7.3 Notes

The main result of this chapter is due to McNeal [76]. It can also be found in [91]. Inequality (7.2) corresponds to the basic estimates (4.38) for original $\bar{\partial}$-complex. Besides the important applications carried out in the following chapter, the twisted basic estimates also yield an elegant proof of the Ohsawa–Takegoshi extension theorem: let $\Omega \subset \mathbb{C}^n$ be a bounded pseudoconvex domain, and let H be a complex hyperplane. There exists a constant C, which depends only on the diameter of Ω, such that for each holomorphic function $f \in L^2(H \cap \Omega)$ (with the induced Euclidean measure), there exists a holomorphic function $F \in L^2(\Omega)$, such that $F|_{H\cap\Omega} = f$ and

$$\int_{\Omega} |F(z)|^2 \, d\lambda(z) \leq C \int_{H\cap\Omega} |f(z)|^2 \, d\lambda(z).$$

For a proof of this result, see [76, 60, 91].

8 Applications

Here we first prove the basic estimates, and then we apply the basic estimates to get the Hörmander L^2-estimates for the $\overline{\partial}$-equation (see [57]) and a result of Shigekawa on the dimension of weighted spaces of entire functions [89]. Some of these results have interesting operator theoretic aspects; they are related to the Brascamp–Lieb inequality. For this interpretation, we need some general results about invertible operators (see Chapter 9).

8.1 Hörmander's L^2-estimates

We derive the basic estimates (4.38) from the theorem on the twisted $\overline{\partial}$-complex in the last chapter and the density result in Proposition 5.14. Applying Theorem 7.2, it becomes clear why the geometric condition of pseudoconvexity is crucial.

Theorem 8.1. *Let $\Omega \subset \mathbb{C}^n$ be a smoothly bounded pseudoconvex domain such that $\Omega \subseteq B(0,R)$. Then for each $u \in \operatorname{dom}(\overline{\partial}) \cap \operatorname{dom}(\overline{\partial}^*)$, we have*

$$\|\overline{\partial}^* u\|^2 + \|\overline{\partial} u\|^2 \geq \frac{1}{4R^2}\|u\|^2. \tag{8.1}$$

Proof. Since Ω is pseudoconvex, the boundary integral in Theorem 7.2 is ≥ 0. Take $\varphi = 0$ and $\tau(z) = R^2 - |z|^2$. Then $-i\partial\overline{\partial}\tau(u,u) = |u|^2$ and $|\langle \partial\tau, u\rangle|^2 = |\sum_{j=1}^n \overline{z}_j u_j|^2 \leq |z|^2 |u|^2$. We choose $A = 2|z|^2$. Then $|\langle \partial\tau, u\rangle|^2/A \leq \frac{1}{2}|u|^2$. Hence from Theorem 7.2 we get

$$\left\|\sqrt{R^2 + |z|^2}\,\overline{\partial}^* u\right\|^2 + \left\|\sqrt{R^2 - |z|^2}\,\overline{\partial} u\right\|^2 \geq \frac{1}{2}\|u\|^2 \tag{8.2}$$

for $u \in \mathcal{D}^{0,1}$. Now the result follows from Proposition 5.14. □

Remark 8.2. In a similar way, we can show that for $(0,q)$-forms $u \in \operatorname{dom}(\overline{\partial}) \cap \operatorname{dom}(\overline{\partial}^*)$, $1 \leq q \leq n$, the corresponding basic estimate reads as

$$\|\overline{\partial}^* u\|^2 + \|\overline{\partial} u\|^2 \geq \frac{q}{4R^2}\|u\|^2;$$

the factor q arises because each strictly increasing multi-index of length q appears q times in the corresponding term on the right-hand side of (8.2). We can also dispense with the smoothness requirement on $b\Omega$; see [91].

Corollary 8.3. *Let Ω be as in Theorem 8.1, and let $\alpha \in L^2_{(0,1)}(\Omega)$ be such that $\overline{\partial}\alpha = 0$. Then there exists $s \in L^2(\Omega)$ such that $\overline{\partial}s = \alpha$ and*

$$\int_\Omega |s|^2\, d\lambda \leq 4R^2 \int_\Omega |\alpha|^2\, d\lambda. \tag{8.3}$$

https://doi.org/10.1515/9783111182926-008

Proof. Apply Proposition 7.1 to $T = \overline{\partial} \circ \sqrt{R^2 + |z|^2}$ and $S = \sqrt{R^2 - |z|^2} \circ \overline{\partial}$, and set $P = 1/\sqrt{2}\,\mathrm{Id}$. Then we have $T^* = \sqrt{R^2 + |z|^2} \circ \overline{\partial}^*$, and Theorem 8.1 gives

$$\|Pu\|^2 \leq \|T^* u\|^2 + \|Su\|^2.$$

By Proposition 7.1 we obtain $\sigma \in L^2(\Omega)$ such that $T\sigma = \overline{\partial}(\sqrt{R^2 + |z|^2}\,\sigma) = \alpha$ and $\|\sigma\|^2 \leq 2\|\alpha\|^2$. Now set $s = \sqrt{R^2 + |z|^2}\,\sigma$; then $\overline{\partial} s = \alpha$, and

$$\int_\Omega \frac{|s|^2}{R^2 + |z|^2}\, d\lambda \leq 2 \int_\Omega |\alpha|^2\, d\lambda,$$

so we get

$$\frac{1}{2R^2} \int_\Omega |s|^2\, d\lambda \leq 2 \int_\Omega |\alpha|^2\, d\lambda. \qquad \square$$

Theorem 8.4. *Let $\varphi : \mathbb{C}^n \longrightarrow \mathbb{R}$ be a real-valued function in $\mathcal{C}^2(\mathbb{C}^n)$ such that*

$$c(z) \sum_{j=1}^n |w_j|^2 \leq \sum_{j,k=1}^n \frac{\partial^2 \varphi(z)}{\partial z_j \partial \overline{z}_k}\, w_j \overline{w}_k, \quad z \in \mathbb{C}^n,\ w \in \mathbb{C}^n,$$

where c is a positive continuous function in $\mathbb{C}^n$. If $g \in L^2_{(0,1)}(\mathbb{C}^n, e^{-\varphi})$ and $\overline{\partial} g = 0$, then we can find $f \in L^2(\mathbb{C}^n, e^{-\varphi})$ with $\overline{\partial} f = g$ and

$$\int_{\mathbb{C}^n} |f(z)|^2 e^{-\varphi(z)}\, d\lambda(z) \leq \int_{\mathbb{C}^n} |g(z)|^2 \frac{e^{-\varphi(z)}}{c(z)}\, d\lambda(z), \tag{8.4}$$

provided that the right-hand side is finite.

Proof. From Proposition 6.3 we have

$$\int_{\mathbb{C}^n} \sum_{j,k=1}^n \frac{\partial^2 \varphi}{\partial z_j \partial \overline{z}_k}\, u_j \overline{u}_k\, e^{-\varphi}\, d\lambda \leq \|\overline{\partial} u\|_\varphi^2 + \|\overline{\partial}_\varphi^* u\|_\varphi^2$$

for each $(0,1)$-form $u = \sum_{k=1}^n u_k\, d\overline{z}_k$ with coefficients $u_k \in \mathcal{C}_0^\infty(\mathbb{C}^n)$, $k = 1, \dots, n$.

Let $P : L^2_{(0,1)}(\mathbb{C}^n, e^{-\varphi}) \longrightarrow L^2_{(0,1)}(\mathbb{C}^n, e^{-\varphi})$ be the multiplication operator by the function $\sqrt{c}$. Then from the assumption it follows that

$$\|Pu\|_\varphi^2 \leq \|\overline{\partial} u\|_\varphi^2 + \|\overline{\partial}_\varphi^* u\|_\varphi^2.$$

By Proposition 7.1 we get a function $f \in L^2(\mathbb{C}^n, e^{-\varphi})$ with $\overline{\partial} f = g$ and $\|f\|_\varphi \leq \|P^{-1} g\|_\varphi$. $\square$

Theorem 8.5. *Let $\varphi : \mathbb{C}^n \longrightarrow \mathbb{R}$ be a plurisubharmonic function in $\mathcal{C}^2(\mathbb{C}^n)$. If $g \in L^2_{(0,1)}(\mathbb{C}^n, e^{-\varphi})$ and $\overline{\partial} g = 0$, then we can find a solution u of $\overline{\partial} u = g$ such that*

$$2\int_{\mathbb{C}^n} |u(z)|^2 e^{-\varphi(z)} \left(1+|z|^2\right)^{-2} d\lambda(z) \le \int_{\mathbb{C}^n} |g(z)|^2 e^{-\varphi(z)} d\lambda(z). \tag{8.5}$$

Proof. We apply Theorem 8.4 with φ replaced by $\varphi + 2\log(1+|z|^2)$ and use that

$$\sum_{j,k=1}^n w_j\overline{w}_k \frac{\partial^2}{\partial z_j\partial\overline{z}_k}\log(1+|z|^2) = \left(1+|z|^2\right)^{-2}\left(|w|^2(1+|z|^2) - \left|(w,z)\right|^2\right)$$
$$\ge \left(1+|z|^2\right)^{-2}|w|^2,$$

so we can take $c(z) = 2(1+|z|^2)^{-2}$ to obtain the desired result. □

Theorem 8.6. *Let Ω be a smoothly bounded pseudoconvex domain in $\mathbb{C}^n$, and let $\varphi : \Omega \longrightarrow \mathbb{R}$ be a real-valued function in $C^2(\Omega)$ such that*

$$c(z)\sum_{j=1}^n |w_j|^2 \le \sum_{j,k=1}^n \frac{\partial^2\varphi(z)}{\partial z_j\partial\overline{z}_k} w_j\overline{w}_k, \quad z\in\Omega,\ w\in\mathbb{C}^n,$$

where c is a positive continuous function on Ω. If $g \in L^2_{(0,1)}(\Omega, e^{-\varphi})$ and $\overline{\partial}g = 0$, then we can find $f \in L^2(\Omega, e^{-\varphi})$ such that $\overline{\partial}f = g$ and

$$\int_\Omega |f(z)|^2 e^{-\varphi(z)} d\lambda(z) \le 2\int_\Omega |g(z)|^2 \frac{e^{-\varphi(z)}}{c(z)} d\lambda(z), \tag{8.6}$$

provided that the right-hand side is finite.

Proof. We use Theorem 7.2 for $\tau = A = 1$ and get

$$\int_\Omega \sum_{j,k=1}^n \frac{\partial^2\varphi}{\partial z_j\partial\overline{z}_k} u_j\overline{u}_k e^{-\varphi} d\lambda \le \|\overline{\partial}u\|_\varphi^2 + 2\|\overline{\partial}_\varphi^* u\|_\varphi^2,$$

for $u \in \mathcal{D}^{0,1}$.

Let $P : L^2_{(0,1)}(\Omega, e^{-\varphi}) \longrightarrow L^2_{(0,1)}(\Omega, e^{-\varphi})$ be the multiplication operator by the function $\sqrt{c}$. Then from the assumption it follows that

$$\|Pu\|_\varphi^2 \le \|\overline{\partial}u\|_\varphi^2 + 2\|\overline{\partial}_\varphi^* u\|_\varphi^2.$$

By Proposition 7.1 we get a function $f \in L^2(\Omega, e^{-\varphi})$ such that $\overline{\partial}f = g$ and $\|f\|_\varphi \le \sqrt{2}\|P^{-1}g\|_\varphi$. □

For a positive $\psi = \sum_{j,k=1}^n \psi_{j,k}\, dz_j \wedge d\overline{z}_k \in \Lambda^{1,1}(\Omega)$ and $\alpha = \sum_{j=1}^n \alpha_j\, d\overline{z}_j \in \Lambda^{0,1}(\Omega)$, we set

$$|\alpha|_\psi^2 := \sum_{j,k=1}^n \psi^{j,k}\,\alpha_j\overline{\alpha}_k, \tag{8.7}$$

where $\psi^{j,k} = (\psi_{\ell,m})^{-1}_{j,k}$. Then we get a more general result.

Theorem 8.7. *Let Ω be a smoothly bounded pseudoconvex domain in $\mathbb{C}^n$, and let $\varphi : \Omega \longrightarrow \mathbb{R}$ be a strictly plurisubharmonic function belonging to $C^2(\Omega)$. If $\alpha \in L^2_{(0,1)}(\Omega, e^{-\varphi})$ satisfies $\overline{\partial}\alpha = 0$, then we can find $u \in L^2(\Omega, e^{-\varphi})$ such that $\overline{\partial}u = \alpha$ and*

$$\int_\Omega |u(z)|^2 e^{-\varphi(z)}\, d\lambda(z) \leq 2 \int_\Omega |\alpha(z)|^2_{i\partial\overline{\partial}\varphi} e^{-\varphi(z)}\, d\lambda(z), \tag{8.8}$$

provided that the right-hand side is finite.

8.2 Weighted spaces of entire functions

In this part, we use Hörmander's L^2-estimates to construct nontrivial entire functions with given growth behavior at infinity.

Theorem 8.8. *Let Ω be a smoothly bounded pseudoconvex domain in $\mathbb{C}^n$, and let $\varphi : \Omega \longrightarrow \mathbb{R}$ be a plurisubharmonic function in $C^2(\Omega)$. For every $g \in L^2_{(0,1)}(\Omega, e^{-\varphi})$ with $\overline{\partial}g = 0$, there is a solution $u \in L^2(\Omega, \mathrm{loc})$ of the equation $\overline{\partial}u = g$ such that*

$$\int_\Omega |u(z)|^2 e^{-\varphi(z)} (1+|z|^2)^{-2}\, d\lambda(z) \leq \int_\Omega |g(z)|^2 e^{-\varphi(z)}\, d\lambda(z). \tag{8.9}$$

Proof. We apply Theorem 8.6 with φ replaced by $\varphi + 2\log(1+|z|^2)$ and use that

$$\begin{aligned}\sum_{j,k=1}^n w_j \overline{w}_k \frac{\partial^2}{\partial z_j \partial \overline{z}_k} \log(1+|z|^2) &= (1+|z|^2)^{-2}\left(|w|^2(1+|z|^2) - |(w,z)|^2\right) \\ &\geq (1+|z|^2)^{-2}|w|^2,\end{aligned}$$

so we can take $c(z) = 2(1+|z|^2)^{-2}$ to obtain the desired result. □

Theorem 8.9. *Let $\Omega \subseteq \mathbb{C}^n$ be a smoothly bounded pseudoconvex domain in $\mathbb{C}^n$, and let $\varphi : \Omega \longrightarrow \mathbb{R}$ be a plurisubharmonic function in $C^2(\Omega)$. If $z_0 \in \Omega$, then we can find a holomorphic function u in Ω such that $u(z_0) = 1$ and*

$$\int_\Omega |u(z)|^2 e^{-\varphi(z)} (1+|z|^2)^{-3n}\, d\lambda(z) < \infty. \tag{8.10}$$

Proof. We may assume that $z_0 = 0$. Choose a polydisc

$$D = \{z : |z_j| < r, j = 1, \ldots, n\} \subset \Omega.$$

The function $e^{-\varphi}$ is integrable there. Define

$$\Omega_k := \{z \in \Omega : |z_j| < r \text{ for } j > k\}$$

for $k = 0, 1, \ldots, n$.

We will prove inductively that for every k, there is a holomorphic function u_k in Ω_k such that $u_k(z_0) = 1$ and

$$\int_{\Omega_k} |u_k(z)|^2 e^{-\varphi(z)} (1 + |z|^2)^{-3k} \, d\lambda(z) < \infty.$$

When $k = 0$, we can take $u_0(z) \equiv 1$, and u_n will have the desired properties.

Assume that $0 < k \leq n$ and that u_{k-1} has already been constructed. Choose $\psi \in C_0^\infty(\mathbb{C})$ such that $\psi(z_k) = 0$ when $|z_k| > r/2$ and $\psi(z_k) = 1$ when $|z_k| < r/3$, and set

$$u_k(z) := \psi(z_k) u_{k-1}(z) - z_k v(z).$$

Notice that $\psi(z_k) u_{k-1}(z) = 0$ in $\Omega_k \setminus \Omega_{k-1}$. To make u_k holomorphic, we must choose v as a solution of the equation $\overline{\partial} v = z_k^{-1} u_{k-1} \overline{\partial} \psi = f$. By the inductive hypothesis we have

$$\int_{\Omega_k} |f(z)|^2 e^{-\varphi(z)} (1 + |z|^2)^{-3(k-1)} \, d\lambda(z) < \infty.$$

Hence it follows from Theorem 8.8 that we can find v such that

$$\int_{\Omega_k} |v(z)|^2 e^{-\varphi(z)} (1 + |z|^2)^{1-3k} \, d\lambda(z) < \infty.$$

Together with the inductive hypothesis on u_{k-1}, this implies that

$$\int_{\Omega_k} |u_k(z)|^2 e^{-\varphi(z)} (1 + |z|^2)^{-3k} \, d\lambda(z) < \infty.$$

Since $\overline{\partial} v = 0$ in a neighborhood of 0, v is a C^∞ function there, and we have $u_k(0) = u_{k-1}(0) = 1$, so u_k has the required properties. □

Lemma 8.10. *Let $\zeta \in \mathbb{C}^n$ and $K > 0$ and define $g(z) = \log(1 + K|z - \zeta|^2)$. Then for each $w \in \mathbb{C}^n$, we have*

$$\frac{K}{(1 + K|z - \zeta|^2)^2} |w|^2 \leq i\partial\overline{\partial} g(w, w)(z) \leq \frac{K}{1 + K|z - \zeta|^2} |w|^2, \tag{8.11}$$

where we use notation (7.5).

Proof. An easy computation shows

$$\frac{\partial^2 g}{\partial z_j \partial \overline{z}_k}(z) = -\frac{K^2(\overline{z}_j - \overline{\zeta}_j)(z_k - \zeta_k)}{(1 + K|z - \zeta|^2)^2} + \frac{K\delta_{jk}}{1 + K|z - \zeta|^2}$$

$$= \frac{K}{(1+K|z-\zeta|^2)^2}\left[(1+K|z-\zeta|^2)\,\delta_{jk} - K(\overline{z}_j-\overline{\zeta}_j)(z_k-\zeta_k)\right].$$

This implies

$$i\partial\overline{\partial}g(w,w)(z) = \frac{K}{(1+K|z-\zeta|^2)^2}\left[(1+K|z-\zeta|^2)\,|w|^2 - K\left|(w,z-\zeta)\right|^2\right],$$

and hence

$$\frac{K}{(1+K|z-\zeta|^2)^2}\left[(1+K|z-\zeta|^2)\,|w|^2 - K|w|^2|z-\zeta|^2\right] \le i\partial\overline{\partial}g(w,w)(z)$$
$$\le \frac{K}{1+K|z-\zeta|^2}\,|w|^2. \qquad \square$$

We are now able to show that weighted spaces of entire functions are of infinite dimension if the weight function has an appropriate behavior at infinity.

Theorem 8.11. *Let $W : \mathbb{C}^n \longrightarrow \mathbb{R}$ be a $\mathcal{C}^\infty$-function, and let $\mu(z)$ denote the lowest eigenvalue of the Levi matrix*

$$i\partial\overline{\partial}W(z) = \left(\frac{\partial^2 W(z)}{\partial z_j \partial \overline{z}_k}\right)_{j,k=1}^n.$$

Suppose that

$$\lim_{|z|\to\infty} |z|^2\mu(z) = \infty. \tag{8.12}$$

Then the Hilbert space $A^2(\mathbb{C}^n, e^{-W})$ of all entire functions f such that

$$\int_{\mathbb{C}^n} |f(z)|^2 \exp(-W(z))\,d\lambda(z) < \infty$$

is of infinite dimension.

Proof. Assumption (8.12) implies that there exists a constant $K > 0$ such that

$$i\partial\overline{\partial}W(w,w)(z) \ge -K|w|^2$$

for all $z, w \in \mathbb{C}^n$ and that $i\partial\overline{\partial}W(z)$ is strictly positive for large $|z|$.

From Lemma 8.10 we have

$$i\partial\overline{\partial}g(w,w)(z) \ge \frac{4K}{(1+4K|z-\zeta|^2)^2}\,|w|^2,$$

where $g(z) = \log(1+4K|z-\zeta|^2)$.

Hence, for $|z-\zeta| \le 1/\sqrt{4K}$, we have

$$i\partial\bar{\partial}g(w,w)(z) \ge K|w|^2.$$

Since $i\partial\bar{\partial}W(w,w)(z)$ is negative at most on a compact set in $\mathbb{C}^n$, there exist finitely many points $\zeta_1,\ldots,\zeta_M \in \mathbb{C}^n$ such that this compact set is covered by the balls $\{z : |z-\zeta_l| < 1/\sqrt{4K}\}$. Hence

$$\tilde{\varphi}(z) := W(z) + \sum_{l=1}^{M} g_l(z)$$

is strictly plurisubharmonic, where $g_l(z) = \log(1+4K|z-\zeta_l|^2)$, $l = 1,\ldots,M$.

Let $\tilde{\mu}(z)$ be the least eigenvalue of $i\partial\bar{\partial}\tilde{\varphi}$. Then by assumption (8.12) we have

$$\lim_{|z|\to\infty} |z|^2\tilde{\mu}(z) = \infty.$$

For each $N \in \mathbb{N}$, there exists $R > 0$ such that

$$\tilde{\mu}(z) \ge \frac{N+M+1}{|z|^2} \quad \text{for } |z| > R.$$

Let $\tilde{\mu}_0 := \inf\{\tilde{\mu}(z) : |z| \le R\}$. Then $\tilde{\mu}_0 > 0$. Set

$$\kappa = \frac{\tilde{\mu}_0}{2(N+M)}$$

and

$$\varphi(z) := W(z) + \sum_{l=1}^{M} g_l(z) - (N+M)\log(1+\kappa|z|^2).$$

It follows that $e^{-\varphi}$ is locally integrable.

Next, we claim that φ is strictly plurisubharmonic. Notice that

$$i\partial\bar{\partial}\varphi(w,w)(z) \ge |w|^2\left(\tilde{\mu}(z) - \frac{(N+M)\kappa}{1+\kappa|z|^2}\right).$$

For $|z| \le R$, we have

$$\tilde{\mu}(z) - \frac{(N+M)\kappa}{1+\kappa|z|^2} \ge \tilde{\mu}_0 - (N+M)\kappa = \tilde{\mu}_0 - \frac{(N+M)\tilde{\mu}_0}{2(N+M)} = \frac{\tilde{\mu}_0}{2} > 0,$$

and for $|z| > R$, we have

$$\tilde{\mu}(z) - \frac{(N+M)\kappa}{1+\kappa|z|^2} \ge \frac{N+M+1}{|z|^2} - \frac{N+M}{|z|^2} = \frac{1}{|z|^2},$$

which implies that φ is strictly plurisubharmonic.

Therefore we can apply Theorem 8.9 and get an entire function f such that $f(0) = 1$ and

$$\int_{\mathbb{C}^n} |f(z)|^2 (1+|z|^2)^{-3n} e^{-\varphi(z)}\, d\lambda(z) < \infty.$$

Now we set $\tilde{N} = N - 3n$ and get

$$\begin{aligned}
&\int_{\mathbb{C}^n} |f(z)|^2 (1+|z|^2)^{\tilde{N}} e^{-W(z)}\, d\lambda(z) \\
&= \int_{\mathbb{C}^n} \frac{\prod_{l=1}^M (1+4K|z-\zeta_l|^2)}{(1+\kappa|z|^2)^{N+M}} |f(z)|^2 (1+|z|^2)^{\tilde{N}} e^{-\varphi(z)}\, d\lambda(z) \\
&\le \sup_{z\in\mathbb{C}^n} \left\{ \frac{(1+|z|^2)^N \prod_{l=1}^M (1+4K|z-\zeta_l|^2)}{(1+\kappa|z|^2)^{N+M}} \right\} \int_{\mathbb{C}^n} |f(z)|^2 (1+|z|^2)^{-3n} e^{-\varphi(z)}\, d\lambda(z) < \infty.
\end{aligned}$$

Hence $fp \in A^2(\mathbb{C}^n, e^{-W})$ for any polynomial p of degree $< \tilde{N}$, and since $N = \tilde{N} + 3n$ was arbitrary, we are done. □

The following example in $\mathbb{C}^2$ shows that Theorem 8.11 is not sharp.

Let $\varphi(z,w) = |z|^2|w|^2 + |w|^4$. In this case, we have that $A^2(\mathbb{C}^2, e^{-\varphi})$ contains all the functions $f_k(z,w) = w^k$ for $k \in \mathbb{N}$, since

$$\begin{aligned}
\int_0^\infty\int_0^\infty r_2^{2k} e^{-(r_1^2 r_2^2 + r_2^4)} r_1 r_2\, dr_1\, dr_2 &= \int_0^\infty \left(\int_0^\infty r_1 r_2^2 e^{-r_1^2 r_2^2}\, dr_1 \right) r_2^{2k-1} e^{-r_2^4}\, dr_2 \\
&= \int_0^\infty \left(\frac{1}{2} \int_0^\infty e^{-s}\, ds \right) r_2^{2k-1} e^{-r_2^4}\, dr_2 = \frac{1}{2} \int_0^\infty r_2^{2k-1} e^{-r_2^4}\, dr_2 < \infty.
\end{aligned}$$

The Levi matrix of φ has the form

$$i\partial\bar{\partial}\varphi = \begin{pmatrix} |w|^2 & \bar{z}w \\ \bar{w}z & |z|^2 + 4|w|^2 \end{pmatrix},$$

and hence φ is plurisubharmonic, and the lowest eigenvalue has the form

$$\begin{aligned}
\mu_\varphi(z,w) &= \frac{1}{2}\left(5|w|^2 + |z|^2 - (9|w|^4 + 10|z|^2|w|^2 + |z|^4)^{1/2}\right) \\
&= \frac{16|w|^4}{2(5|w|^2 + |z|^2 + (9|w|^4 + 10|z|^2|w|^2 + |z|^4)^{1/2})},
\end{aligned}$$

and hence

$$\lim_{|z|\to\infty} |z|^2 \mu_\varphi(z,0) = 0,$$

which implies that condition (8.12) of Theorem 8.11 is not satisfied.

8.3 Notes

The original proof of Hörmander's L^2-estimates uses a density lemma in weighted spaces. The L^2-estimates turned out to be an important method in different problems of analysis such as approximation and interpolation. The method is known as a real method in complex analysis (see [60]). It can also be used to prove the Levi problem that domains of holomorphy coincide with pseudoconvex domains [57]. The basic estimates (4.38) also hold for general bounded pseudoconvex domains (see [91]).

We simplified the proof of Shikegawa's result, Theorem 8.11. In the following, we will show that the infinite dimension of weighted spaces of entire functions is closely related to the compactness of the $\overline{\partial}$-Neumann operator on the corresponding weighted L^2-spaces, a phenomenon that also appears in applications to Dirac and Pauli operators.

Rozenblum and Shirokov [82] showed that $A^2(\mathbb{C}, e^{-\varphi})$ is of infinite dimension if φ is a subharmonic $\mathcal{C}^2$-function such that $\int_{\mathbb{C}} \triangle\varphi(z)\, d\lambda(z) = \infty$; see also Chapter 13. Borichev, Van An, and Youssfi [11] extended and completed these results.

9 Spectral analysis

In this chapter, we discuss the spectral theorem for self-adjoint unbounded operators on a separable Hilbert space and important consequences of it. For the general approach, we use the basic notions of measure theory such as σ-algebras and Borel sets. Moreover, we prove Ruelle's lemma and show how it can be used to interpret Hörmander's L^2-estimates. In the following, we will discuss applications of these results to the $\overline{\partial}$-Neumann operator, Schrödinger operators with magnetic fields, Dirac operators, and the Witten Laplacian.

9.1 Resolutions of the identity

We start with some definitions and basic facts.

Definition 9.1. Let Ω be a subset of $\mathbb{C}$, let M be a σ-algebra in Ω, and let H be a Hilbert space. A resolution of the identity is a mapping

$$E : \mathrm{M} \longrightarrow \mathcal{L}(H)$$

of M to the algebra $\mathcal{L}(H)$ of bounded linear operators on H with the following properties:
(a) $E(\emptyset) = 0$ (zero operator), $E(\Omega) = I$ (identity on H).
(b) For each $\omega \in \mathrm{M}$, the image $E(\omega)$ is an orthogonal projection on H.
(c) $E(\omega' \cap \omega'') = E(\omega')E(\omega'')$.
(d) If $\omega' \cap \omega'' = \emptyset$, then $E(\omega' \cup \omega'') = E(\omega') + E(\omega'')$.
(e) For all $x \in H$ and $y \in H$, the set function $E_{x,y}$ defined by

$$E_{x,y}(\omega) = (E(\omega)x, y)$$

is a complex measure on M.

We collect some immediate consequences of these properties.

Since each $E(\omega)$ is an orthogonal (i. e., self-adjoint) projection, for $x \in H$, we have

$$E_{x,x}(\omega) = (E(\omega)x, x) = (E(\omega)^2 x, x) = (E(\omega)x, E(\omega)x) = \|E(\omega)x\|^2, \tag{9.1}$$

and hence each $E_{x,x}$ is a positive measure on M with total variation

$$E_{x,x}(\Omega) = \|x\|^2. \tag{9.2}$$

By (c) any two of the projections $E(\omega)$ commute with each other; if $\omega \cap \omega' = \emptyset$, then (a) and (c) show that $\mathrm{im}(E(\omega)) \perp \mathrm{im}(E(\omega'))$, which follows from

$$(E(\omega)x, E(\omega')y) = (E(\omega)^2 x, E(\omega')y) = (E(\omega)x, E(\omega)E(\omega')y) = 0.$$

https://doi.org/10.1515/9783111182926-009

By (d), E is finitely additive. Concerning countable additivity, we have the following result.

Proposition 9.2. *If E is a resolution of the identity, and if $x \in H$, then*

$$\omega \mapsto E(\omega)x$$

is a countably additive H-valued measure on M.

If $\omega_n \in \mathrm{M}$ and $E(\omega_n) = 0$ for $n \in \mathbb{N}$, and if $\omega = \bigcup_{n=1}^{\infty} \omega_n$, then $E(\omega) = 0$.

Proof. By (d), $\omega \mapsto (E(\omega)x, y)$ is a complex measure, and hence

$$\sum_{n=1}^{\infty} (E(\omega_n)x, y) = (E(\omega)x, y) \tag{9.3}$$

for every $y \in H$.

For $n \neq m$, we have $E(\omega_n)x \perp E(\omega_m)x$. Let

$$\Lambda_N(y) = \sum_{n=1}^{N} (y, E(\omega_n)x).$$

By (9.3) the sequence $(\Lambda_N(y))_N$ converges for every $y \in H$. The uniform boundedness principle (Theorem 4.14) implies that $(\|\Lambda_N\|)_N$ is bounded, where

$$\|\Lambda_N\| = \|E(\omega_1)x + \cdots + E(\omega_N)x\| = \left(\|E(\omega_1)x\|^2 + \cdots + \|E(\omega_N)x\|^2\right)^{1/2},$$

and hence by the orthogonality the partial sums

$$\sum_{n=1}^{N} E(\omega_n)x$$

form a Cauchy sequence in H, so

$$\sum_{n=1}^{\infty} E(\omega_n)x = E(\omega)x,$$

and $\omega \mapsto E(\omega)x$ is countably additive and therefore a complex measure on M.

For the second claim, observe that $E(\omega_n) = 0$ implies $E_{x,x}(\omega_n) = 0$ for every $x \in H$. Since $E_{x,x}$ is countably additive, it follows that $E_{x,x}(\omega) = 0$. Since $\|E(\omega)x\|^2 = E_{x,x}(\omega)$, $E(\omega) = 0$. □

Definition 9.3. Let E be a resolution of the identity on M, and let f be a complex M-measurable function on Ω. There is a countable family $(D_k)_k$ of open discs forming a base for the topology of $\mathbb{C}$. Let V be the union of those D_k for which $E(f^{-1}(D_k)) = 0$.

By Proposition 9.2, $E(f^{-1}(V)) = 0$. Also, V is the largest open subset of $\mathbb{C}$ with this property. The essential range of f is, by definition, the complement of V. It is the smallest closed subset of $\mathbb{C}$ that contains $f(z)$ for all $z \in \Omega$ except those that lie in some set $\omega \in \mathrm{M}$ with $E(\omega) = 0$. We say that f is essentially bounded if its essential range is bounded, hence compact. The largest value of $|\lambda|$ as λ runs through the essential range of f is called the essential supremum $\|f\|_\infty$ of f.

Let B be the algebra of all bounded complex M-measurable functions on Ω with the norm

$$\|f\| = \sup_{z\in\Omega}|f(z)|,$$

and let

$$N = \{f \in B : \|f\|_\infty = 0\},$$

which by Proposition 9.2 is a closed ideal. Hence B/N is a Banach algebra, which is denoted by $L^\infty(E)$. The norm of a coset $[f] = f + N$ is $\|f\|_\infty$, the spectrum $\sigma([f])$ is the essential range of f, and the spectrum of an element g in a Banach algebra is the set of all complex numbers λ such that $\lambda e - g$ is not invertible.

In the next step, we describe that a resolution of the identity induces an isometric isomorphism of the Banach algebra $L^\infty(E)$ onto a closed normal subalgebra $\mathcal{A}$ of $\mathcal{L}(H)$, the algebra of all bounded linear operators from H to H (a normal subalgebra is a commutative one that contains T^* for every $T \in \mathcal{A}$).

For this purpose, let $\{\omega_1, \dots, \omega_n\}$ be a partition of Ω with $\omega_j \in \mathrm{M}$, and let s be a simple function, such that $s = a_j$ on ω_j. Define $\Psi(s) \in \mathcal{L}(H)$ by

$$\Psi(s) = \sum_{j=1}^{n} a_j E(\omega_j).$$

Since each $E(\omega_j)$ is self-adjoint, $\Psi(s)^* = \Psi(\overline{s})$. If t is another simple function and $\alpha, \beta \in \mathbb{C}$, then we have

$$\Psi(s)\Psi(t) = \Psi(st) \quad \text{and} \quad \Psi(\alpha s + \beta t) = \alpha\Psi(s) + \beta\Psi(t).$$

For $x, y \in H$, we get

$$(\Psi(s)x, y) = \sum_{j=1}^{n} a_j (E(\omega_j)x, y) = \sum_{j=1}^{n} a_j E_{x,y}(\omega_j) = \int_\Omega s\, dE_{x,y}.$$

In addition, we have

$$\Psi(s)^*\Psi(s) = \Psi(|s|^2) \quad \text{and} \quad \|\Psi(s)x\|^2 = \int_\Omega |s|^2\, dE_{x,x}.$$

By (9.1) this implies

$$\|\Psi(s)x\| \le \|s\|_\infty \|x\|, \tag{9.4}$$

and if $x \in \operatorname{im}(E(\omega_j))$, then

$$\Psi(s)x = \alpha_j E(\omega_j)x = \alpha_j x,$$

since the projections $E(\omega_j)$ have mutually orthogonal ranges. If j is chosen so that $|\alpha_j| = \|s\|_\infty$, then by (9.4) it follows that

$$\|\Psi(s)\| = \sup_{\|x\|\le 1} \|\Psi(s)x\| = \|s\|_\infty. \tag{9.5}$$

Now suppose that $f \in L^\infty(E)$. There is a sequence of simple measurable functions s_k that converges to f in the norm of $L^\infty(E)$. By (9.5) the corresponding operators $\Psi(s_k)$ form a Cauchy sequence in $\mathcal{L}(H)$, which is therefore norm-convergent to an operator that we call $\Psi(f)$. By (9.5) we get

$$\|\Psi(f)\| = \|f\|_\infty. \tag{9.6}$$

Thus Ψ is an isometric isomorphism of $L^\infty(E)$ into $\mathcal{L}(H)$. Since $L^\infty(E)$ is complete, $\mathcal{A} = \Psi(L^\infty(E))$ is closed in $\mathcal{L}(H)$. In addition, we have

$$(\Psi(f)x, y) = \int_\Omega f\, dE_{x,y} \quad \text{and} \quad \|\Psi(f)x\|^2 = \int_\Omega |f|^2\, dE_{x,x},$$

which justifies the notation

$$\Psi(f) = \int_\Omega f\, dE.$$

9.2 Spectral decomposition of bounded normal operators

We will now relate a resolution of the identity to a certain operator in $\mathcal{L}(H)$.

Definition 9.4. The spectrum $\sigma(T)$ of an operator $T \in \mathcal{L}(H)$ is the set of all $\lambda \in \mathbb{C}$ such that $\lambda I - T$ has no inverse in $\mathcal{L}(H)$. The complement $\rho(T) = \mathbb{C}\backslash\sigma(T)$ is called the resolvent set. If $\lambda \in \rho(T)$, then the operator $(\lambda I - T)^{-1} \in \mathcal{L}(H)$ is called the resolvent of T at λ and is denoted by $R_T(\lambda)$. We have an operator-valued function

$$R_T : \rho(T) \longrightarrow \mathcal{L}(H).$$

An operator $T \in \mathcal{L}(H)$ is called normal if $TT^* = T^*T$.

Proposition 9.5. *Let $T \in \mathcal{L}(H)$. Then the spectrum $\sigma(T)$ of T is a compact subset of $\mathbb{C}$, and $|\lambda| \leq \|T\|$ for every $\lambda \in \sigma(T)$.*

Proof. First, we show that $\rho(T)$ is open. Let $\lambda \in \rho(T)$. Then

$$\alpha = \|R_T(\lambda)\|^{-1} > 0.$$

Let $\mu \in \mathbb{C}$ with $|\mu| < \alpha$. We will show that $(\lambda + \mu)I - T$ has a bounded inverse. Then we prove that $\rho(T)$ is open. We have

$$\begin{aligned}(\lambda + \mu)I - T &= \lambda I - T + \mu I \\ &= (\lambda I - T)[I + \mu(\lambda I - T)^{-1}] \\ &= (\lambda I - T)(I + \mu R_T(\lambda)).\end{aligned}$$

Formally,

$$(I + \mu R_T(\lambda))^{-1} = I + \sum_{k=1}^{\infty}(-1)^k(\mu R_T(\lambda))^k,$$

but as $|\mu| < \alpha$, we have

$$\left\|\sum_{k=1}^{\infty}(-1)^k(\mu R_T(\lambda))^k\right\| \leq \sum_{k=1}^{\infty}|\mu|^k\,\|R_T(\lambda)\|^k = \sum_{k=1}^{\infty}(|\mu|/\alpha)^k < \infty,$$

and therefore the partial sums of $\sum_{k=1}^{\infty}(-1)^k(\mu R_T(\lambda))^k$ form a Cauchy sequence in $\mathcal{L}(H)$. Since $\mathcal{L}(H)$ is complete, we obtain that

$$\sum_{k=1}^{\infty}(-1)^k(\mu R_T(\lambda))^k \in \mathcal{L}(H),$$

and $\rho(T)$ is open.

If $\eta \in \mathbb{C}$ with $|\eta| > \|T\|$, then $I - T/\eta$ has a bounded inverse, since

$$(I - T/\eta)^{-1} = I + \sum_{k=1}^{\infty}(T/\eta)^k.$$

This implies that $\eta I - T$ has a bounded inverse. Hence, if $\lambda \in \sigma(T)$, then $|\lambda| \leq \|T\|$, and $\sigma(T)$ is a bounded set. □

The resolvent has the following properties:

Lemma 9.6. *If $\lambda, \mu \in \rho(T)$, then*

$$R_T(\lambda) - R_T(\mu) = (\mu - \lambda)R_T(\lambda)R_T(\mu). \tag{9.7}$$

If $\lambda \in \rho(T)$ and $|\lambda - \mu| < \|R_T(\lambda)\|^{-1}$, then

$$R_T(\mu) = \sum_{k=0}^{\infty} (\lambda - \mu)^k [R_T(\lambda)]^{k+1}, \tag{9.8}$$

and therefore we say that R_T is a holomorphic operator-valued function.

Proof. Relation (9.7) follows from

$$\begin{aligned} R_T(\lambda) = R_T(\lambda)(\mu I - T)R_T(\mu) &= R_T(\lambda)[(\lambda I - T) + (\mu - \lambda)I]R_T(\mu) \\ &= R_T(\mu) + (\mu - \lambda)R_T(\lambda)R_T(\mu). \end{aligned}$$

Equality (9.8) follows immediately from the proof of Proposition 9.5. □

The spectral theorem indicates that every bounded normal operator T on a Hilbert space induces a resolution E of the identity on the Borel subsets of its spectrum $\sigma(T)$ and that T can be reconstructed from E by an integral of the type discussed before.

Using Banach algebra techniques such as the Gelfand transform (see [85]), we obtain the spectral decomposition for a single normal operator.

Proposition 9.7. *If $T \in \mathcal{L}(H)$ and T is normal, then there exists a uniquely determined resolution of the identity E on the Borel subsets of the spectrum $\sigma(T)$ that satisfies*

$$T = \int_{\sigma(T)} \lambda \, dE(\lambda) \text{ and } (Tx, y) = \int_{\sigma(T)} \lambda \, dE_{x,y}(\lambda). \tag{9.9}$$

Furthermore, every projection $E(\omega)$ commutes with every $S \in \mathcal{L}(H)$ that commutes with T.

We will refer to this E as the spectral decomposition of T.

We list a few consequences of the spectral decomposition.
- If $\omega \subseteq \sigma(T)$ is a nonempty open set, then $E(\omega) \neq 0$.
- If f is a bounded Borel function on $\sigma(T)$, then it is customary to denote the operator

$$\Psi(f) = \int_{\sigma(T)} f \, dE$$

by $f(T)$.

The mapping $f \mapsto f(T)$ establishes a homomorphism of the algebra of all bounded Borel functions on $\sigma(T)$ into $\mathcal{L}(H)$, which carries the function 1 to I, the identity function on $\sigma(T)$ to T, and satisfies

$$\bar{f}(T) = f(T)^* \quad \text{and} \quad \|f(T)\| \leq \sup\{|f(\lambda)| : \lambda \in \sigma(T)\}.$$

The procedure explained above is also called symbolic calculus.

If $f \in C(\sigma(T))$, then $f \mapsto f(T)$ is an isomorphism on $C(\sigma(T))$ satisfying

$$\|f(T)x\|^2 = \int_{\sigma(T)} |f|^2 \, dE_{x,x}.$$

The eigenvalues of a normal operator can be characterized in terms of the spectral decomposition. For this purpose, we mention the following applications of the symbolic calculus.

Proposition 9.8. *Let $T \in \mathcal{L}(H)$ be a normal operator, and let E be its spectral decomposition. If $f \in C(\sigma(T))$ and $\omega_0 = f^{-1}(0)$, then*

$$\ker(f(T)) = \operatorname{im}(E(\omega_0)).$$

Proof. We set $h(\lambda) = 1$ on ω_0 and $h(\lambda) = 0$ on $\tilde{\omega} = \sigma(T) \setminus \omega_0$. Then $fh = 0$, and by the symbolic calculus, $f(T)h(T) = 0$. Since $h(T) = E(\omega_0)$, it follows that

$$\operatorname{im}(E(\omega_0)) \subseteq \ker(f(T)).$$

For the opposite inclusion, we define for $n \in \mathbb{N}$ the set

$$\omega_n = \{\lambda \in \sigma(T) : 1/n \leq |f(\lambda)| < 1/(n-1)\}.$$

Then $\tilde{\omega}$ is the union of the disjoint Borel sets ω_n. We define $f_n(\lambda) = 1/f(\lambda)$ on ω_n and $f_n(\lambda) = 0$ on $\sigma(T) \setminus \omega_n$. Then each f_n is a bounded Borel function on $\sigma(T)$, and

$$f_n(T)f(T) = E(\omega_n), \quad n \in \mathbb{N}.$$

If $f(T)x = 0$, then it follows that $E(\omega_n)x = 0$. Since the mapping $\omega \mapsto E(\omega)x$ is countably additive (Proposition 9.2), we obtain $E(\tilde{\omega})x = 0$. We also have that

$$E(\tilde{\omega}) + E(\omega_0) = I.$$

Hence $E(\omega_0)x = x$, and therefore

$$\ker(f(T)) \subseteq \operatorname{im}(E(\omega_0)).$$ □

Proposition 9.9. *Let $T \in \mathcal{L}(H)$ be a normal operator, and let E be its spectral decomposition. Let $\lambda_0 \in \sigma(T)$ and $E_0 = E(\{\lambda_0\})$. Then*
(a) $\ker(T - \lambda_0 I) = \operatorname{im}(E_0)$;
(b) *λ_0 is an eigenvalue of T if and only if $E_0 \neq 0$;*
(c) *Every isolated point of $\sigma(T)$ is an eigenvalue of T;*

(d) *If* $\sigma(T) = \{\lambda_1, \lambda_2, \dots\}$ *is a countable set, then every* $x \in H$ *has a unique expansion of the form*

$$x = \sum_{j=1}^{\infty} x_j,$$

where $Tx_j = \lambda_j x_j$, *and* $x_j \perp x_k$ *for* $j \neq k$;

(e) *If* $\sigma(T)$ *has no limit point except possibly* 0 *and if* $\dim \ker(T - \lambda I) < \infty$ *for* $\lambda \neq 0$, *then* T *is compact (compare with Section* 2.1*).*

Proof. (a) is an immediate consequence of Proposition 9.8 with $f(\lambda) = \lambda - \lambda_0$.

(b) follows from (a).

(c) If λ_0 is an isolated point of $\sigma(T)$, then $\{\lambda_0\}$ is a nonempty open subset of $\sigma(T)$, and by the properties of the spectral decomposition listed above we get $E_0 \neq 0$, and therefore (c) follows from (b).

(d) Let $E_j = E(\{\lambda_j\})$ for $j \in \mathbb{N}$. The projections E_j have pairwise orthogonal ranges, and the mapping $\omega \mapsto E(\omega)x$ is countably additive (Proposition 9.2). Hence for each $x \in H$, we have

$$\sum_{j=1}^{\infty} E_j x = E(\sigma(T))x = x,$$

and the series converges in the norm of H. Now set $x_j = E_j x$ and observe that the uniqueness follows from the orthogonality of the vectors x_j and that $Tx_j = \lambda_j x_j$ follows from (a).

(e) Let $\{\lambda_j\}$ be an enumeration of the nonzero points of $\sigma(T)$ such that $|\lambda_1| \geq |\lambda_2| \geq \cdots$. Define $f_n(\lambda) = \lambda$ if $\lambda = \lambda_j$ and $j \leq n$, and put $f_n(\lambda) = 0$ at the other points of $\sigma(T)$. We set again $E_j = E(\{\lambda_j\})$ and obtain

$$f_n(T) = \lambda_1 E_1 + \cdots + \lambda_n E_n.$$

Since $\dim \operatorname{im}(E_j) = \dim \ker(T - \lambda_j I) < \infty$, each $f_n(T)$ is a compact operator. We have $|\lambda - f_n(\lambda)| \leq |\lambda_n|$ for all $\lambda \in \sigma(T)$ and, by the symbolic calculus,

$$\|T - f_n(T)\| \leq |\lambda_n| \to 0 \quad \text{as } n \to \infty.$$

By Proposition 2.2, T is compact. □

The symbolic calculus is a powerful tool in operator theory. Finally, we mention important applications to positive operators:

Definition 9.10. An operator $T \in \mathcal{L}(H)$ is called positive if $(Tx, x) \geq 0$ for every $x \in H$. We write $T \geq 0$.

Proposition 9.11.

(a) *$T \in \mathcal{L}(H)$ is positive if and only if $T = T^*$ and $\sigma(T) \subset [0, \infty)$.*

(b) *Every positive $T \in \mathcal{L}(H)$ has a unique positive square root $S \in \mathcal{L}(H)$, i. e., $S^2 = T$.*

(c) *If $T \in \mathcal{L}(H)$, then T^*T is positive, and the positive square root P of T^*T is the only positive operator in $\mathcal{L}(H)$ satisfying $\|Px\| = \|Tx\|$ for every $x \in H$.*

(d) *If $T \in \mathcal{L}(H)$ is normal, then T has a polar decomposition $T = UP$, where U is unitary, and P is positive.*

Proof. (a) (Tx, x) and (x, Tx) are complex conjugates of each other. If T is positive, then (Tx, x) is real, so that

$$(x, T^*x) = (Tx, x) = (x, Tx)$$

for every $x \in H$. Hence $T = T^*$ (by the proof of Theorem 2.7). Let $\lambda = \alpha + i\beta \in \sigma(T)$ and put $T_\lambda = T - \lambda I$. Then

$$\|T_\lambda x\|^2 = \|Tx - \alpha x\|^2 + \beta^2\|x\|^2,$$

so that $\|T_\lambda x\| \geq |\beta|\,\|x\|$. If $\beta \neq 0$, then it follows that T_λ is invertible, which means that $\lambda \notin \sigma(T)$. So we get that $\sigma(T)$ lies in the real axis. If $\lambda > 0$, then we obtain

$$\lambda\|x\|^2 = (\lambda x, x) \leq ((T + \lambda I)x, x) \leq \|(T + \lambda I)x\|\,\|x\|,$$

so that $\|(T + \lambda I)x\| \geq \lambda\|x\|$, which implies that $T + \lambda I$ is invertible in $\mathcal{L}(H)$ and $-\lambda \notin \sigma(T)$, and hence $\sigma(T) \subset [0, \infty)$.

Now assume that $T = T^*$ and $\sigma(T) \subset [0, \infty)$. Let E be the spectral decomposition of T. We have

$$(Tx, x) = \int_{\sigma(T)} \lambda\, dE_{x,x}(\lambda).$$

Since $E_{x,x}$ is a positive measure and $\lambda \geq 0$ on $\sigma(T)$, we obtain that T is positive.

(b) By (a), Proposition 9.5, and the symbolic calculus, $\sigma(T)$ is a compact subset of $\mathbb{R}^+$, and there exists a uniquely determined spectral decomposition E such that

$$T = \int_{\sigma(T)} \lambda\, dE(\lambda).$$

Define

$$S = \int_{\sigma(T)} \lambda^{1/2}\, dE(\lambda).$$

Then S is a positive self-adjoint operator with $S^2 = T$. In addition, there is a sequence of polynomials p_n such that $p_n(\lambda) \to \lambda^{1/2}$ uniformly on $\sigma(T)$ (see Corollary 12.5) and

$$\lim_{n\to\infty} \|p_n(T) - S\| = 0.$$

Let $\tilde{S}$ be an arbitrary positive self-adjoint operator with $\tilde{S}^2 = T$. Since $T\tilde{S} = \tilde{S}^3$ and $\tilde{S}T = \tilde{S}^3$, the operator $\tilde{S}$ commutes with T, and so with polynomials of T. Hence also with $S = \lim_{n\to\infty} p_n(T)$. Let $x \in H$ and put $y = (S - \tilde{S})x$. Using that $\tilde{S}S = S\tilde{S}$ and $S^2 = \tilde{S}^2$, we obtain

$$(Sy, y) + (\tilde{S}y, y) = ((S + \tilde{S})(S - \tilde{S})x, y) = ((S^2 - \tilde{S}^2)x, y) = 0.$$

Since S and $\tilde{S}$ are positive, $(Sy, y) = (\tilde{S}y, y) = 0$. Hence $Sy = \tilde{S}y = 0$, because $(S\cdot, \cdot)$ is a positive semidefinite sesquilinear form, so we can apply the Cauchy–Schwarz inequality

$$|(Sy, z)|^2 \leq (Sy, y)(Sz, z)$$

for all $z \in H$. Now we get

$$\|(S - \tilde{S})x\|^2 = ((S - \tilde{S})^2 x, x) = ((S - \tilde{S})y, x) = 0,$$

which yields $Sx = \tilde{S}x$ and $S = \tilde{S}$.

(c) Note first that

$$(T^*Tx, x) = (Tx, Tx) = \|Tx\|^2 \geq 0 \quad \text{for } x \in H,$$

so that T^*T is positive. If $P \in \mathcal{L}(H)$ and $P^* = P$, then

$$(P^2 x, x) = (Px, Px) = \|Px\|^2.$$

Then by the proof of Theorem 2.7 it follows that $\|Px\| = \|Tx\|$ for every $x \in H$ if and only if $P^2 = T^*T$.

(d) Put $p(\lambda) = |\lambda|$ and $u(\lambda) = \lambda/|\lambda|$ for $\lambda \neq 0$ and $u(0) = 1$. Then p and u are bounded Borel functions on $\sigma(T)$. Put $P = p(T)$ and $U = u(T)$. As $p \geq 0$, we get from (a) that $P \geq 0$. Since $u\bar{u} = 1$, we get $UU^* = U^*U = I$, and since $\lambda = u(\lambda)p(\lambda)$, the relation $T = UP$ follows from the symbolic calculus. □

Remark 9.12. We apply the last results to the $\overline{\partial}$-Neumann operator N to prove the following:

Suppose that

$$\square : \mathrm{dom}(\square) \longrightarrow L^2_{(0,1)}(\Omega)$$

is bijective and has a bounded inverse N. Then we have the basic estimate

$$\|u\|^2 \leq C\,(\|\overline{\partial} u\|^2 + \|\overline{\partial}^* u\|^2), \quad u \in \operatorname{dom}(\Box). \quad (4.38)$$

The operator N is self-adjoint and bounded and therefore, by Proposition 9.11, has a bounded self-adjoint root $N^{1/2}$, which is again injective. By Lemma 4.28, $N^{1/2}$ has a self-adjoint inverse, which we denote by $N^{-1/2}$. Let $u \in \operatorname{dom}(\Box)$. Then there exists $w \in L^2_{(0,1)}(\Omega)$ such that $Nw = u$. Hence we have $N^{1/2}v = u$, where $v = N^{1/2}w$, and $N^{-1/2}v = w = N^{-1/2}N^{-1/2}u$ is well defined. Now we get

$$\begin{aligned}\|u\|^2 = \|N^{1/2}v\|^2 &\leq C\|v\|^2 = C\,(N^{-1/2}u, N^{-1/2}u)\\ &= C\,(N^{-1/2}N^{-1/2}u, u) = C\,(N^{-1/2}N^{-1/2}Nw, Nw)\\ &= C\,(w, Nw) = C\,(\Box u, u)\\ &= C\,(\|\overline{\partial} u\|^2 + \|\overline{\partial}^* u\|^2),\end{aligned}$$

which is the basic estimate (4.38).

9.3 Spectral decomposition of unbounded self-adjoint operators

Let Ω be a subset of $\mathbb{C}$, let M be a σ-algebra in Ω, and let H be a Hilbert space. Let

$$E : \mathrm{M} \longrightarrow \mathcal{L}(H)$$

be a resolution of the identity. The symbolic calculus associates with every $f \in L^\infty(E)$ an operator $\Psi(f) \in \mathcal{L}(H)$ by the formula

$$(\Psi(f)x, y) = \int_\Omega f\, dE_{x,y}, \quad x, y \in H.$$

Now we will extend this for unbounded measurable functions f.

Lemma 9.13. *Let $f : \Omega \longrightarrow \mathbb{C}$ be a measurable function. Put*

$$\mathcal{D}_f = \left\{x \in H : \int_\Omega |f|^2\, dE_{x,x} < \infty\right\}.$$

Then $\mathcal{D}_f$ is a dense subspace of H. If $x, y \in H$, then

$$\int_\Omega |f|\, d|E_{x,y}| \leq \|y\| \left[\int_\Omega |f|^2\, dE_{x,x}\right]^{1/2}. \tag{9.10}$$

If f is bounded and $u = \Psi(f)v$ for $v \in H$, then

$$dE_{x,u} = \overline{f}\, dE_{x,v}, \quad x \in H. \tag{9.11}$$

Proof. Let $z = x + y$ and $\omega \in \mathrm{M}$. Then

$$\|E(\omega)z\|^2 \leq (\|E(\omega)x\| + \|E(\omega)y\|)^2 \leq 2(\|E(\omega)x\|^2 + \|E(\omega)y\|^2).$$

Recall that $E_{x,x}(\omega) = (E(\omega)x, x) = (E(\omega)^2 x, x) = \|E(\omega)x\|^2$, so we get from the above

$$E_{z,z}(\omega) \leq 2(E_{x,x}(\omega) + E_{y,y}(\omega)),$$

which implies that $\mathcal{D}_f$ is closed under addition. It is clear that $\mathcal{D}_f$ is also closed under scalar multiplication. Therefore $\mathcal{D}_f$ is a subspace of H.

For $n \in \mathbb{N}$, let ω_n be the subset of Ω where $|f| < n$. If $x \in \operatorname{im}(E(\omega_n))$, then

$$E(\omega)x = E(\omega)E(\omega_n)x = E(\omega \cap \omega_n)x, \quad \omega \in \mathrm{M}.$$

Hence

$$E_{x,x}(\omega) = E_{x,x}(\omega \cap \omega_n),$$

and therefore

$$\int_{\Omega} |f|^2 \, dE_{x,x} = \int_{\omega_n} |f|^2 \, dE_{x,x} \leq n^2 \|x\|^2 < \infty.$$

Thus $\operatorname{im}(E(\omega_n)) \subset \mathcal{D}_f$. Since $\Omega = \bigcup_{n=1}^{\infty} \omega_n$, the countable additivity of $\omega \mapsto E(\omega)y$ implies that $y = \lim_{n\to\infty} E(\omega_n)y$ for every $y \in H$. Hence y lies in the closure of $\mathcal{D}_f$, and $\mathcal{D}_f$ is dense in H.

If $x, y \in H$ and f is bounded and measurable, the Radon–Nikodym theorem (see, for instance, [83]) implies that there is a measurable function g on Ω such that $|g| = 1$ on Ω and

$$gf \, dE_{x,y} = |f| \, d|E_{x,y}|.$$

Hence

$$\int_{\Omega} |f| \, d|E_{x,y}| = (\Psi(gf)x, y) \leq \|\Psi(gf)x\| \, \|y\|. \tag{9.12}$$

As in Section 9.1, we get

$$\|\Psi(gf)x\|^2 = \int_{\Omega} |gf|^2 \, dE_{x,x} = \int_{\Omega} |f|^2 \, dE_{x,x},$$

which implies (9.10) for a bounded function f. The general case is done as in the first statement of this proposition.

To show (9.11), for an arbitrary bounded measurable function h, we have

$$\begin{aligned}\int_\Omega h\,dE_{x,u} &= (\Psi(h)x, u) = (\Psi(h)x, \Psi(f)v)\\ &= (\Psi(\bar f)\Psi(h)x, v) = (\Psi(\bar f h)x, v)\\ &= \int_\Omega h\bar f\,dE_{x,v}.\end{aligned}$$

□

In the next step, we carry over the results of Section 9.1 (symbolic calculus) for unbounded measurable functions.

Proposition 9.14. *Let E be a resolution of identity on Ω.*

(a) *To every measurable $f : \Omega \longrightarrow \mathbb{C}$, there corresponds a densely defined closed operator $\Psi(f)$ on H with domain $\operatorname{dom}(\Psi(f)) = \mathcal{D}_f$, which is characterized by*

$$(\Psi(f)x, y) = \int_\Omega f\,dE_{x,y}, \quad x \in \mathcal{D}_f, \quad y \in H, \tag{9.13}$$

and satisfies

$$\|\Psi(f)x\|^2 = \int_\Omega |f|^2\,dE_{x,x}, \quad x \in \mathcal{D}_f. \tag{9.14}$$

(b) *If f and g are measurable, then*

$$\Psi(f)\Psi(g) \subset \Psi(fg),$$

which means that $\operatorname{dom}(\Psi(f)\Psi(g)) \subset \operatorname{dom}(\Psi(fg))$ and $\Psi(f)\Psi(g) = \Psi(fg)$ on $\operatorname{dom}(\Psi(f)\Psi(g))$, and

$$\operatorname{dom}(\Psi(f)\Psi(g)) = \mathcal{D}_g \cap \mathcal{D}_{fg}.$$

Hence $\Psi(f)\Psi(g) = \Psi(fg)$ if and only if $\mathcal{D}_{fg} \subseteq \mathcal{D}_g$.

(c) *For every measurable $f : \Omega \longrightarrow \mathbb{C}$,*

$$\Psi(f)^* = \Psi(\bar f) \quad \text{and} \quad \Psi(f)\Psi(f)^* = \Psi(|f|^2) = \Psi(f)^*\Psi(f).$$

Proof. Fix $x \in \mathcal{D}_f$. Then the conjugate-linear functional $y \mapsto \int_\Omega f\,dE_{x,y}$ is bounded on H (Lemma 9.13). Hence there is a unique element $\Psi(f)x \in H$ satisfying (9.13) and

$$\|\Psi(f)x\|^2 \le \int_\Omega |f|^2\,dE_{x,x}, \quad x \in \mathcal{D}_f. \tag{9.15}$$

The linearity of $\Psi(f)$ on $\mathcal{D}_f$ follows from (9.13) and the fact that $E_{x,y}$ is linear in x.

Now we associate with each f its truncations $f_n = f\phi_n$, where $\phi_n(p) = 1$ if $|f(p)| \le n$ and $\phi_n(p) = 0$ if $|f(p)| > n$.

Then $\mathcal{D}_{f-f_n} = \mathcal{D}_f$, since each f_n is bounded, and therefore (9.15) shows, using the dominated convergence theorem, that

$$\|\Psi(f)x - \Psi(f_n)x\|^2 \le \int_\Omega |f - f_n|^2 \, dE_{x,x} \to 0 \quad \text{as } n \to \infty \tag{9.16}$$

for every $x \in \mathcal{D}_f$. Since f_n is bounded, (9.14) holds for f_n. Hence (9.16) implies (9.14) for f.

This proves (a), except for the statement that $\Psi(f)$ is closed. This will follow from (c) (to be proved) and Lemma 4.5 with $\overline{f}$ instead of f.

(b) First, assume that f is bounded. Then $\mathcal{D}_{fg} \subset \mathcal{D}_g$. If $v \in H$ and $u = \Psi(\overline{f})v$, then we get from Section 9.1 and (9.11) that

$$\begin{aligned}(\Psi(f)\Psi(g)x, v) &= (\Psi(g)x, \Psi(\overline{f})v) = (\Psi(g)x, u)\\ &= \int_\Omega g \, dE_{x,u} = \int_\Omega fg \, dE_{x,v}\\ &= (\Psi(fg)x, v).\end{aligned}$$

So we have shown that

$$\Psi(f)\Psi(g)x = \Psi(fg)x, \quad x \in \mathcal{D}_g, \quad f \in L^\infty.$$

The last line implies that for $y = \Psi(g)x$,

$$\int_\Omega |f|^2 \, dE_{y,y} = \int_\Omega |fg|^2 \, dE_{x,x}, \quad x \in \mathcal{D}_g, \quad f \in L^\infty. \tag{9.17}$$

Using truncation, we see that (9.17) also holds for arbitrary f (possibly unbounded). Since $\operatorname{dom}(\Psi(f)\Psi(g))$ consists of all $x \in \mathcal{D}_g$ such that $y = \Psi(g)x \in \mathcal{D}_f$ and since (9.17) shows that $y \in \mathcal{D}_f$ if and only if $x \in \mathcal{D}_{fg}$, we see that

$$\operatorname{dom}(\Psi(f)\Psi(g)) = \mathcal{D}_g \cap \mathcal{D}_{fg}.$$

If $x \in \mathcal{D}_g \cap \mathcal{D}_{fg}$ and $y = \Psi(g)x$ and if the truncations f_n are defined as above, we obtain $f_n \to f$ in $L^2(E_{x,x})$ and $f_n g \to fg$ in $L^2(E_{y,y})$ and finally

$$\Psi(f)\Psi(g)x = \Psi(f)y = \lim_{n\to\infty} \Psi(f_n)y = \lim_{n\to\infty} \Psi(f_n g)x = \Psi(fg)x.$$

This proves (b).

(c) Suppose that $x \in \mathcal{D}_f$ and $y \in \mathcal{D}_{\overline{f}} = \mathcal{D}_f$. It follows from (9.16) and Section 9.1 that

$$(\Psi(f)x, y) = \lim_{n\to\infty} (\Psi(f_n)x, y) = \lim_{n\to\infty} (x, \Psi(\overline{f}_n)y) = (x, \Psi(\overline{f})y).$$

Hence $y \in \operatorname{dom}(\Psi(f)^*)$, and $\operatorname{dom}(\Psi(\bar{f})) \subseteq \operatorname{dom}(\Psi(f)^*)$. If we can show that each $u \in \operatorname{dom}(\Psi(f)^*)$ lies in $\mathcal{D}_f$, then we obtain $\Psi(f)^* = \Psi(\bar{f})$. Fix u for this purpose and put $v = \Psi(f)^* u$. Since $f_n = f\phi_n$, the multiplication theorem yields

$$\Psi(f_n) = \Psi(f)\Psi(\phi_n).$$

Since $\Psi(\phi_n)$ is self-adjoint and bounded, by Lemma 4.3 and Section 9.1 we have that

$$\Psi(\phi_n)\Psi(f)^* = [\Psi(f)\Psi(\phi_n)]^* = \Psi(f_n)^* = \Psi(\bar{f}_n).$$

Hence

$$\Psi(\phi_n)v = \Psi(\bar{f}_n)u, \quad n \in \mathbb{N}.$$

Since $|\phi_n| \leq 1$, we now have

$$\int_\Omega |f_n|^2 \, dE_{u,u} = \int_\Omega |\phi_n|^2 \, dE_{v,v} \leq E_{v,v}(\Omega), \quad n \in \mathbb{N}.$$

Hence $u \in \mathcal{D}_f$.

Finally, since $\mathcal{D}_{f\bar{f}} \subset \mathcal{D}_f$, another application of the multiplication theorem gives the last statement of (c). □

Definition 9.15. The resolvent set of a linear operator $T : \operatorname{dom}(T) \longrightarrow H$ is the set of all $\lambda \in \mathbb{C}$ such that $\lambda I - T$ is an injective mapping of $\operatorname{dom}(T)$ onto H whose inverse belongs to $\mathcal{L}(H)$. The spectrum $\sigma(T)$ of T is the complement of the resolvent set of T.

First, we collect some information about the spectrum of an unbounded operator.

Lemma 9.16. *If the spectrum $\sigma(T)$ of an operator T does not coincide with the whole complex plane $\mathbb{C}$, then T must be a closed operator. The spectrum of a linear operator is always closed. Moreover, if $\zeta \notin \sigma(T)$ and $c := \|R_T(\zeta)\| = \|(\zeta I - T)^{-1}\|$, then the spectrum $\sigma(T)$ does not intersect the ball $\{w \in \mathbb{C} : |\zeta - w| < c^{-1}\}$. The resolvent operator R_T is a holomorphic operator-valued function.*

Proof. For $\zeta \notin \sigma(T)$, let $S = (\zeta I - T)^{-1}$, which is a bounded operator. Let $x_n \in \operatorname{dom}(T)$ with $x = \lim_{n\to\infty} x_n = x$ and $\lim_{n\to\infty} Tx_n = y$ and set $u_n = (\zeta I - T)x_n$. Then

$$\lim_{n\to\infty} u_n = \lim_{n\to\infty} (\zeta x_n - Tx_n) = \zeta x - y,$$

and therefore

$$S(\zeta x - y) = \lim_{n\to\infty} Su_n = \lim_{n\to\infty} x_n = x.$$

This implies $x \in \operatorname{dom}(T)$ and $(\zeta I - T)x = \zeta x - y$, and so $Tx = y$. Hence T is closed.

The remainder of the proof is similar to the case where T is bounded; see Lemma 9.6. □

In the next proposition, we refer to the notion of the essential range of a function with respect to a given resolution of the identity (Definition 9.3).

Proposition 9.17. *Let E be a resolution of the identity on Ω, and let $f : \Omega \longrightarrow \mathbb{C}$ be a measurable function. For $\alpha \in \mathbb{C}$, put*

$$\omega_\alpha = \{p \in \Omega : f(p) = \alpha\}.$$

(a) *If α is in the essential range of f and $E(\omega_\alpha) \neq 0$, then $\alpha I - \Psi(f)$ is not injective.*
(b) *If α is in the essential range of f but $E(\omega_\alpha) = 0$, then $\alpha I - \Psi(f)$ is an injective mapping of $\mathcal{D}_f$ onto a proper dense subspace of H, and there exist vectors $x_n \in H$ with $\|x_n\| = 1$ such that*

$$\lim_{n\to\infty} [\alpha x_n - \Psi(f)x_n] = 0.$$

(c) *$\sigma(\Psi(f))$ is the essential range of f.*

We say that α lies in the point spectrum of $\Psi(f)$ in case (a) and in the continuous spectrum of $\Psi(f)$ in case (b).

Proof. Without loss of generality, we can assume that $\alpha = 0$.

(a) If $E(\omega_0) \neq 0$, then there exists $x_0 \in \operatorname{im}(E(\omega_0))$ with $\|x_0\| = 1$. Let ϕ_0 be the characteristic function of ω_0. Then $f\phi_0 = 0$, and $\Psi(f)\Psi(\phi_0) = 0$. Since $\Psi(\phi_0) = E(\omega_0)$, it follows that

$$\Psi(f)x_0 = \Psi(f)E(\omega_0)x_0 = \Psi(f)\Psi(\phi_0)x_0 = 0.$$

(b) Now we have $E(\omega_0) = 0$ but $E(\omega_n) \neq 0$ for $n \in \mathbb{N}$, where

$$\omega_n = \{p \in \Omega : |f(p)| < 1/n\}.$$

Let $x_n \in \operatorname{im}(E(\omega_n))$ with $\|x_n\| = 1$, and let ϕ_n be the characteristic functions of ω_n. As in (a), we obtain

$$\|\Psi(f)x_n\| = \|\Psi(f\phi_n)x_n\| \le \|\Psi(f\phi_n)\| = \|f\phi_n\|_\infty \le 1/n.$$

Thus $\Psi(f)x_n \to 0$ although $\|x_n\| = 1$.

If $\Psi(f)x = 0$ for some $x \in \mathcal{D}_f$, then

$$\int_\Omega |f|^2\, dE_{x,x} = \|\Psi(f)x\|^2 = 0.$$

Since $|f| > 0$ almost everywhere ($E_{x,x}$), we must have $E_{x,x}(\Omega) = 0$. However, $E_{x,x}(\Omega) = \|x\|^2$. Hence $\Psi(f)$ is injective. Similarly, $\Psi(f)^* = \Psi(\bar{f})$ is injective. If $y \perp \operatorname{im}(\Psi(f))$, then $x \mapsto (\Psi(f)x, y) = 0$ is continuous in $\mathcal{D}_f$; hence $y \in \operatorname{dom}(\Psi(f)^*)$, and

$$(x, \Psi(\bar{f})y) = (\Psi(f)x, y) = 0, \quad x \in \mathcal{D}_f.$$

Hence $\Psi(\bar{f})y = 0$ and $y = 0$. Therefore $\operatorname{im}(\Psi(f))$ is dense in H.

Since $\Psi(f)$ is closed, so is $\Psi(f)^{-1}$. If $\operatorname{im}(\Psi(f)) = H$, then the closed graph theorem would imply that $\Psi(f)^{-1} \in \mathcal{L}(H)$. This is impossible in view of the sequence $(x_n)_n$ constructed above. Hence (b) is proved.

(c) It follows from (a) and (b) that the essential range of f is a subset of $\sigma(\Psi(f))$. Now assume that 0 is not in the essential range of f. Then $g = 1/f \in L^\infty(E)$, and $fg = 1$, hence $\Psi(f)\Psi(g) = \Psi(1) = I$, which proves that $\operatorname{im}(\Psi(f)) = H$. Since $|f| > 0$, we have that $\Psi(f)$ is injective, as in the proof of (b). By the closed graph theorem, $\Psi(f)^{-1} \in \mathcal{L}(H)$. Therefore $0 \notin \sigma(\Psi(f))$, and (c) is proved. □

In the following proposition, we describe the change of measure principle.

Proposition 9.18. *Let* M *and* M′ *be σ-algebras in the sets* $\Omega, \Omega' \subseteq \mathbb{C}$, *let* $E : \mathrm{M} \longrightarrow \mathcal{L}(H)$ *be a resolution of identity, and suppose that* $\Phi : \Omega \longrightarrow \Omega'$ *has the property that* $\Phi^{-1}(\omega') \in \mathrm{M}$ *for every* $\omega' \in \mathrm{M}'$.

If $E'(\omega') = E(\Phi^{-1}(\omega'))$, *then* $E' : \mathrm{M}' \longrightarrow \mathcal{L}(H)$ *is a resolution of the identity, and*

$$\int_{\Omega'} f \, dE'_{x,y} = \int_{\Omega} (f \circ \Phi) \, dE_{x,y} \tag{9.18}$$

for every M′*-measurable* $f : \Omega' \longrightarrow \mathbb{C}$ *for which either of these integrals exists.*

Proof. A straightforward verification gives that E' is again a resolution of the identity. For characteristic functions, (9.18) is just the definition of E'. So (9.18) follows for simple functions and also in the general case. □

To derive the general spectral theorem for unbounded self-adjoint operators, we will use the Cayley transform.

The mapping

$$t \mapsto \frac{t - i}{t + i}$$

sets up a bijection between the real line and the unit circle minus the point 1. The symbolic calculus developed in Section 9.2 therefore shows that every self-adjoint operator $T \in \mathcal{L}(H)$ gives rise to a unitary operator

$$U = (T - iI)(T + iI)^{-1}$$

and that every unitary U whose spectrum does not contain the point 1 is obtained in this way. This relation will now be extended to unbounded symmetric operators.

If T is a symmetric operator, then we have

$$\|Tx + ix\|^2 = (Tx + ix, Tx + ix) = \|x\|^2 + \|Tx\|^2 = \|Tx - ix\|^2$$

for $x \in \text{dom}(T)$. This implies that $(T + iI)$ is injective and that there is an isometry U with $\text{dom}(U) = \text{im}(T + iI)$ and $\text{im}(U) = \text{im}(T - iI)$ defined by

$$U(Tx + ix) = T - ix, \quad x \in \text{dom}(T).$$

Since $(T + iI)^{-1}$ maps $\text{dom}(U)$ onto $\text{dom}(T)$, we can write

$$U = (T - iI)(T + iI)^{-1}.$$

This operator U is called the Cayley transform of T.

Lemma 9.19. *Let U be an isometry, i. e., $\|Ux\| = \|x\|$ for all $x \in \text{dom}(U)$.*
(a) *For $x, y \in \text{dom}(U)$, we have $(Ux, Uy) = (x, y)$.*
(b) *If $\text{im}(I - U)$ is dense in H, then $I - U$ is injective.*
(c) *If any of the three spaces $\text{dom}(U), \text{im}(U)$, and $\mathcal{G}(U)$ is closed, so are the other two.*

Proof. (a) follows from the polarization identity:

$$(x, y) = \frac{1}{4}(\|x + y\|^2 - \|x - y\|^2 + i\|x + iy\|^2 - i\|x - iy\|^2). \tag{9.19}$$

(b) Let $x \in \text{dom}(U)$ and $(I - U)x = 0$. Then $x = Ux$, and

$$(x, (I - U)y) = (x, y) - (x, Uy) = (Ux, Uy) - (x, Uy) = 0$$

for every $y \in \text{dom}(U)$. This implies $x \perp \text{im}(I - U)$, so that $x = 0$ if $\text{im}(I - U)$ is dense in H.

(c) follows from

$$\|Ux - Uy\| = \|x - y\| = \frac{1}{\sqrt{2}}\|(x, Ux) - (y, Uy)\|, \quad x, y \in \text{dom}(U),$$

where in the last term, $(x, Ux), (y, Uy) \in \mathcal{G}(U)$ are elements of the graph of U, and

$$\|(x, Ux) - (y, Uy)\| = (\|x - y\|^2 + \|Ux - Uy\|^2)^{1/2}. \qquad \square$$

Proposition 9.20. *Let T be a symmetric operator on H (not necessarily densely defined), and let U be its Cayley transform. The following statements are true:*

(a) *U is closed if and only if T is closed.*
(b) $\operatorname{im}(I - U) = \operatorname{dom}(T)$, *$I - U$ is injective, and T can be reconstructed from U by*

$$T = i(I + U)(I - U)^{-1}.$$

The Cayley transforms of distinct symmetric operators are distinct.
(c) *U is unitary if and only if T is self-adjoint.*

Conversely, if V is an operator in H that is an isometry and if $I - V$ is injective, then V is the Cayley transform of a symmetric operator in H.

Proof. (a) The identity $\|Tx + ix\|^2 = \|x\|^2 + \|Tx\|^2$ implies that $(T + iI)x \leftrightarrow (x, Tx)$ is an isometric one-to-one correspondence between $\operatorname{im}(T + iI)$ and the graph $\mathcal{G}(T)$ of T. Hence T is closed if and only if $\operatorname{im}(T + iI)$ is closed. By Lemma 9.19, U is closed if and only if $\operatorname{dom}(U)$ is closed. However, by the definition of the Cayley transform, $\operatorname{dom}(U) = \operatorname{im}(T + iI)$, which proves (a).

(b) The one-to-one correspondence $x \leftrightarrow z$ between $\operatorname{dom}(T)$ and $\operatorname{dom}(U) = \operatorname{im}(T + iI)$ given by

$$z = Tx + ix, \quad Uz = Tx - ix$$

can be written in the form

$$(I - U)z = 2ix, \quad (I + U)z = 2Tx.$$

Hence $(I - U)$ is injective, and $\operatorname{im}(I - U) = \operatorname{dom}(T)$, and therefore $(I - U)^{-1}$ maps $\operatorname{dom}(T)$ onto $\operatorname{dom}(U)$, and

$$2Tx = (I + U)z = (I + U)(I - U)^{-1}(2ix), \quad x \in \operatorname{dom}(T).$$

This proves (b).

(c) Assume that T is self-adjoint. Then by Lemma 4.29

$$\operatorname{im}(I + T^* T) = \operatorname{im}(I + T^2) = H. \tag{9.20}$$

We have

$$(T + iI)(T - iI) = T^2 + I = (T - iI)(T + iI),$$

where all operators have domain $\operatorname{dom}(T^2)$. Hence (9.20) implies that

$$\operatorname{dom}(U) = \operatorname{im}(T + iI) = H \tag{9.21}$$

and

$$\operatorname{im}(U) = \operatorname{im}(T - iI) = H. \tag{9.22}$$

Now $(U^*Ux, x) = (Ux, Ux) = (x, x)$ for every $x \in H$, which implies $U^*U = I$, and U is unitary.

Now assume that U is unitary. Then

$$(\operatorname{im}(I - U))^\perp = \ker(I - U)^* = \{0\},$$

and $\operatorname{dom}(T) = \operatorname{im}(I - U)$ is dense in H. Thus T^* is defined, and $T \subset T^*$. Fix $y \in \operatorname{dom}(T^*)$. Since $\operatorname{im}(T + iI) = \operatorname{dom}(U) = H$, there exists $y_0 \in \operatorname{dom}(T)$ such that

$$(T^* + iI)y = (T + iI)y_0 = (T^* + iI)y_0.$$

Set $y_1 = y - y_0$. Then $y_1 \in \operatorname{dom}(T^*)$, and for every $x \in \operatorname{dom}(T)$, we have

$$((T - iI)x, y_1) = (x, (T^* + iI)y_1) = (x, 0) = 0.$$

Thus $y_1 \perp \operatorname{im}(T - iI) = \operatorname{im}(U) = H$, so $y_1 = 0$ and $y = y_0 \in \operatorname{dom}(T)$. Hence $\operatorname{dom}(T) = \operatorname{dom}(T^*)$, and (c) is proved.

Finally, let V be as in the statement of the converse. Then there is a one-to-one correspondence $z \leftrightarrow x$ between $\operatorname{dom}(V)$ and $\operatorname{im}(I - V)$ given by

$$x = z - Vz.$$

Define S on $\operatorname{dom}(S) = \operatorname{im}(I - V)$ by

$$Sx = i(z + Vz) \text{ if } x = z - Vz. \tag{9.23}$$

If $x, y \in \operatorname{dom}(S)$, then $x = z - Vz$ and $y = u - Vu$ for some $z, u \in \operatorname{dom}(V)$. Since V is an isometry, it follows from Lemma 9.19 that

$$\begin{aligned}(Sx, y) &= i(z + Vz, u - Vu) = i(Vz, u) - i(z, Vu) \\ &= (z - Vz, iu + iVu) = (x, Sy).\end{aligned}$$

Hence S is symmetric. For $z \in \operatorname{dom}(V)$, (9.23) can be written in the form

$$2iVz = Sx - ix, \quad 2iz = Sx + ix,$$

and hence if $x \in \operatorname{dom}(S)$, then

$$V(Sx + ix) = Sx - ix,$$

and $\operatorname{dom}(V) = \operatorname{im}(S + iI)$. Therefore V is the Cayley transform of S. □

At this point, we use the methods developed above to prove a key result about the spectrum of an unbounded self-adjoint operator. It transfers properties of unbounded self-adjoint operators to the bounded resolvent operators.

Let $T \in \mathcal{L}(H)$ be a self-adjoint operator. If $\Im\lambda \neq 0$, then

$$|\Im\lambda|\,\|u\|^2 = \big|\Im((T-\lambda I)u,u)\big| \leq \big\|(T-\lambda I)u\big\|\,\|u\|$$

for all $u \in H$, where we used that $(Tu,u)=(u,Tu)$ is real.

This implies that $T-\lambda I$ is injective and has a closed range. As

$$(\operatorname{im}(T-\lambda I))^{\perp} = \ker(T-\bar{\lambda}I)$$

and this kernel reduces to $\{0\}$, we obtain that $T-\lambda I$ is bijective.

This follows also from Proposition 4.35, once we have observed that

$$\big|((T-\lambda I)u,u)\big| \geq |\Im\lambda|\,\|u\|^2.$$

Proposition 9.21. *Let T be a bounded self-adjoint operator. The spectrum of T is contained in the interval $[m,M]$, where*

$$m = \inf_{u\neq 0}\frac{(Tu,u)}{(u,u)} \quad \text{and} \quad M = \sup_{u\neq 0}\frac{(Tu,u)}{(u,u)},$$

and m and M belong to the spectrum of T.

Proof. We have already shown that the spectrum is real (Proposition 9.23). Let λ be a real number with $\lambda > M$. We consider the scalar product

$$\langle u,v\rangle := M(u,v) - (Tu,v).$$

Notice that $\langle u,u\rangle \geq 0$. By the Cauchy–Schwarz inequality we have

$$\big|(Mu-Tu,v)\big| \leq (Mu-Tu,u)^{1/2}(Mv-Tv,v)^{1/2};$$

in particular, setting $v=(M-T)u$, we have

$$\begin{aligned}\big\|(M-T)u\big\|^2 &\leq (Mu-Tu,u)^{1/2}\big((M-T)^2u,(M-T)u\big)^{1/2}\\ &\leq (Mu-Tu,u)^{1/2}\big(\|M-T\|\|(M-T)u\|^2\big)^{1/2},\end{aligned}$$

which implies that

$$\big\|(M-T)u\big\| \leq \|M-T\|^{1/2}\,(Mu-Tu,u)^{1/2}. \tag{9.24}$$

Now let $(u_n)_n$ be a sequence such that $\|u_n\| = 1$ for all n and $(Tu_n,u_n) \to M$ as $n \to \infty$. By (9.24) we obtain that $(M-T)u_n \to 0$. Hence $M \in \sigma(T)$; otherwise, we would obtain that

$$u_n = (MI - T)^{-1}(MI - T)u_n \to 0,$$

in contradiction with $\|u_n\| = 1$. □

Corollary 9.22. *Let $T \in \mathcal{L}(H)$ be a self-adjoint operator such that $\sigma(T) = \{0\}$. Then $T = 0$.*

Proof. By Proposition 9.21 we have $m = M = 0$, and hence $(Tu, u) = 0$ for each $u \in H$. As (Tu, v) can be written as a linear combination of terms of the form (Tw, w), we obtain $T = 0$. □

Proposition 9.23. *The spectrum $\sigma(T)$ of any self-adjoint operator T is real and nonempty. If $\zeta \notin \sigma(T)$, then*

$$\|(\zeta I - T)^{-1}\| \leq |\Im\zeta|^{-1}. \tag{9.25}$$

Moreover,

$$(\overline{\zeta} I - T)^{-1} = ((\zeta I - T)^{-1})^*. \tag{9.26}$$

Proof. Let $\zeta = \xi + i\eta$ and $\eta \neq 0$ and set $K = \frac{1}{\eta}(T - \xi I)$. Using Lemma 4.4, it follows that $K^* = K$. Let $f \in \operatorname{dom}(K)$ be such that $Kf = K^*f = if$. Then $i(f, f) = (Kf, f) = (f, Kf) = -i(f, f)$, which implies that $f = 0$ and $K - iI$ is injective. In a similar way, we show that $K + iI$ is injective. The proof of Proposition 9.20(a) implies that $\operatorname{im}(K \pm iI)$ is closed. Now we obtain from Lemma 4.8 that $\operatorname{im}(K \pm iI)^\perp = \ker(K \pm iI) = \{0\}$. Therefore $(K \pm iI)^{-1}$ is defined on the whole of H. Since

$$\|Kx \pm ix\|^2 = \|Kx\|^2 + \|x\|^2, \quad x \in \operatorname{dom}(K),$$

we get

$$\|(K \pm iI)^{-1}y\| = \|(K \pm iI)^{-1}(K \pm iI)x\| = \|x\| \leq \|(K \pm iI)x\| = \|y\|$$

for each $y \in H$, which implies that

$$\|(K \pm iI)^{-1}\| \leq 1. \tag{9.27}$$

Thus $\pm i \notin \sigma(K)$, and hence $\zeta \notin \sigma(T)$. In addition, (9.27) implies (9.25).

Now let $x_1, x_2 \in \operatorname{dom}(T)$. Then

$$((T - \zeta I)x_1, x_2) = (x_1, (T - \overline{\zeta} I)x_2).$$

Putting $y_1 = (T - \zeta I)x_1$ and $y_2 = (T - \overline{\zeta} I)x_2$ and rewriting the last equation in terms of y_1 and y_2 yield (9.26).

Finally, suppose T has the empty spectrum. Then T^{-1} is a bounded self-adjoint operator. We claim that $\sigma(T^{-1}) = \{0\}$. For $\lambda \neq 0$, we can write the inverse of $T^{-1} - \lambda I$ in the form

$$(T^{-1} - \lambda I)^{-1} = ((\lambda^{-1} I - T)\lambda T^{-1})^{-1} = \lambda^{-1} T(\lambda^{-1} I - T)^{-1},$$

and

$$\lambda^{-1} T(\lambda^{-1} I - T)^{-1} - \lambda^{-2}(\lambda^{-1} I - T)^{-1} = -\lambda^{-1}(\lambda^{-1} I - T)(\lambda^{-1} I - T)^{-1} = -\lambda^{-1} I;$$

hence

$$(T^{-1} - \lambda I)^{-1} = -\lambda^{-1} I + \lambda^{-2}(\lambda^{-1} I - T)^{-1},$$

which is a bounded operator. So $\sigma(T^{-1}) = \{0\}$. The spectrum of the bounded self-adjoint operator T is contained in the interval $[m, M]$, where

$$m = \inf_{u \neq 0} \frac{(Tu, u)}{(u, u)} \quad \text{and} \quad M = \sup_{u \neq 0} \frac{(Tu, u)}{(u, u)},$$

and m and M belong to the spectrum of T; see Proposition 9.21. So we get $T^{-1} = 0$, which contradicts to $TT^{-1} = I$. □

The Cayley transform is now used to reduce the construction of the spectral decomposition of an unbounded self-adjoint operator to the spectral decomposition of a unitary operator.

Proposition 9.24. *Let T be an unbounded self-adjoint operator ($T = T^*$ and $\mathrm{dom}(T) = \mathrm{dom}(T^*)$). Then there exists a uniquely determined resolution of the identity E on the Borel subsets of $\mathbb{R}$ such that*

$$(Tx, y) = \int_{-\infty}^{\infty} t \, dE_{x,y}(t), \quad x \in \mathrm{dom}(T), \quad y \in H. \tag{9.28}$$

Moreover, E is concentrated on the spectrum $\sigma(T) \subset \mathbb{R}$ of T in the sense that $E(\sigma(T)) = I$.

Proof. Let U be the Cayley transform of T, and let Ω be the unit circle with the point 1 removed. Let E' be the spectral decomposition of U (Proposition 9.7). By Proposition 9.20, $I - U$ is injective, and $E'(\{1\}) = 0$ by Proposition 9.9. Hence

$$(Ux, y) = \int_{\Omega} \lambda \, dE'_{x,y}(\lambda), \quad x, y \in H. \tag{9.29}$$

Define

$$f(\lambda) = \frac{i(1+\lambda)}{(1-\lambda)}, \quad \lambda \in \Omega.$$

We define $\Psi(f)$ as in Proposition 9.14 with E' in place of E.

$$(\Psi(f)x, y) = \int_{\Omega} f\, dE'_{x,y}, \quad x \in \mathcal{D}_f, \quad y \in H. \tag{9.30}$$

Since f is real-valued, $\Psi(f)$ is self-adjoint (Proposition 9.14). Since

$$f(\lambda)(1-\lambda) = i(1+\lambda),$$

the symbolic calculus gives

$$\Psi(f)(I-U) = i(I+U). \tag{9.31}$$

This in particular implies

$$\operatorname{im}(I-U) \subset \operatorname{dom}(\Psi(f)). \tag{9.32}$$

By Proposition 9.20

$$T(I-U) = i(I+U), \tag{9.33}$$

and $\operatorname{dom}(T) = \operatorname{im}(I-U) \subset \operatorname{dom}(\Psi(f))$. Relations (9.32) and (9.33) imply that $\Psi(f)$ is a self-adjoint extension of the self-adjoint operator T. Thus we have $T \subset \Psi(f)$ and $\Psi(f) = \Psi(f)^* \subset T^* = T$, and hence $\Psi(f) = T$. In addition, we get

$$(Tx, y) = \int_{\Omega} f\, dE'_{x,y}, \quad x \in \operatorname{dom}(T), \quad y \in H. \tag{9.34}$$

By Proposition 9.17(c), $\sigma(T)$ is the essential range of f. Thus $\sigma(T) \subset \mathbb{R}$. Since f is injective in Ω, we can define $E(f(\omega)) = E'(\omega)$ for every Borel set $\omega \subset \Omega$ to obtain the desired resolution E, which converts (9.34) into (9.28). The uniqueness of E follows from the uniqueness of representation (9.29). □

The symbolic calculus is now used to prove the following statements.

Proposition 9.25. *Let T be a self-adjoint operator on H.*
(a) *$(Tx, x) \geq 0$ for every $x \in \operatorname{dom}(T)$ (briefly $T \geq 0$) if and only if $\sigma(T) \subset [0, \infty)$.*
(b) *If $T \geq 0$, then there exists a unique self-adjoint operator $S \geq 0$ such that $\operatorname{dom}(S) \supseteq \operatorname{dom}(T)$ and $S^2 = T$ on $\operatorname{dom}(T)$. The symbolic calculus implies that $x \in \operatorname{dom}(T)$ if and only if $x \in \operatorname{dom}(S)$ and also $Sx \in \operatorname{dom}(S)$. In addition, $\operatorname{dom}(T)$ is a core of S.*

Proof. (a) see Proposition 9.11(a).
(b) Assume that $T \geq 0$, so that $\sigma(T) \subset [0, \infty)$, and

$$(Tx, y) = \int_0^{\infty} t\, dE_{x,y}(t), \quad x \in \operatorname{dom}(T), \quad y \in H, \tag{9.35}$$

where $\operatorname{dom}(T) = \{x \in H : \int_0^\infty t^2\, dE_{x,x}(t) < \infty\}$. Let $s(t)$ be the nonnegative square root of $t \geq 0$ and put $\Psi(s) = S$; explicitly,

$$(Sx, y) = \int_0^\infty s(t)\, dE_{x,y}(t), \quad x \in \mathcal{D}_s, \quad y \in H. \tag{9.36}$$

By Proposition 9.14 we obtain that $S^2 = T$ on $\operatorname{dom}(T)$ and $S \geq 0$.

To prove the uniqueness, suppose R is self-adjoint, $R \geq 0$, and $R^2 = T$, and let E^R be its spectral decomposition:

$$(Rx, y) = \int_0^\infty t\, dE^R_{x,y}(t), \quad x \in \operatorname{dom}(R), \quad y \in H. \tag{9.37}$$

We apply Proposition 9.18 with $\Omega = [0, \infty)$, $\phi(t) = t^2$, $f(t) = t$, and

$$E'(\phi(\omega)) = E^R(\omega) \quad \text{for } \omega \subset [0, \infty) \tag{9.38}$$

to obtain

$$(Tx, y) = (R^2 x, y) = \int_0^\infty t^2\, dE^R_{x,y(t)} = \int_0^\infty t\, dE'_{x,y}(t). \tag{9.39}$$

Relations (9.35) and (9.39) and the uniqueness statement in Proposition 9.24 show that $E' = E$. By (9.38), E determines E^R and hence R.

The statement about the domains of the operators S and T follows from the symbolic calculus and the definition of the domains there (see Proposition 9.14). As $S^* = S$, Lemma 4.32 immediately implies that $\operatorname{dom}(T)$ is a core of S. □

As applications of these results, we prove a useful characterization of self-adjoint and essentially self-adjoint operators.

Proposition 9.26. *Let T be a closed symmetric operator. Then the following statements are equivalent:*
(i) *T is self-adjoint;*
(ii) $\ker(T^* + iI) = \{0\}$ *and* $\ker(T^* - iI) = \{0\}$;
(iii) $\operatorname{im}(T + iI) = H$ *and* $\operatorname{im}(T - iI) = H$.

Proof. (i) implies (ii): By Proposition 9.23, $\pm i \notin \sigma(T)$.

(ii) implies (iii): Notice that $\ker(T^* \pm iI) = \{0\}$ if and only if $\operatorname{im}(T \mp iI)$ is dense in H. This follows easily from

$$(Tu \pm iu, v) = (u, T^* \mp iv)$$

for $u, v \in \mathrm{dom}(T)$. So it remains to show that $\mathrm{im}(T \mp iI)$ is closed. The symmetry of T implies that

$$\|(T \mp iI)u\|^2 = \|Tu\|^2 + \|u\|^2 \tag{9.40}$$

for $u \in \mathrm{dom}(T)$. Now since T is closed, we easily obtain that $\mathrm{im}(T \mp iI)$ is closed.

(iii) implies (i): Let $u \in \mathrm{dom}(T^*)$. By (iii) there exists $v \in \mathrm{dom}(T)$ such that

$$(T - iI)v = (T^* - iI)u.$$

Since T is symmetric, we have also $(T^* - iI)(v - u) = 0$. However, if $(T + iI)$ is surjective, then $(T^* - iI)$ is injective (Lemma 4.8), and we obtain $u = v$. This proves that $u \in \mathrm{dom}(T)$ and that T is self-adjoint. □

We proved during the statement that (ii) implies (iii) the following:

Lemma 9.27. *If T is closed and symmetric, then* $\mathrm{im}(T \pm iI)$ *is closed.*

In a similar way, we obtain a characterization for essentially self-adjoint operators.

Proposition 9.28. *Let A be a symmetric operator. Then the following statements are equivalent:*
(i) *A is essentially self-adjoint;*
(ii) $\ker(A^* + iI) = \{0\}$ *and* $\ker(A^* - iI) = \{0\}$;
(iii) $\mathrm{im}(A + iI)$ *and* $\mathrm{im}(A - iI)$ *are dense in H.*

Proof. We apply Proposition 9.26 to $\overline{A}$ and notice that $\overline{A}$ is symmetric and that Lemma 4.5 implies that $A^* = (\overline{A})^*$. In addition, we use Lemma 9.27. □

If A is also a positive operator, then we get the following:

Proposition 9.29. *Let A be a positive symmetric operator. Then the following statements are equivalent:*
(i) *A is essentially self-adjoint;*
(ii) $\ker(A^* + bI) = \{0\}$ *for some $b > 0$;*
(iii) $\mathrm{im}(A + bI)$ *is dense in H.*

Proof. We proceed in a similar way as before and notice that for a positive symmetric operator A, we have

$$((A + bI)u, u) \geq b\|u\|^2 \tag{9.41}$$

for $u \in \mathrm{dom}(A)$, which is a good substitute for (9.40).

By Lemma 4.8, (ii) and (iii) are equivalent. Since the closure of a positive symmetric operator is again positive and symmetric, it remains to show that a closed positive symmetric operator T is self-adjoint if and only if $\ker(T^* + bI) = \{0\}$ for some $b > 0$.

We can suppose that $b = 1$. If T is self-adjoint, then the spectrum $\sigma(T) \subseteq \mathbb{R}^+$, and hence $\ker(T + I) = \ker(T^* + I) = \{0\}$.

For the converse, we first show that $\operatorname{im}(T + I)$ is closed: let $(y_k)_k \subset \operatorname{im}(T + I)$ be a convergent sequence. There exists a sequence $(x_k)_k \subset \operatorname{dom}(T)$ such that $y_k = (T + I)x_k$. Then

$$(x_k, y_k) = (x_k, Tx_k) + \|x_k\|^2 \geq \|x_k\|^2,$$

and, by the Cauchy–Schwarz inequality,

$$\|x_k\| \leq \|y_k\|. \tag{9.42}$$

Since $(y_k)_k$ is convergent, $\sup_k \|y_k\| < \infty$, and, by (9.42), $\sup_k \|x_k\| < \infty$. Now the positivity implies

$$\begin{aligned}\|x_k - x_\ell\|^2 &\leq (x_k - x_\ell, (T + I)(x_k - x_\ell)) \\ &\leq (\|x_k\| + \|x_\ell\|)\|y_k - y_\ell\| \\ &\leq C\|y_k - y_\ell\|.\end{aligned}$$

Hence $(x_k)_k$ is a Cauchy sequence. Since we supposed that T is closed, there exists $x \in \operatorname{dom}(T)$ such that $x = \lim_{k\to\infty} x_k$ and $(T+I)x = y = \lim_{k\to\infty} y_k$. Hence $\operatorname{im}(T+I)$ is closed.

The assumption $\ker(T^* + I) = \{0\}$ now gives $\operatorname{im}(T + I) = H$. To show that T is self-adjoint, it suffices to show that $\operatorname{dom}(T^*) \subseteq \operatorname{dom}(T)$. Let $x \in \operatorname{dom}(T^*)$. There exists $y \in \operatorname{dom}(T)$ such that

$$(T + I)y = (T^* + I)y = (T^* + I)x,$$

since $\operatorname{dom}(T) \subseteq \operatorname{dom}(T^*)$. This implies $(T^* + I)(x - y) = 0$, and hence $x = y \in \operatorname{dom}(T)$. □

Finally, we mention that every self-adjoint operator is unitarily equivalent to a multiplication operator, which is important for applications to the solution of other spectral problems. We omit the proof, as we will not use this version of the spectral theorem in the sequel. Using the Riesz representation theorem in measure theory (see, for instance, [83]), this version follows easily from Proposition 9.24.

Theorem 9.30. *Let T be a self-adjoint operator on H with spectrum $\sigma(T)$. Then there exist a finite measure μ on $\sigma(T) \times \mathbb{N}$ and a unitary operator*

$$U : H \longrightarrow L^2(\sigma(T) \times \mathbb{N}, d\mu)$$

with the following properties: if

$$g : \sigma(T) \times \mathbb{N} \longrightarrow \mathbb{R}$$

is the function $g(t,n) = t$, *then* $x \in H$ *lies in* $\operatorname{dom}(T)$ *if and only if* $g \cdot Ux \in L^2(\sigma(T)\times\mathbb{N}, d\mu)$. *In addition, we have*

$$UTU^{-1}h = gh$$

for all $h \in U(\operatorname{dom}(T))$, *and*

$$Uf(T)U^{-1}h = f(g)h$$

for all bounded Borel functions f *on* $\sigma(T)$.

In particular, $f(T)$ *is a bounded operator, and*

$$\|f(T)\| = \|f\|_\infty. \tag{9.43}$$

In the proof of this version of the spectral theorem, we start with a function $f \in \mathcal{C}_0(\sigma(T))$, a continuous function vanishing at infinity. Then we consider the linear functional

$$\Lambda(f) := (f(T)x, x),$$

where $x \in H$ is an appropriate vector (a cyclic vector). The Riesz representation theorem (see [83]) implies that there exists a finite countably additive measure μ on $\mathbb{R}$ such that

$$(f(T)x, x) = \int_{\sigma(T)} f(t)\, d\mu(t).$$

This is the main step. The rest of the proof consists of an application of the symbolic calculus (see [24] for all details).

9.4 Determination of the spectrum

In this part, we prove some results of the spectral theory of unbounded self-adjoint operators, which are used later on for the applications to the $\Box$-operator, to Schrödinger operators with magnetic field, and to Pauli and Dirac operators. These results enable us to determine the spectrum of the $\Box$-operator in some particular cases, and they yield methods to decide whether the corresponding differential operators are with compact resolvent.

First, we prove some general results about the spectrum of an unbounded self-adjoint operator.

Lemma 9.31. *Let* T *be an unbounded self-adjoint operator. Then* $\lambda \in \sigma(T)$ *if and only if there exists a sequence* $(x_k)_k$ *in* $\operatorname{dom}(T)$ *such that* $\|x_k\| = 1$ *for each* $k \in \mathbb{N}$ *and*

$$\lim_{k\to\infty} \|(T - \lambda I)x_k\| = 0.$$

If T and S are self-adjoint operators such that $T = U^{-1}SU$ for some unitary operator U, where $\operatorname{dom}(T) = U^{-1}(\operatorname{dom}(S))$, then $\sigma(T) = \sigma(S)$.

Proof. If $\lambda \in \sigma(T)$, then $T - \lambda I$ is not injective, and we can find $x \in \operatorname{dom}(T)$ such that $\|x\| = 1$ and $(T - \lambda I)x = 0$. So the constant sequence $x_k = x$ has the desired property.

Conversely, suppose that $\lambda \notin \sigma(T)$. Then $(T - \lambda I)^{-1}$ is a bounded operator. If there exists a sequence $(x_k)_k$ in $\operatorname{dom}(T)$ such that $\|x_k\| = 1$ for each $k \in \mathbb{N}$ and $\lim_{k\to\infty} \|(T - \lambda I)x_k\| = 0$, then

$$\lim_{k\to\infty} (T - \lambda I)^{-1}(T - \lambda I)x_k = \lim_{k\to\infty} x_k = 0$$

yields a contradiction to $\|x_k\| = 1$ for each $k \in \mathbb{N}$.

To prove the second statement, take $\lambda \in \sigma(S)$ and a sequence $(y_k)_k$ in $\operatorname{dom}(S)$ such that $\|y_k\| = 1$ for each $k \in \mathbb{N}$ and

$$\lim_{k\to\infty} \|(S - \lambda I)y_k\| = 0.$$

Then for $U^{-1}y_k = z_k \in \operatorname{dom}(T)$, we have $\|z_k\| = 1$, since U is unitary and

$$\|(T - \lambda I)z_k\| = \|U^{-1}(Sy_k - \lambda y_k)\| = \|(S - \lambda I)y_k\|.$$

This shows that $\lambda \in \sigma(T)$. □

Lemma 9.32. *Let T be an unbounded self-adjoint operator, and let E be the uniquely determined resolution of the identity E on the Borel subsets of $\mathbb{R}$ such that*

$$(Tx, y) = \int_{-\infty}^{\infty} t\, dE_{x,y}(t), \quad x \in \operatorname{dom}(T), \quad y \in H$$

(see Proposition 9.24).

Then

$$\sigma(T) = \{\lambda \in \mathbb{R} : E((\lambda - \epsilon, \lambda + \epsilon)) \neq 0, \forall \epsilon > 0\}. \tag{9.44}$$

Proof. Let $\lambda \in \sigma(T)$. Suppose that there exists $\epsilon_0 > 0$ such that $E((\lambda - \epsilon_0, \lambda + \epsilon_0)) = 0$. Then there exists a continuous function f on $\mathbb{R}$ such that $f(t) = (t - \lambda)^{-1}$ on the support of the measure $dE(t)$. By the symbolic calculus we obtain a bounded operator $(T - \lambda I)^{-1}$ and hence a contradiction.

Conversely, let λ belong to the right-hand side of (9.44). For each $k \in \mathbb{N}$, we can now find $x_k \in \operatorname{dom}(T)$ such that $\|x_k\| = 1$ and

$$E((\lambda - 1/k, \lambda + 1/k))x_k = x_k.$$

Consider the bounded Borel functions

$$f_k(t) = (t - \lambda)\chi_{(\lambda-1/k,\lambda+1/k)}(t), \quad t \in \mathbb{R}, \quad k \in \mathbb{N},$$

where $\chi_{(\lambda-1/k,\lambda+1/k)}$ is the characteristic function of the interval $(\lambda - 1/k, \lambda + 1/k)$. Then by (9.43)

$$\|(T - \lambda I)x_k\| = \|(T - \lambda I)\, E((\lambda - 1/k, \lambda + 1/k))x_k\| \le 1/k\, \|x_k\| = 1/k.$$

Using Lemma 9.31, we get $\lambda \in \sigma(T)$. □

Lemma 9.33. *Let T be a symmetric operator on H with domain* dom(T), *and suppose that $\{x_k\}_k$ is a complete orthonormal system in H. If each x_k lies in* dom(T), *and there exist $\lambda_k \in \mathbb{R}$ such that*

$$Tx_k = \lambda_k x_k$$

for every $k \in \mathbb{N}$, then T is essentially self-adjoint. Moreover, the spectrum of $\overline{T}$ is the closure in $\mathbb{R}$ of the set of all λ_k.

Proof. If $x = \sum_{k=1}^{\infty} \alpha_k x_k$ belongs to dom(T) and

$$y = Tx = \sum_{k=1}^{\infty} \beta_k x_k,$$

then

$$\beta_k = (y, x_k) = (Tx, x_k) = (x, Tx_k) = \lambda_k (x, x_k) = \lambda_k \alpha_k.$$

Since $x, y \in H$, we have

$$\sum_{k=1}^{\infty} |\alpha_k|^2 < \infty, \quad \sum_{k=1}^{\infty} |\beta_k|^2 < \infty,$$

and hence

$$\sum_{k=1}^{\infty} (1 + \lambda_k^2)|\alpha_k|^2 < \infty.$$

We define an operator $\tilde{T}$ as follows: let

$$\operatorname{dom}(\tilde{T}) = \left\{ x \in H : x = \sum_{k=1}^{\infty} \alpha_k x_k \text{ with } \sum_{k=1}^{\infty} (1 + \lambda_k^2)|\alpha_k|^2 < \infty \right\}$$

and define

$$\tilde{T}x = \sum_{k=1}^{\infty} \alpha_k \lambda_k x_k$$

for $x \in \mathrm{dom}(\tilde{T})$. It follows that $\tilde{T}$ is an extension of T.

Let Σ be the closure of the set $\{\lambda_k : k \in \mathbb{N}\}$. Each λ_k is an eigenvalue of $\tilde{T}$, and, by Lemma 9.16, $\sigma(\tilde{T})$ is closed, so $\Sigma \subseteq \sigma(\tilde{T})$. For $z \notin \Sigma$ and $x = \sum_{k=1}^{\infty} \alpha_k x_k$, we define the operator S on H by

$$Sx = S\left(\sum_{k=1}^{\infty} \alpha_k x_k\right) := \sum_{k=1}^{\infty} \alpha_k (z - \lambda_k)^{-1} x_k.$$

It follows that S is injective and that

$$\left\| S\left(\sum_{k=1}^{\infty} \alpha_k x_k\right)\right\| \le \sup_k |z - \lambda_k|^{-1} \left(\sum_{k=1}^{\infty} |\alpha_k|^2\right)^{1/2} = C\,\|x\|,$$

which implies that the operator S is bounded. Its range is precisely $\mathrm{dom}(\tilde{T})$, and $(zI - \tilde{T})Sx = x$ for all $x \in H$. Thus $z \notin \sigma(\tilde{T})$ and $S = (zI - \tilde{T})^{-1}$. This implies $\Sigma = \sigma(\tilde{T})$.

Now we claim that $\tilde{T}$ is the closure of T. Since $\sigma(\tilde{T})$ is not equal to $\mathbb{C}$, Lemma 9.16 implies that $\tilde{T}$ is a closed operator. Let

$$u = \sum_{k=1}^{\infty} \alpha_k x_k \in \mathrm{dom}(\tilde{T})$$

and put $u_m = \sum_{k=1}^{m} \alpha_k x_k$. Then $\lim_{m\to\infty} u_m = u$, and

$$\lim_{m\to\infty} Tu_m = \lim_{m\to\infty} \sum_{k=1}^{m} \alpha_k \lambda_k x_k = \sum_{k=1}^{\infty} \alpha_k \lambda_k x_k = \tilde{T}u.$$

Hence $\tilde{T} = \overline{T}$.

Finally, we prove that $\tilde{T}$ is self-adjoint. For $x \in \mathrm{dom}(\tilde{T}^*)$ and $\tilde{T}^* x = y$, we have

$$(y, x_k) = (\tilde{T}^* x, x_k) = (x, \tilde{T} x_k) = \lambda_k (x, x_k).$$

If $x = \sum_{k=1}^{\infty} \alpha_k x_k$, then the above implies that $y = \sum_{k=1}^{\infty} \alpha_k \lambda_k x_k$. Hence $x \in \mathrm{dom}(\tilde{T})$, thus $\mathrm{dom}(\tilde{T}^*) = \mathrm{dom}(\tilde{T})$, and $\tilde{T} = \tilde{T}^*$. □

Definition 9.34. Let T be a self-adjoint operator on H. The discrete spectrum $\sigma_d(T)$ of T is the set of all eigenvalues λ of finite multiplicity that are isolated in the sense that the intervals $(\lambda - \epsilon, \lambda)$ and $(\lambda, \lambda + \epsilon)$ are disjoint from the spectrum for some $\epsilon > 0$. The nondiscrete part of the spectrum of T is called the essential spectrum of T and is denoted by $\sigma_e(T)$.

A closed linear subspace L of H is called invariant if

$$(\zeta I - T)^{-1}(L) \subseteq L$$

for all $\zeta \notin \mathbb{R}$.

Lemma 9.35. *Let T be a self-adjoint operator. Then*

$$\sigma_d(T) = \{\lambda \in \sigma(T) : \exists \epsilon > 0,\ \dim \operatorname{im}(E((\lambda - \epsilon, \lambda + \epsilon))) < \infty\}. \tag{9.45}$$

Proof. Let λ belong to the right-hand side of (9.45). Then there exists $\epsilon_0 > 0$ such that for each $\epsilon \in (0, \epsilon_0)$, the projection $E((\lambda - \epsilon, \lambda + \epsilon))$ becomes a projection with finite range independent of ϵ. This is in fact the projection $E(\{\lambda\})$, and we observe that $E((\lambda - \epsilon_0, \lambda)) = 0$ and $E((\lambda, \lambda + \epsilon_0)) = 0$. This shows that $\lambda \in \sigma_d(T)$.

If $\lambda \in \sigma_d(T)$, then λ is an eigenvalue of finite multiplicity, and there exists $\epsilon > 0$ such that the intervals $(\lambda - \epsilon, \lambda)$ and $(\lambda, \lambda + \epsilon)$ are disjoint from the spectrum, which means that $\dim \operatorname{im}(E((\lambda - \epsilon, \lambda + \epsilon))) < \infty$. □

As the whole spectrum of a self-adjoint operator is closed, it now follows that the essential spectrum is closed. The following characterization of the essential spectrum of a self-adjoint operator is similar to the characterization of the whole spectrum; compare with Lemma 9.31.

Lemma 9.36. *Let T be a self-adjoint operator. Then λ belongs to the essential spectrum $\sigma_e(T)$ if and only if there exists a sequence $(x_k)_k$ in $\operatorname{dom}(T)$ such that $\|x_k\| = 1$ for each $k \in \mathbb{N}$, x_k converges weakly to 0, and*

$$\lim_{k\to\infty} \|(T - \lambda I)x_k\| = 0.$$

Proof. The sequence appearing in the lemma is called a Weyl sequence. We say that λ belongs to the Weyl spectrum $W(T)$ if there exists an associated Weyl sequence. By Lemma 9.31 we already know that

$$W(T) \subseteq \sigma(T).$$

Let $\lambda \in W(T)$ and suppose that $\lambda \in \sigma_d(T)$. Then the spectral projection $E(\{\lambda\})$ has a finite-dimensional range and hence is compact. So, by Proposition 4.17,

$$\lim_{k\to\infty} E(\{\lambda\})x_k = 0$$

in H. Now let $y_k := (I - E(\{\lambda\}))x_k$. Then, as $E(\{\lambda\})T = TE(\{\lambda\})$ (Proposition 9.24), we get $\lim_{k\to\infty} \|y_k\| = 1$ and

$$\lim_{k\to\infty} (T - \lambda I)y_k = \lim_{k\to\infty} (I - E(\{\lambda\}))(T - \lambda I)x_k = 0.$$

Since $(T - \lambda I)$ is invertible on $\operatorname{im}(I - E(\{\lambda\}))$, we obtain $\lim_{k\to\infty} y_k = 0$, which is a contradiction. Hence $\lambda \in \sigma_e(T)$.

Conversely, let $\lambda \in \sigma_e(T)$. Then, by Lemma 9.35,

$$\dim \operatorname{im}(E((\lambda - \epsilon, \lambda + \epsilon))) = \infty$$

for any $\epsilon > 0$. Let ϵ_k be a decreasing sequence of positive numbers such that $\lim_{k\to\infty} \epsilon_k = 0$. We can now choose an orthonormal system $(x_k)_k$ such that

$$x_k \in \operatorname{im}(E((\lambda - \epsilon_k, \lambda + \epsilon_k))).$$

Then, by Bessel's inequality (Proposition 1.9), x_k converges weakly to 0, and the same reasoning as in Lemma 9.32 yields

$$\lim_{k\to\infty} \|(T - \lambda I)x_k\| = 0. \qquad \square$$

Next, we will characterize the situation where $\sigma_e(T) = \emptyset$. For this purpose, we need some preparations.

Lemma 9.37. *Let T be a self-adjoint operator on H, and let $\lambda \in \mathbb{R}$ be an eigenvalue of T. Then*

$$L_\lambda = \{x \in \operatorname{dom}(T) : Tx = \lambda x\}$$

is a closed invariant subspace of H.

If L is an invariant subspace of T, then $L^\perp$ is also invariant.

Proof. Since $L_\lambda = \ker(T - \lambda I)$ and $T - \lambda I$ is a closed operator, it follows by Lemma 4.4 that L_λ is a closed subspace. Now let $x \in L_\lambda$ and $\zeta \notin \mathbb{R}$. Then

$$\begin{aligned}(\zeta I - T)^{-1}x &= (\lambda - \zeta)^{-1}(\zeta I - T)^{-1}(\lambda - \zeta)x \\ &= (\lambda - \zeta)^{-1}(\zeta I - T)^{-1}(T - \zeta I)x \\ &= -(\lambda - \zeta)^{-1}x,\end{aligned}$$

and hence $(\zeta I - T)^{-1}(L_\lambda) \subseteq L_\lambda$.

Let L be an invariant subspace and $y \in L^\perp$. Then by (9.26)

$$((\zeta I - T)^{-1}y, x) = (y, (\overline{\zeta} I - T)^{-1}x) = 0$$

for all $x \in L$ and $\zeta \notin \mathbb{R}$. Therefore $(\zeta I - T)^{-1}y \in L^\perp$, and $L^\perp$ is invariant. $\square$

For the following results, we always suppose that the underlying Hilbert space H is separable and infinite-dimensional, i. e., each complete orthonormal system is countably infinite.

Proposition 9.38. *Let T be a self-adjoint operator on H. The essential spectrum $\sigma_e(T)$ of T is empty if and only if there exists a complete orthonormal system of eigenvectors $\{x_n\}_n$ of T such that the corresponding eigenvalues λ_n converge in absolute value to ∞ as $n \to \infty$.*

Proof. If $\sigma_e(T) = \emptyset$, then the spectrum $\sigma(T)$ consists of a set $\{r_n\}_n$ of isolated eigenvalues of finite multiplicity, which can only converge to $\pm\infty$. We enumerate the eigenvalues in order of increasing absolute values and repeat each eigenvalue according to its multiplicity. In this way, we get an associated orthonormal system of eigenvectors $\{x_n\}_n$ of T. Suppose that this system is not complete. Then by Lemma 9.37 the subspace

$$L = \{x \in H : (x, x_n) = 0, n \in \mathbb{N}\}$$

is invariant with respect to T in the sense of Definition 9.33. If $x \in \text{dom}(T) \cap L$, then

$$(Tx, x_n) = (x, Tx_n) = r_n(x, x_n) = 0,$$

and hence $T(\text{dom}(T) \cap L) \subseteq L$. The essential spectrum of the restriction of T to L is also empty. Since the spectrum of this restriction is nonempty (Proposition 9.23), T has further eigenvalues and eigenvectors not accounted for in the above list.

Conversely, suppose that a sequence of eigenvalues and eigenvectors with the stated properties exists, and let $\{s_n\}_n$ be the set of distinct eigenvalues. By the assumption that λ_n converge in absolute value to ∞ as $n \to \infty$ we deduce that s_n are isolated eigenvalues of finite multiplicity. It follows from Lemma 9.33 that

$$\sigma(T) = \bigcup_{n=1}^{\infty} \{s_n\}.$$

Thus $\sigma_e(T) = \emptyset$. □

Proposition 9.39. *Let T be an unbounded self-adjoint operator on H that is nonnegative in the sense that $\sigma(T) \subseteq [0, \infty)$. Then the following conditions are equivalent:*
(i) *The resolvent operator $(I + T)^{-1}$ is compact.*
(ii) $\sigma_e(T) = \emptyset$.
(iii) *There exists a complete orthonormal system of eigenvectors $\{x_n\}_n$ of T with corresponding eigenvalues $\mu_n \geq 0$ that converge to $+\infty$ as $n \to \infty$.*

Proof. (i) $\Rightarrow$ (iii): The operator $(I + T)^{-1} : H \longrightarrow \text{dom}(T)$ is compact and self-adjoint and has the dense image. By Proposition 2.7 there exists a complete orthonormal system of eigenvectors x_n of $(I + T)^{-1}$ and eigenvalues λ_n (of finite multiplicity) tending to 0 such that

$$(I + T)^{-1}x = \sum_{n=1}^{\infty} \lambda_n(x, x_n)x_n.$$

Since all $x_n \in \mathrm{dom}(T)$, we have $T(I+T)^{-1}x_n = \lambda_n T x_n$. Adding $(I+T)^{-1}x_n$ to this equality, we get

$$\lambda_n T x_n + (I+T)^{-1}x_n = T(I+T)^{-1}x_n + (I+T)^{-1}x_n = (I+T)(I+T)^{-1}x_n = x_n,$$

which implies that

$$\lambda_n T x_n + \lambda_n x_n = x_n,$$

and therefore

$$T x_n = \frac{1-\lambda_n}{\lambda_n} x_n.$$

Setting $\mu_n = \frac{1-\lambda_n}{\lambda_n}$, we get (iii).

Now suppose that (iii) holds. We rearrange the eigenvectors x_n so that the sequence $\{\mu_n\}_n$ is nondecreasing. Each $x \in H$ can be written in the form $x = \sum_{n=1}^{\infty}(x, x_n)x_n$, and we obtain

$$(I+T)^{-1}x = \sum_{n=1}^{\infty}(x, x_n)(I+T)^{-1}x_n = \sum_{n=1}^{\infty}\frac{1}{1+\mu_n}(x, x_n)x_n.$$

Let

$$A_N = \sum_{n=1}^{N}\frac{1}{1+\mu_n}(x, x_n)x_n.$$

Then the operators A_N are of finite rank, and by Bessel's inequality (1.6) we obtain

$$\|((I+T)^{-1} - A_N)x\| = \left\| \sum_{n=N+1}^{\infty}\frac{1}{1+\mu_n}(x, x_n)x_n \right\| \le \frac{1}{1+\mu_N}\|x\|,$$

from which we see that A_N converges in operator norm to $(I+T)^{-1}$ as $N \to \infty$. Now, by Proposition 2.6, $(I+T)^{-1}$ is a compact operator.

The equivalence of (ii) and (iii) is a consequence of Proposition 9.38. □

The following general result explains the approach to the $\overline{\partial}$-Neumann operator (4.42) by means of the embedding

$$j : \mathrm{dom}(\overline{\partial}) \cap \mathrm{dom}(\overline{\partial}^*) \hookrightarrow L^2_{(0,q)}(\Omega),$$

where $\mathrm{dom}(\overline{\partial}) \cap \mathrm{dom}(\overline{\partial}^*)$ is endowed with the graph norm

$$u \mapsto (\|\overline{\partial}u\|^2 + \|\overline{\partial}^* u\|^2)^{1/2}.$$

It will also be crucial for the question of compactness of the $\overline{\partial}$-Neumann operator, which will be discussed in the following chapters.

Proposition 9.40. *Let A be a nonnegative self-adjoint operator (i. e., $\sigma(A)$ is contained in $[0,\infty)$). There exists a unique self-adjoint square root $A^{1/2}$ of A, and* $\mathrm{dom}(A^{1/2}) \supseteq \mathrm{dom}(A)$. *In addition,* $\mathrm{dom}(A)$ *endowed with the norm*

$$\|f\|_{\mathcal{D}} := \left(\|A^{1/2}f\|^2 + \|f\|^2\right)^{1/2}$$

becomes a Hilbert space; the norm $\|\cdot\|_{\mathcal{D}}$ stems from the inner product

$$(f,g)_{\mathcal{D}} = (A^{1/2}f, A^{1/2}g) + (f,g).$$

Let $\mathrm{dom}(A)$ *be endowed with the norm $\|\cdot\|_{\mathcal{D}}$. Then A has a compact resolvent if and only if the canonical embedding*

$$j : \mathrm{dom}(A) \hookrightarrow H$$

is a compact linear operator.

Furthermore, A has a compact resolvent if and only if $A^{1/2}$ has a compact resolvent.

Proof. In Proposition 9.25, we proved the existence and uniqueness of the square root of A.

For $n \in \mathbb{N}$, we define the functions

$$q_n(t) = \frac{nt}{n+t}, \quad t \in [0,\infty).$$

These are continuous functions such that $q_n(t) \le q_{n+1}(t)$ and $\lim_{n\to\infty} q_n(t) = t$. Moreover, $q_n(t) \le n$ for each $t \in [0,\infty)$. By Theorem 9.30 the operator $nA(nI+A)^{-1}$ is bounded on H, and the function

$$Q_n(x) := (nA(nI+A)^{-1}x, x), \quad x \in H,$$

is bounded on the unit ball of H and continuous. The functional calculus implies that $Q_n(x)$ increases monotonically to $Q(x)$, where

$$Q(x) = \begin{cases} (A^{1/2}x, A^{1/2}x) & \text{for } x \in \mathrm{dom}(A), \\ +\infty & \text{otherwise.} \end{cases}$$

A function $\Theta : H \longrightarrow (-\infty,+\infty]$ is said to be lower semicontinuous if for every convergent sequence $x_n \longrightarrow x$ in H, we have

$$\Theta(x) \le \liminf_{n\to\infty} \Theta(x_n).$$

It is easily seen that a function Θ is lower semicontinuous if and only if

$$\{x : \Theta(x) > a\}$$

is open for every real a and that the pointwise limit of an increasing sequence of continuous functions is a lower semicontinuous function. Therefore the function Q is lower semicontinuous.

Now let $\{x_n\}_n$ be a Cauchy sequence with respect to $\|\cdot\|_{\mathcal{D}}$. Then $\{x_n\}_n$ is also a Cauchy sequence with respect to $\|\cdot\|$ and therefore converges to $x \in H$. Given $\epsilon > 0$, there exists $N \in \mathbb{N}$ such that

$$Q(x_m - x_n) + \|x_m - x_n\|^2 < \epsilon^2$$

for all $m, n \geq N$. Letting $m \to \infty$ and using the lower semicontinuity of Q, we deduce that $x \in \mathrm{dom}(A)$ and

$$Q(x - x_n) + \|x - x_n\|^2 \leq \epsilon^2$$

for all $n > N$. Hence $\|x - x_n\|_{\mathcal{D}} \leq \epsilon$ and $\mathrm{dom}(A)$ endowed with the norm $\|\cdot\|_{\mathcal{D}}$ is complete.

Since $-1 \notin \sigma(A)$, we know that $(I + A)^{-1}$ is a bounded operator on H. From (9.7) we get that $R_A(-1) = (I + A)^{-1}$ is compact if and only if $R_A(z)$ is compact for all $z \notin \sigma(A)$.

Let $u \in H$ and $v \in \mathrm{dom}(A)$. Then

$$\begin{aligned}
(j^* u, v)_{\mathcal{D}} &= (u, jv) = (u, v) = ((I + A)(I + A)^{-1}u, v) \\
&= ((I + A)^{-1}u, (I + A)v) \\
&= ((I + A)^{-1}u, Av) + ((I + A)^{-1}u, v) \\
&= (A^{1/2}(I + A)^{-1}u, A^{1/2}v) + ((I + A)^{-1}u, v) \\
&= ((I + A)^{-1}u, v)_{\mathcal{D}}.
\end{aligned}$$

This implies that $j^* = (I + A)^{-1}$ as an operator on $\mathrm{dom}(A)$ and $j \circ j^* = (I + A)^{-1}$ as an operator on H. So we deduce the desired conclusion by the fact that j is compact if and only if $j \circ j^*$ is compact (Theorem 2.5).

The last statement follows from $(iI + A^{1/2})^* = -iI + A^{1/2}$ and

$$(I + A) = (iI + A^{1/2})(-iI + A^{1/2}). \qquad \square$$

Our next aim is to compare two self-adjoint strictly positive operators and to prove Ruelle's lemma.

Remark 9.41. Let S be a self-adjoint operator. Suppose that $(Sx, x) \geq 0$ for each $x \in \mathrm{dom}(S)$. We say that S is strictly positive if $(Sx, x) > 0$ for each $x \in \mathrm{dom}(S) \setminus \{0\}$. Using the fact that $\mathrm{dom}(S)$ is a core of $S^{1/2}$ (Proposition 9.25), we can easily show that S is strictly positive if and only if $\|S^{1/2}x\| > 0$ for each $x \in \mathrm{dom}(S^{1/2}) \setminus \{0\}$. We also indicate that a strictly positive self-adjoint operator S is injective and $\mathrm{im}(S)$ is dense in H and S^{-1} is self-adjoint (Lemma 4.28).

Lemma 9.42. *Let S and T be strictly positive self-adjoint operators. Then the following statements are equivalent.*

(a) $\operatorname{dom}(T^{1/2}) \subset \operatorname{dom}(S^{1/2})$ *and* $\|S^{1/2}x\| \le \|T^{1/2}x\|$ *for* $x \in \operatorname{dom}(T^{1/2})$;

(b) $\operatorname{dom}(T^{1/2}) \subset \operatorname{dom}(S^{1/2})$ *and* $\|S^{1/2}T^{-1/2}x\| \le \|x\|$ *for* $x \in \operatorname{dom}(T^{-1/2})$;

(c) $\operatorname{dom}(S^{-1/2}) \subset \operatorname{dom}(T^{-1/2})$ *and* $\|T^{-1/2}S^{1/2}x\| \le \|x\|$ *for* $x \in \operatorname{dom}(S^{1/2})$;

(d) $\operatorname{dom}(S^{-1/2}) \subset \operatorname{dom}(T^{-1/2})$ *and* $\|T^{-1/2}x\| \le \|S^{-1/2}x\|$ *for* $x \in \operatorname{dom}(S^{-1/2})$.

Proof. We show that (a) and (b) are equivalent and that (b) implies (c). The remaining implications follow by symmetry and by replacing S and T by S^{-1} and T^{-1}.

Suppose that (a) holds. If $x \in \operatorname{dom}(T^{-1/2})$, then $T^{-1/2}x \in \operatorname{dom}(T^{1/2})$, so we get $\|S^{1/2}T^{-1/2}x\| \le \|x\|$.

If (b) holds and $x \in \operatorname{dom}(T^{1/2})$, then we have $T^{1/2}x \in \operatorname{dom}(T^{-1/2})$, and hence

$$\|S^{1/2}x\| = \|S^{1/2}T^{-1/2}T^{1/2}x\| \le \|T^{1/2}x\|.$$

Suppose that (b) holds. We observe that $\operatorname{dom}(S^{-1/2}) = \operatorname{im}(S^{1/2})$ and that we have to show

$$\operatorname{im}(S^{1/2}) \subset \operatorname{dom}(T^{-1/2}) \quad \text{and} \quad \|T^{-1/2}S^{1/2}x\| \le \|x\| \quad \text{for } x \in \operatorname{dom}(S^{1/2}).$$

If $x \in \operatorname{dom}(S^{1/2})$, then $S^{1/2}x \in \operatorname{im}(S^{1/2})$. We consider the linear functional

$$\psi_x(y) := (S^{1/2}x, T^{-1/2}y) \quad \text{for } y \in \operatorname{dom}(T^{-1/2}).$$

Then

$$|\psi_x(y)| = |(S^{1/2}x, T^{-1/2}y)| = |(x, S^{1/2}T^{-1/2}y)| \le \|x\|\,\|y\|.$$

This implies that $S^{1/2}x \in \operatorname{dom}((T^{-1/2})^*) = \operatorname{dom}(T^{-1/2})$. Finally, we get

$$|(T^{-1/2}S^{1/2}x, y)| \le \|x\|\,\|y\|, \quad \text{for } y \in \operatorname{dom}(T^{-1/2}).$$

As $\operatorname{dom}(T^{-1/2})$ is dense in H, we obtain $\|T^{-1/2}S^{1/2}x\| \le \|x\|$. □

For Ruelle's lemma, we consider strictly positive self-adjoint operators S and T, and we write $S \le T$ if and only if $\operatorname{dom}(T) \subseteq \operatorname{dom}(S)$ and $(Sx, x) \le (Tx, x)$ for each $x \in \operatorname{dom}(T)$. By Proposition 9.25 the square roots of S and T exist and are themselves positive self-adjoint operators.

Lemma 9.43 (Ruelle's lemma). *Let S and T be strictly positive self-adjoint operators. Suppose that $S \le T$ and $0 \in \rho(S)$. Then $T^{-1} \le S^{-1}$.*

Proof. We have $\operatorname{dom}(T) \subseteq \operatorname{dom}(S)$ and

$$\|S^{1/2}x\|^2 = (Sx, x) \le (Tx, x) = \|T^{1/2}x\|^2 \tag{9.46}$$

for all $x \in \operatorname{dom}(T)$.

Next, we show that $\operatorname{dom}(T^{1/2}) \subset \operatorname{dom}(S^{1/2})$. Let $x \in \operatorname{dom}(T^{1/2})$. By Proposition 9.25, $\operatorname{dom}(T)$ is a core of $\operatorname{dom}(T^{1/2})$. Hence there exists a sequence $(x_k)_k$ in $\operatorname{dom}(T)$ such that $x_k \to x$ and $T^{1/2}x_k \to T^{1/2}x$. By (9.46) we have

$$\|S^{1/2}(x_m - x_k)\| \leq \|T^{1/2}(x_m - x_k)\|,$$

which implies that $(S^{1/2}x_k)_k$ is a Cauchy sequence. Since $S^{1/2}$ is also a closed operator, we have that $x \in \operatorname{dom}(S^{1/2})$ and $S^{1/2}x_k \to S^{1/2}x$. In addition, by (9.46) we have

$$\|S^{1/2}x\| = \lim_{k\to\infty} \|S^{1/2}x_k\| \leq \lim_{k\to\infty} \|T^{1/2}x_k\| = \|T^{1/2}x\|$$

for each $x \in \operatorname{dom}(T^{1/2})$.

Now we can apply Lemma 9.42 (d) and get $\operatorname{dom}(S^{-1/2}) \subset \operatorname{dom}(T^{-1/2})$ and $\|T^{-1/2}x\| \leq \|S^{-1/2}x\|$ for $x \in \operatorname{dom}(S^{-1/2})$. Our assumption $0 \in \rho(S)$ implies that $H = \operatorname{dom}(S^{-1}) = \operatorname{dom}(S^{-1/2})$, which proves the lemma. □

Example 9.44. We return to the $\overline{\partial}$-Neumann operator on $L^2_{(0,1)}(\mathbb{C}^n, e^{-\varphi})$ and remark that the Kohn–Morrey formula from Proposition 6.3 can be written in the form

$$(M_\varphi u, u)_\varphi \leq (\Box_\varphi u, u)_\varphi$$

for a $(0,1)$-form $u \in \operatorname{dom}(\overline{\partial}) \cap \operatorname{dom}(\overline{\partial}^*_\varphi)$. So, under the assumptions of Theorem 8.4, we obtain using Ruelle's Lemma 9.43 that

$$(N_\varphi u, u)_\varphi \leq (M_\varphi^{-1} u, u)_\varphi,$$

and setting $\overline{\partial} v = u$, we get

$$\|v\|^2_\varphi = (v, v)_\varphi = (v, \overline{\partial}^*_\varphi N_\varphi u)_\varphi = (\overline{\partial} v, N_\varphi u)_\varphi = (u, N_\varphi u)_\varphi \leq (M_\varphi^{-1}\overline{\partial} v, \overline{\partial} v)_\varphi$$

for each $v \in \operatorname{dom}(\overline{\partial})$ orthogonal to $\ker(\overline{\partial})$.

This gives a different proof of Hörmander's L^2-estimates similar to the Brascamp–Lieb inequality (see [46] and [61]):

$$\int_{\mathbb{C}^n} |v(z)|^2 e^{-\varphi(z)}\, d\lambda(z) \leq \int_{\mathbb{C}^n} |\overline{\partial} v(z)|^2_{i\partial\overline{\partial}\varphi}\, e^{-\varphi(z)}\, d\lambda(z) \tag{9.47}$$

for each $v \in \operatorname{dom}(\overline{\partial})$ orthogonal to $\ker(\overline{\partial})$, where we used the notation in formula (8.7).

Let $1 \leq q \leq n$. If u is a $(0,q)$-form in $\operatorname{dom}(\overline{\partial}) \cap \operatorname{dom}(\overline{\partial}^*_\varphi)$, then by (6.11) we get

$$\sum_{|K|=q-1}{}' \sum_{j,k=1}^{n} \int_{\mathbb{C}^n} \frac{\partial^2 \varphi}{\partial z_j \partial \overline{z}_k}\, u_{jK}\, \overline{u}_{kK}\, e^{-\varphi}\, d\lambda \leq (\Box_\varphi u, u)_\varphi.$$

The left-hand side can be written in the form $(\tilde{M}_\varphi u, u)_\varphi$. We suppose that $\tilde{M}_\varphi$ is invertible and get as above

$$\|v\|_\varphi^2 \leq (\tilde{M}_\varphi^{-1}\overline{\partial}v, \overline{\partial}v)_\varphi \tag{9.48}$$

for each $(0, q-1)$-form $v \in \operatorname{dom}(\overline{\partial})$ orthogonal to $\ker(\overline{\partial})$.

9.5 Variational characterization of the discrete spectrum

Here we explain the max-min principle to describe the lowest part of the spectrum of a self-adjoint operator when it is discrete. This is done in the setting of semibounded operators.

Definition 9.45. Let T be a symmetric unbounded operator with $\operatorname{dom}(T)$. We say that T is semibounded (from below) if there exists a constant $C > 0$ such that

$$(Tu, u) \geq -C\|u\|^2 \ \forall u \in \operatorname{dom}(T).$$

See the next chapter for examples of semibounded operators.

Using Proposition 4.36, we can show that a symmetric semibounded operator T admits a self-adjoint extension. For this purpose, set $a(u, v) = (Tu, v)$ for the sesquilinear form from there. The resulting self-adjoint extension is called the Friedrichs extension.

Proposition 9.46. *Let A be a self-adjoint semibounded operator. Let*

$$\Sigma = \inf \sigma_e(A).$$

The set $\sigma(A) \cap (-\infty, \Sigma)$ can be described as a sequence (finite or infinite) of eigenvalues λ_j ordered increasingly. Then

$$\lambda_1 = \inf\{(A\phi, \phi)\, \|\phi\|^{-2} : \phi \in \operatorname{dom}(A),\ \phi \neq 0\}, \tag{9.49}$$

and for $k \geq 2$,

$$\lambda_k = \inf\{(A\phi, \phi)\, \|\phi\|^{-2} : \phi \in \operatorname{dom}(A) \cap \mathcal{K}_{k-1}^{\perp},\ \phi \neq 0\}, \tag{9.50}$$

where

$$\mathcal{K}_j = \bigoplus_{m \leq j} \ker(A - \lambda_m I).$$

Proof. Let μ_1 denote the right-hand side of (9.49). If ϕ_1 is an eigenfunction for the eigenvalue λ_1, then we get $\mu_1 \leq \lambda_1$. On the other side, we have $\sigma(A - \lambda_1 I) \subseteq [0, \infty)$, and hence, by Proposition 9.25,

$$(A\phi - \lambda_1\phi, \phi) \geq 0$$

for each $\phi \in \operatorname{dom}(A)$, which implies $\lambda_1 \leq \mu_1$.

In fact, we have shown that if $\mu_1 < \Sigma$, then the spectrum below Σ is not empty. In particular, the bottom of the spectrum is an eigenvalue equal to μ_1.

For $k = 2, 3, \ldots$, we apply the first step of the proof to $A\ |_{\operatorname{dom}(A)\cap\mathcal{K}_{k-1}^{\perp}}$ and use the spectral theorem. □

Our next aim is to generalize the following min-max lemma for positive compact operators.

Lemma 9.47. *Let $A : H \longrightarrow H$ be a positive compact operator (i. e., $(Ax, x) \geq 0$ for all $x \in H$) with spectral decomposition*

$$Ax = \sum_{n=0}^{\infty} \lambda_n (x, x_n) x_n,$$

where $\lambda_0 \geq \lambda_1 \geq \cdots$. Then

(i)

$$\lambda_0 = \max_{x\in H} \frac{(Ax, x)}{(x, x)},$$

where the maximum is attained by an eigenvector x_0 with eigenvalue λ_0.

(ii)

$$\lambda_j = \min_{L\in N_j} \max_{x\in L^{\perp}} \frac{(Ax, x)}{(x, x)}, \quad j \geq 1,$$

where N_j denotes the set of all j-dimensional subspaces of H. The minimum is attained by the subspace $L = L_j = \langle x_0, \ldots, x_{j-1}\rangle$, i. e.,

$$\lambda_j = \max_{x\in L_j^{\perp}} \frac{(Ax, x)}{(x, x)}.$$

Proof. (i) follows directly from the proof of the spectral Theorem 2.7.

For $j \geq 1$, we have $\lambda_j = (Ax_j, x_j)/(x_j, x_j)$. The statement follows if we can show that for each j-dimensional subspace L, there exists $z_0 \perp L$ with $z_0 \neq 0$ and $z_0 = \sum_{k=0}^{j} (z_0, x_k) x_k$. In that case, we then have

$$\frac{(Az_0, z_0)}{(z_0, z_0)} = \frac{\sum_{k=0}^{j} \lambda_k |(z_0, x_k)|^2}{\sum_{k=0}^{j} |(z_0, x_k)|^2} \geq \lambda_j,$$

as $\lambda_j \leq \lambda_i$ for $0 \leq i \leq j$ and

$$\max_{x\in L^\perp}\frac{(Ax,x)}{(x,x)} \geq \lambda_j.$$

The existence of z_0 follows from the fact that for a basis $\{y_k : k = 0,\dots,j-1\}$ of L, the system of linear equations

$$\sum_{i=0}^{j} a_i(x_i, y_k) = 0, \quad k = 0,\dots,j-1,$$

has a nontrivial solution. Set $z_0 = \sum_{i=0}^{j} a_i x_i$; then $z_0 \perp L$. □

So if we have positive compact operators A and B such that $A \leq B$, which means $(Ax,x) \leq (Bx,x)$, then the eigenvalues of A and B satisfy

$$\lambda_j(A) \leq \lambda_j(B), \quad j = 0,1,2,\dots.$$

The corresponding result for unbounded operators yields information about the bottom of the spectrum and of the essential spectrum.

Proposition 9.48. *Let H be a Hilbert space of infinite dimension. Let A be a self-adjoint semibounded operator with domain* $\operatorname{dom}(A)$. *Let*

$$\mu_1(A) = \inf\{(A\phi,\phi) : \phi \in \operatorname{dom}(A), \|\phi\| = 1\}$$

and, for $n \geq 2$,

$$\mu_n(A) = \sup_{L\in N_{n-1}} \inf\{(A\phi,\phi) : \phi \in L^\perp \cap \operatorname{dom}(A), \|\phi\| = 1\}, \tag{9.51}$$

where N_{n-1} denotes the set of all subspaces of H of dimension $\leq n-1$. Then either
(a) *$\mu_n(A)$ is the nth eigenvalue when the eigenvalues are increasingly ordered (counting the multiplicities) and A has a discrete spectrum in $(-\infty,\mu_n(A)]$, or*
(b) *$\mu_n(A)$ corresponds to the bottom of the essential spectrum. In this case, we have $\mu_j(A) = \mu_n(A)$ for all $j \geq n$.*

Proof. Let E be the uniquely determined resolution of identity on the Borel subsets of $\mathbb{R}$. First, we show that

$$\dim \operatorname{im} E((-\infty,a)) < n \quad \text{if } a < \mu_n(A), \tag{9.52}$$

$$\dim \operatorname{im} E((-\infty,a)) \geq n \quad \text{if } a > \mu_n(A). \tag{9.53}$$

To prove (9.52), let $a < \mu_n(A)$ and suppose that

$$\dim \operatorname{im} E((-\infty,a)) \geq n.$$

If $y \in \operatorname{im} E((-\infty,a))$, then we get $y = E((-\infty,a))x$ for some $x \in H$, and therefore

$$E_{y,y}(\omega) = (E(\omega)y, y) = (E(\omega)E((-\infty, a))x, E((-\infty, a))x) = 0 \tag{9.54}$$

for each Borel subset ω of $\mathbb{R}$ such that $(-\infty, a) \cap \omega = \emptyset$. Since A is bounded from below, we have $(Ax, x) \geq -C\|x\|^2$ for all $x \in \operatorname{dom}(A)$, and Proposition 9.25 and Lemma 9.13 imply that

$$\operatorname{dom}(A) = \left\{ u \in H : \int_{-C}^{\infty} t^2 \, dE_{u,u}(t) < \infty \right\},$$

so we obtain for $y \in \operatorname{im} E((-\infty, a))$ by (9.54) that

$$\int_{-C}^{\infty} t^2 \, dE_{y,y}(t) \leq C' \max(C^2, a^2) < \infty,$$

which implies that $\operatorname{im} E((-\infty, a)) \subseteq \operatorname{dom}(A)$. So we can find an n-dimensional subspace $L \subset \operatorname{dom}(A)$ such that

$$(Au, u) = \int_{-C}^{a} t \, dE_{u,u} \leq a(u, u) \quad \forall u \in L. \tag{9.55}$$

Then, given any $\psi_1, \dots, \psi_{n-1} \in H$, we can find

$$\phi \in L \cap \langle \psi_1, \dots, \psi_{n-1} \rangle^{\perp}$$

such that $\|\phi\| = 1$ and $(A\phi, \phi) \leq a$. Returning to the definition of $\mu_n(A)$, we would have $\mu_n(A) \leq a$, which is a contradiction. Hence we have shown (9.52).

To prove (9.53), let $a > \mu_n(A)$ and suppose that

$$\dim \operatorname{im} E((-\infty, a)) \leq n - 1.$$

Then we can find $(n-1)$ generators $\psi_1, \dots, \psi_{n-1}$ of this space, and any

$$\phi \in \operatorname{dom}(A) \cap \langle \psi_1, \dots, \psi_{n-1} \rangle^{\perp}$$

is in $\operatorname{im} E([a, +\infty))$, so

$$(A\phi, \phi) \geq a\|\phi\|^2,$$

which is again a contradiction, and we get (9.53).

In the next step, we will show that $\mu_n(A) < +\infty$ for each $n \in \mathbb{N}$. Since A is semi-bounded from below, $\mu_n(A)$ has a uniform lower bound. Suppose that $\mu_n(A) = +\infty$. By (9.52) this means that

$$\dim \operatorname{im} E((-\infty, a)) < n$$

for all $a \in \mathbb{R}$. Hence H must be of finite dimension, and we arrive at a contradiction. (If H is finite-dimensional, then we have $\mu_n(A) \geq \|A\|$.)

For the rest of the proof, we distinguish between the following two cases:

$$\dim \operatorname{im} E((-\infty, \mu_n(A) + \epsilon)) = \infty \quad \forall \epsilon > 0 \tag{9.56}$$

and

$$\dim \operatorname{im} E((-\infty, \mu_n(A) + \epsilon_0)) < \infty \quad \text{for some } \epsilon_0 > 0. \tag{9.57}$$

Assuming (9.56), we claim that statement (b) of the proposition holds: using (9.52) in this case, we obtain

$$\dim \operatorname{im} E((\mu_n(A) - \epsilon, \mu_n(A) + \epsilon)) = \infty \quad \forall \epsilon > 0.$$

By Lemma 9.35, this shows that $\mu_n(A) \in \sigma_e(A)$.

Using (9.52) once more, we see that the interval $(-\infty, \mu_n(A))$ does not contain any point of the essential spectrum. Hence

$$\mu_n(A) = \inf\{\lambda : \lambda \in \sigma_e(A)\}.$$

From (9.51) we obtain that $\mu_{n+1}(A) \geq \mu_n(A)$. If we had $\mu_{n+1}(A) > \mu_n(A)$, then (9.52) would also be satisfied for $\mu_{n+1}(A)$. This is a contradiction to (9.56). Hence statement (b) is proved.

Finally, suppose that (9.57) holds. By Lemma 9.35 it is clear that the spectrum of A is discrete in $(-\infty, \mu_n(A) + \epsilon_0)$. Then for $\epsilon_1 > 0$ small enough,

$$\operatorname{im} E((-\infty, \mu_n(A)]) = \operatorname{im} E((-\infty, \mu_n(A) + \epsilon_1)),$$

and by (9.53)

$$\dim \operatorname{im} E((-\infty, \mu_n(A)]) \geq n.$$

So there are at least n eigenvalues

$$\lambda_1 \leq \lambda_2 \leq \cdots \leq \lambda_n \leq \mu_n(A)$$

for A. If λ_n were strictly less than $\mu_n(A)$, then we would have

$$\dim \operatorname{im} E((-\infty, \lambda_n]) = n,$$

which yields a contradiction to (9.52). So $\lambda_n = \mu_n(A)$, and $\mu_n(A)$ is an eigenvalue. This proves statement (a). □

We note that the proof of (9.52) gives the following:

Proposition 9.49. *Suppose that there exist a and an n-dimensional subspace $L \subset \mathrm{dom}(A)$ such that* (9.55) *is satisfied. Then $\mu_n(A) \leq a$.*

Using Proposition 9.48, we now get the following:

Corollary 9.50. *Under the same assumptions as in Proposition* 9.49, *if a is below the bottom of the essential spectrum of A, then A has at least n eigenvalues (counted with multiplicities).*

The last results allow us to compare the spectra of two operators. If $(Au, u) \leq (Bu, u)$ for all $u \in \mathrm{dom}(B) \subseteq \mathrm{dom}(A)$, then $\lambda_n(A) \leq \lambda_n(B)$.

9.6 Notes

Most of the material on spectral theory is taken from [85] and [24]. It would be beyond the scope of this book to provide a complete proof of the spectral theorem. The notion of the spectral decomposition is explained in detail for both the bounded and the unbounded self-adjoint operators. For the complete proof, some aspects of Banach algebras and the Gelfand transform are needed, which appear to be not important enough for a thorough understanding of the spectral properties of the $\overline{\partial}$-Neumann operator. See [85] for a complete proof of the spectral theorem and [24] for a different approach using the Helffer–Sjöstrand formula. The use of the square root of the $\overline{\partial}$-Neumann operator and of the $\Box$-operator, as done in Remark 9.12 and Example 9.44, gives a better understanding of the basic estimate (4.38) and other fundamental results such as Theorem 8.6. Lemma 9.33 is taken from [24]. It is very useful for operators whose eigenvectors can be determined explicitly; it will be of importance in Chapter 14. In the next chapters on operators of mathematical physics, certain aspects of spectral theory developed in this chapter will be crucial. Variational methods in spectral analysis (see [51, 82]) are important to explain special properties of Schrödinger operators like diamagnetism and paramagnetism.

10 Schrödinger operators and Witten Laplacians

In this chapter, we collect basic information about Schrödinger operators with magnetic field in real dimension 2 and the Witten Laplacian for complex dimension ≥ 2. It turns out that there is a close relationship between these operators and the weighted $\square$-operator of the $\overline{\partial}$-complex. In addition, we include the definition of Dirac and Pauli operators, for which, later on, spectral properties are investigated as applications of corresponding results on the weighted $\square$-operator of the $\overline{\partial}$-complex. First, we will study the regularity of the solutions for elliptic differential equations of order two.

10.1 Difference quotients

To apply Sobolev space theory, we are forced to study difference quotient approximations to weak derivatives.

Definition 10.1. Let $\Omega \subset \mathbb{R}^n$ be a bounded domain, let $u \in L^2_{\text{loc}}(\Omega)$, and suppose that $V \Subset \Omega$.

The jth difference quotient of size h is

$$D_j^h u(x) = \frac{u(x + he_j) - u(x)}{h}$$

for $j = 1, \ldots, n$, where $x \in V$ and $h \in \mathbb{R}$, $0 < |h| < \operatorname{dist}(V, b\Omega)$.

Further, we denote

$$D^h u := (D_1^h u, \ldots, D_n^h u).$$

Proposition 10.2.

(i) *Let $u \in W^1(\Omega)$. Then for each $V \Subset \Omega$, we have*

$$\|D^h u\|_{L^2(V)} \leq C \|\nabla u\|_{L^2(\Omega)} \tag{10.1}$$

for some constant $C > 0$ and all h with $0 < |h| < \frac{1}{2} \operatorname{dist}(V, b\Omega)$.

(ii) *Assume that $u \in L^2(V)$ and that there exists a constant $C > 0$ such that*

$$\|D^h u\|_{L^2(V)} \leq C \tag{10.2}$$

for all h with $0 < |h| < \frac{1}{2} \operatorname{dist}(V, b\Omega)$. Then $u \in W^1(V)$, and

$$\|\nabla u\|_{L^2(V)} \leq C. \tag{10.3}$$

Proof. Suppose first that $u \in C^1(\Omega)$. Then for all $x \in V$, $j = 1, \ldots, n$, and $0 < |h| < \frac{1}{2} \operatorname{dist}(V, b\Omega)$, we have

https://doi.org/10.1515/9783111182926-010

$$u(x + he_j) - u(x) = h \int_0^1 u_{x_j}(x + the_j)\, dt,$$

and hence

$$\left|u(x + he_j) - u(x)\right| \le |h| \int_0^1 \left|\nabla u(x + the_j)\right| dt.$$

So we obtain

$$\begin{aligned}\int_V \left|D^h u\right|^2 d\lambda &\le \sum_{j=1}^n \int_V \int_0^1 \left|\nabla u(x + the_j)\right|^2 dt\, d\lambda(x)\\ &= \sum_{j=1}^n \int_0^1 \int_V \left|\nabla u(x + the_j)\right|^2 d\lambda(x)\, dt.\end{aligned}$$

Thus

$$\int_V \left|D^h u\right|^2 d\lambda \le C \int_\Omega |\nabla u|^2\, d\lambda(x).$$

This estimate is true for a smooth u and by Lemma 5.9 for arbitrary $u \in W^1(\Omega)$, and hence we have shown (i).

Now suppose that (10.2) holds. We choose $j = 1, \ldots, n$ and $\phi \in C_0^\infty(V)$ and note that for small enough h, we have

$$\int_V u(x) \left(\frac{\phi(x + he_j) - \phi(x)}{h}\right) d\lambda(x) = -\int_V \left(\frac{u(x) - u(x - he_j)}{h}\right) \phi(x)\, d\lambda(x),$$

which means that

$$\int_V u\, D_j^h \phi\, d\lambda = -\int_V (D_j^{-h} u)\, \phi\, d\lambda. \tag{10.4}$$

Hence (10.2) implies that

$$\sup_h \|D_j^{-h} u\|_{L^2(V)} < \infty.$$

Using the proof of Proposition 4.17, we conclude that there exist a function $v_j \in L^2(V)$ and a subsequence $h_k \to 0$ such that $D_j^{-h_k} u \to v_j$ weakly in $L^2(V)$. Then we have

$$\int_V u\phi_{x_j}\, d\lambda = \int_\Omega u\phi_{x_j}\, d\lambda = \lim_{h_k \to 0} \int_\Omega u D_j^{h_k} \phi\, d\lambda$$

$$= -\lim_{h_k \to 0} \int_V D_j^{-h_k} u\, \phi\, d\lambda$$
$$= -\int_V v_j \phi\, d\lambda = -\int_\Omega v_j \phi\, d\lambda.$$

Hence $v_j = u_{x_j}$ in the sense of distributions, and so $\nabla u \in L^2(V)$. As $u \in L^2(V)$, we get $u \in W^1(V)$. □

10.2 Interior regularity

We prove a basic result concerning elliptic partial differential equations of order 2 with variable coefficients and the smoothness of their weak solutions.

Definition 10.3. Let $\Omega \subset \mathbb{R}^n$ be a bounded open set, let $a_{jk} \in C^1(\Omega)$ be real-valued, and let $b_j, c \in L^\infty(\Omega)$ for $j, k = 1, \ldots, n$. Define the partial differential operator L by

$$Lu = -\sum_{j,k=1}^{n} (a_{jk}(x)u_{x_j})_{x_k} + \sum_{j=1}^{n} b_j(x)u_{x_j} + c(x)u \tag{10.5}$$

with $a_{jk} = a_{kj}$ for $j, k = 1, \ldots, n$. We say that the operator L is elliptic if there exists a constant $C > 0$ such that

$$\sum_{j,k=1}^{n} a_{jk}(x)t_j\overline{t}_k \geq C|t|^2 \tag{10.6}$$

for almost every $x \in \Omega$ and all $t \in \mathbb{C}^n$.

Ellipticity means that the symmetric $(n \times n)$ matrix $A(x) = (a_{jk}(x))_{j,k=1}^n$ is positive definite with the smallest eigenvalue greater than or equal to C.

If $a_{jk} = \delta_{jk}$ and $b_j = c = 0$, then $L = -\triangle$.

Definition 10.4. Let $f \in L^2(\Omega)$. We say that a function $u \in H^1(\Omega)$ is a weak solution to the elliptic partial differential equation

$$Lu = f \quad \text{in } \Omega$$

if for the bilinear form

$$a(u, v) = \int_\Omega \left(\sum_{j,k=1}^{n} a_{jk}u_{x_j}\overline{v_{x_k}} + \sum_{j=1}^{n} b_j u_{x_j}\overline{v} + cu\overline{v} \right) d\lambda,$$

we have

$$a(u, v) = (f, v)$$

for all $v \in H_0^1(\Omega)$, where $(\cdot,\cdot)$ denotes the inner product in $L^2(\Omega)$.

In the next proposition, we show what is called the interior H^2-regularity of the operator L.

Proposition 10.5. *Let L be as in the above definition, and let $f \in L^2(\Omega)$. Suppose that $u \in H^1(\Omega)$ is a weak solution of $Lu = f$. Then $u \in H^2_{\mathrm{loc}}(\Omega)$, and for each open $V \Subset \Omega$, we have*

$$\|u\|_{H^2(V)} \le \tilde{C}(\|f\|_{L^2(\Omega)} + \|u\|_{H^1(\Omega)}), \tag{10.7}$$

where $\tilde{C} > 0$ is a constant only depending on V, Ω, and the coefficients of L.

Proof. Choose an open set W such that $V \Subset W \Subset \Omega$. Next, select a smooth cut-off function ψ with $0 \le \psi \le 1$, $\psi = 1$ on V, and $\psi = 0$ on $\mathbb{R}^n \setminus W$.

Since u is a weak solution of $Lu = f$, we have $a(u,v) = (f,v)$ for all $v \in H_0^1(\Omega)$, and hence

$$\sum_{j,k=1}^{n} \int_{\Omega} a_{jk} u_{x_j} \overline{v_{x_k}}\, d\lambda = \int_{\Omega} \tilde{f} \bar{v}\, d\lambda, \tag{10.8}$$

where

$$\tilde{f} := f - \sum_{j=1}^{n} b_j u_{x_j} - cu. \tag{10.9}$$

Let $\ell \in \{1,\dots,n\}$, and let $h \in \mathbb{R}$ be such that $|h| > 0$ is small. We substitute

$$v = -D_\ell^{-h}(\psi^2 D_\ell^h u) \tag{10.10}$$

into (10.8), where $D_\ell^h u$ denotes the difference quotient (Definition 10.1). For the left-hand side of (10.8), we get

$$\begin{aligned}
-\sum_{j,k=1}^{n} \int_{\Omega} a_{jk} u_{x_j} [D_\ell^{-h}(\psi^2 D_\ell^h u)]^{-}_{x_k}\, d\lambda &= \sum_{j,k=1}^{n} \int_{\Omega} D_\ell^h(a_{jk} u_{x_j})(\psi^2 D_\ell^h u)^{-}_{x_k}\, d\lambda \\
&= \sum_{j,k=1}^{n} \int_{\Omega} [a_{jk}^h (D_\ell^h u_{x_j})(\psi^2 D_\ell^h u)^{-}_{x_k} + (D_\ell^h a_{jk}) u_{x_j} (\psi^2 D_\ell^h u)^{-}_{x_k}]\, d\lambda,
\end{aligned}$$

where we used the formulas

$$\int_{\Omega} v D_\ell^{-h} w\, d\lambda = -\int_{\Omega} w D_\ell^h v\, d\lambda$$

and

$$D_\ell^h(vw) = v^h D_\ell^h w + w D_\ell^h v$$

for $v^h(x) := v(x + he_\ell)$. We continue the computation of the left-hand side of (10.8) and obtain

$$\sum_{j,k=1}^{n} \int_\Omega a_{jk}^h (D_\ell^h u_{x_j})(D_\ell^h u_{x_k})^- \psi^2 \, d\lambda + \sum_{j,k=1}^{n} \int_\Omega [a_{jk}^h (D_\ell^h u_{x_j})(D_\ell^h u)^- 2\psi\psi_{x_k}$$
$$+ (D_\ell^h a_{jk}) u_{x_j} (D_\ell^h u_{x_k})^- \psi^2 + (D_\ell^h a_{jk}) u_{x_j} (D_\ell^h u)^- 2\psi\psi_{x_k}] \, d\lambda = T_1 + T_2.$$

The first term can be estimated from below using ellipticity (10.6):

$$|T_1| \geq C \int_\Omega \psi^2 |D_\ell^h \nabla u|^2 \, d\lambda. \tag{10.11}$$

For the second term T_2, we have by the assumptions on a_{jk}, b_j, and c that there exists a constant $C' > 0$ such that

$$|T_2| \leq C' \int_\Omega [\psi |D_\ell^h \nabla u| \, |D_\ell^h u| + \psi |D_\ell^h \nabla u| \, |\nabla u| + \psi |D_\ell^h u| \, |\nabla u|] \, d\lambda.$$

Taking into account that $\psi = 0$ on $\mathbb{R}^n \setminus W$ and using the small–large constant trick, we get

$$|T_2| \leq \epsilon \int_\Omega \psi^2 |D_\ell^h \nabla u|^2 \, d\lambda + \frac{C'}{\epsilon} \int_W [|D_\ell^h u|^2 + |\nabla u|^2] \, d\lambda.$$

Now choose $\epsilon = C/2$ and estimate (10.1) to obtain

$$|T_2| \leq \frac{C}{2} \int_\Omega \psi^2 |D_\ell^h \nabla u|^2 \, d\lambda + C'' \int_\Omega |\nabla u|^2 \, d\lambda.$$

Hence by (10.10) we see that the left-hand side of (10.8) can be estimated from below by

$$\frac{C}{2} \int_\Omega \psi^2 |D_\ell^h \nabla u|^2 \, d\lambda - C'' \int_\Omega |\nabla u|^2 \, d\lambda. \tag{10.12}$$

The absolute value of the right-hand side of (10.8) is certainly less than

$$C'' \int_\Omega (|f| + |\nabla u| + |u|)|v| \, d\lambda. \tag{10.13}$$

Using Proposition 10.2(i), we derive

$$\int_\Omega |v|^2 \, d\lambda \leq C'' \int_\Omega |\nabla(\psi^2 D_\ell^h u)|^2 \, d\lambda$$

$$\leq C'' \int_W (|D_\ell^h u|^2 + \psi^4 |D_\ell^h \nabla u|^2)\, d\lambda$$
$$\leq C'' \int_\Omega (|\nabla u|^2 + \psi^4 |D_\ell^h \nabla u|^2)\, d\lambda.$$

Again by the small–large constant trick and by (10.13) we obtain now that the absolute value of the right-hand side of (10.8) can be estimated from above by

$$\epsilon \int_\Omega \psi^2 |D_\ell^h \nabla u|^2\, d\lambda + \frac{C'}{\epsilon} \int_\Omega (|f|^2 + |u|^2 + |\nabla u|^2)\, d\lambda.$$

Let $\epsilon = C/4$. Then the absolute value of the right-hand side of (10.8) can be estimated from above by

$$\frac{C}{4} \int_\Omega \psi^2 |D_\ell^h \nabla u|^2\, d\lambda + C''' \int_\Omega (|f|^2 + |u|^2 + |\nabla u|^2)\, d\lambda. \tag{10.14}$$

Finally, combining (10.8), (10.12), and (10.14), we see that

$$\int_V |D_\ell^h \nabla u|^2\, d\lambda \leq \int_\Omega \psi^2 |D_\ell^h \nabla u|^2\, d\lambda \leq \tilde{C} \int_\Omega (|f|^2 + |u|^2 + |\nabla u|^2)\, d\lambda$$

for some constant $\tilde{C} > 0$, $\ell = 1, \ldots, n$, and all sufficiently small $|h| \neq 0$.

Using Proposition 10.2(ii), we derive that $\nabla u \in H^1_{\mathrm{loc}}(\Omega)$ and hence that $u \in H^2_{\mathrm{loc}}(\Omega)$ with the estimate

$$\|u\|_{H^2(V)} \leq \tilde{C}(\|f\|_{L^2(\Omega)} + \|u\|_{H^1(\Omega)}). \qquad \square$$

Remark 10.6.

(a) It is not difficult to show that even

$$\|u\|_{H^2(V)} \leq \tilde{C}(\|f\|_{L^2(\Omega)} + \|u\|_{L^2(\Omega)})$$

in Proposition 10.5.

(b) The result that $u \in H^2_{\mathrm{loc}}(\Omega)$ implies that $Lu = f$ almost everywhere in Ω. To see this, note that for each $v \in C_0^\infty(\Omega)$, we have

$$a(u, v) = (f, v),$$

and since $u \in H^2_{\mathrm{loc}}(\Omega)$, we can integrate by parts and obtain

$$a(u, v) = (Lu, v).$$

Thus $(Lu - f, v) = 0$ for all $v \in C_0^\infty(\Omega)$, and so $Lu = f$ almost everywhere in Ω.

10.3 Schrödinger operators with magnetic field

We consider differential operators $H(A,V)$ of the form

$$H(A,V) = -\triangle_A + V, \tag{10.15}$$

where $V : \mathbb{R}^n \longrightarrow \mathbb{R}$ is the electric potential, and

$$A = \sum_{j=1}^{n} A_j dx_j, \quad A_j : \mathbb{R}^n \longrightarrow \mathbb{R}, \quad j = 1, \dots, n,$$

is a 1-form, and

$$\triangle_A = \sum_{j=1}^{n} \left(\frac{\partial}{\partial x_j} - iA_j \right)^2.$$

The 2-form

$$B = dA = \sum_{j<k} \left(\frac{\partial A_k}{\partial x_j} - \frac{\partial A_j}{\partial x_k} \right) dx_j \wedge dx_k$$

is the magnetic field, which is responsible for specific spectral properties of the operator $H(A,V)$, as we will see later. It is clear that $H(A,V)$ are second-order elliptic partial differential operators.

Under appropriate assumptions on A and V, the operator $H(A,V)$ acts as an unbounded self-adjoint operator on $L^2(\mathbb{R}^n)$. In many aspects of the spectral theory of the Schrödinger operator with magnetic field $H(A,V)$, it is convenient to compare this operator with the ordinary Schrödinger operator

$$H(0,V) = -\triangle + V$$

and then to employ well-known properties of $H(0,V)$.

Let $X_j = (-i\frac{\partial}{\partial x_j} - A_j)$ for $j = 1, \dots, n$. Then

$$-\triangle_A = \sum_{j=1}^{n} X_j^2, \tag{10.16}$$

and for $u, v \in C_0^\infty(\mathbb{R}^n)$, we have $(X_j u, v) = (u, X_j v)$, $j = 1, \dots, n$, and

$$(-\triangle_A u, u) = \sum_{j=1}^{n} \|X_j u\|^2. \tag{10.17}$$

Proposition 10.7. *Let $A \in C^2(\mathbb{R}^n, \mathbb{R}^n)$, and let V be a continuous real-valued function on $\mathbb{R}^n$ such that*

$$V(x) \geq -C \quad \forall x \in \mathbb{R}^n,$$

where $C > 0$ is a positive constant. Let $\operatorname{dom}(H(A,V)) = C_0^\infty(\mathbb{R}^n)$. *Then $H(A,V)$ is a symmetric semibounded operator on $L^2(\mathbb{R}^n)$.*

Proof. For $u \in C_0^\infty(\mathbb{R}^n)$, we have

$$\begin{aligned}(H(A,V)u,u) &= \int_{\mathbb{R}^n} (-\bigtriangleup_A u + Vu)\,\overline{u}\,d\lambda \\ &= \int_{\mathbb{R}^n} \sum_{j=1}^n |X_j u|^2\,d\lambda + \int_{\mathbb{R}^n} V|u|^2\,d\lambda \\ &\geq -C\,\|u\|^2.\end{aligned}$$

□

Using the Friedrichs extension (Proposition 4.36), we obtain the following:

Proposition 10.8. *Let $H(A,V)$ be as in Proposition* 10.7. *Then $H(A,V)$ admits a self-adjoint extension.*

Proof. Define

$$a(u,v) := (H(A,V)u,v) + (C+1)(u,v),$$

and define $\mathcal{V}$ to be the completion of $C_0^\infty(\mathbb{R}^n)$ with respect to the inner product $a(u,v)$. Then we can apply Proposition 4.36 to get the desired result. □

Let $f, g \in L^1_{\text{loc}}(\mathbb{R}^n)$. We say that $f \geq g$ in the distributional sense if

$$(f,\phi) \geq (g,\phi)$$

for all positive $\phi \in C_0^\infty(\mathbb{R}^n)$.

A useful tool for spectral analysis of Schrödinger operators is Kato's inequality, sometimes also called the diamagnetic inequality.

Proposition 10.9. *Let $A \in C^2(\mathbb{R}^n,\mathbb{R}^n)$. Then for all $f \in L^1_{\text{loc}}(\mathbb{R}^n)$ such that $(-i\nabla - A)^2 f \in L^2_{\text{loc}}(\mathbb{R}^n)$, we have*

$$\bigtriangleup |f| \geq -\Re(\operatorname{sgn}(f)(-i\nabla - A)^2 f) = \Re(\operatorname{sgn}(f)\bigtriangleup_A f) \tag{10.18}$$

in the distributional sense, where sgn is defined as in Chapter 5.

Proof. Let $A_1,\dots,A_n$ be the components of A. Notice that

$$-\bigtriangleup_A f = (-i\nabla - A)^2 f = \sum_{j=1}^n \left(-i\frac{\partial}{\partial x_j} - A_j\right)^2 f.$$

The assumption $(-i\nabla - A)^2 f \in L^2_{\mathrm{loc}}(\mathbb{R}^n)$ and the regularity property of second-order elliptic operators (see Proposition 10.5) imply that $f \in W^2_{\mathrm{loc}}(\mathbb{R}^n)$ and, in particular, $\Delta f, \nabla f \in L^1_{\mathrm{loc}}(\mathbb{R}^n)$.

First, suppose that u is smooth. Then, with $|u|_\epsilon = \sqrt{|u|^2 + \epsilon^2} - \epsilon$, we get

$$\nabla |u|_\epsilon = \frac{\Re(\overline{u}\nabla u)}{\sqrt{|u|^2 + \epsilon^2}} = \frac{\Re(\overline{u}(\nabla - iA)u)}{\sqrt{|u|^2 + \epsilon^2}}. \tag{10.19}$$

A straightforward calculation shows that for a smooth function g, we have

$$g \triangle g = \operatorname{div}(g\nabla g) - |\nabla g|^2.$$

Hence we obtain

$$\begin{aligned}\sqrt{|u|^2 + \epsilon^2} \triangle |u|_\epsilon &= \operatorname{div}\big(\sqrt{|u|^2 + \epsilon^2}\, \nabla |u|_\epsilon\big) - \big|\nabla |u|_\epsilon\big|^2 \\ &= \Re\big[\overline{\nabla u} \cdot (\nabla - iA)u + \overline{u} \operatorname{div}((\nabla - iA)u)\big] - \big|\nabla |u|_\epsilon\big|^2 \\ &= \Re\big[\overline{(\nabla u - iAu)} \cdot (\nabla - iA)u \\ &\quad + (-iA\overline{u}) \cdot (\nabla - iA)u + \overline{u} \operatorname{div}((\nabla - iA)u)\big] - \big|\nabla |u|_\epsilon\big|^2 \\ &= \big|(\nabla - iA)u\big|^2 - \big|\nabla |u|_\epsilon\big|^2 \\ &\quad + \Re\big[(-iA\overline{u}) \cdot (\nabla - iA)u + \overline{u} \operatorname{div}((\nabla - iA)u)\big].\end{aligned}$$

An easy calculation shows that

$$(-iA\overline{u}) \cdot (\nabla - iA)u + \overline{u} \operatorname{div}((\nabla - iA)u) = \overline{u}\, (\nabla - iA)^2 u.$$

From (10.19) we get

$$\big|\nabla |u|_\epsilon\big|^2 \le \frac{|\overline{u}(\nabla - iA)u|^2}{|u|^2 + \epsilon^2} = \frac{|u|^2\, |(\nabla - iA)u|^2}{|u|^2 + \epsilon^2} \le \big|(\nabla - iA)u\big|^2.$$

So we finally see that

$$\triangle |u|_\epsilon \ge \Re \frac{\overline{u}\, (\nabla - iA)^2 u}{\sqrt{|u|^2 + \epsilon^2}}. \tag{10.20}$$

The rest of the proof uses approximative units and follows the same lines as the proof of Proposition 5.7. □

Using Kato's inequality and a criterion for essential self-adjointness, we obtain the following:

Proposition 10.10. *Let $A \in C^2(\mathbb{R}^n, \mathbb{R}^n)$, $V \in L^2_{\mathrm{loc}}(\mathbb{R}^n)$, and $V \ge 0$. Then the Schrödinger operator $H(A, V) = -\triangle_A + V$ is essentially self-adjoint on $C_0^\infty(\mathbb{R}^n)$. In this case the Friedrichs* *extension is the uniquely determined self-adjoint extension (see Remark 4.27(b) and Proposition 10.8).*

Proof. By Proposition 9.29 it is sufficient to show that

$$\ker(H(A,V)^* + I) = \{0\}.$$

Since $\operatorname{dom}(H(A,V)^*) \subseteq L^2(\mathbb{R}^n)$, the triviality of the kernel follows from the following statement: if

$$-\triangle_A u + Vu + u = 0 \tag{10.21}$$

for $u \in L^2(\mathbb{R}^n)$, then $u = 0$.

If $u \in L^2(\mathbb{R}^n)$ and $V \in L^2_{\mathrm{loc}}(\mathbb{R}^n)$, we have $uV \in L^1_{\mathrm{loc}}(\mathbb{R}^n)$. In addition, we have the inclusion

$$L^2(\mathbb{R}^n) \subset L^2_{\mathrm{loc}}(\mathbb{R}^n) \subset L^1_{\mathrm{loc}}(\mathbb{R}^n),$$

which follows from the estimate

$$\int_K |u|\, d\lambda \le |K|^{1/2} \left(\int_K |u|^2\, d\lambda \right)^{1/2}.$$

Hence we have that $u \in L^1_{\mathrm{loc}}(\mathbb{R}^n)$ and, by (10.21), that $\triangle u \in L^1_{\mathrm{loc}}(\mathbb{R}^n)$, where the derivative is taken in the sense of distributions.

From (10.18) and (10.21) we obtain

$$\begin{aligned} \triangle|u| &\ge \Re(\operatorname{sgn}(u)\ \triangle_A\, u) \\ &= \Re(\operatorname{sgn}(u)\,(V+1)u) \\ &= |u|\,(V+1) \ge 0. \end{aligned}$$

If $(\chi_\epsilon)_\epsilon$ is an approximate unit, then we get

$$\triangle\,(\chi_\epsilon * |u|) = \chi_\epsilon * \triangle|u| \ge 0. \tag{10.22}$$

Since $\chi_\epsilon * |u| \in \operatorname{dom}(\triangle)$, we have

$$(\triangle(\chi_\epsilon * |u|), \chi_\epsilon * |u|) = -\|\nabla(\chi_\epsilon * |u|)\|^2 \le 0. \tag{10.23}$$

By (10.22) the left side of (10.23) is nonnegative, so $\nabla(\chi_\epsilon * |u|) = 0$, and hence $\chi_\epsilon * |u| = c \ge 0$. Since $|u| \in L^2(\mathbb{R}^n)$ and $\chi_\epsilon * |u| \to |u|$ in $L^2(\mathbb{R}^n)$, it follows that $c = 0$. Hence $\chi_\epsilon * |u| = 0$, so $|u| = 0$, and thus $u = 0$. □

Before we return to applications of methods from complex analysis, we prove two results on the spectrum of Schrödinger operators:

Proposition 10.11. *Let $A \in C^2(\mathbb{R}^n, \mathbb{R}^n)$, and let $V \in L^2_{\text{loc}}(\mathbb{R}^n)$ with $V \geq 0$. Then*

$$\inf \sigma(H(A, V)) \geq \inf \sigma(H(0, V)). \tag{10.24}$$

Proof. By Kato's inequality (10.18) we have

$$-\triangle |f| \leq \Re(\operatorname{sgn}(f)(-\triangle_A f)),$$

so we get

$$\begin{aligned}(|f|, H(0, V)|f|) &\leq \int_{\mathbb{R}^n} |f| \, [\Re(\operatorname{sgn}(f)(-\triangle_A f)) + V] \, d\lambda \\ &\leq \Re \int_{\mathbb{R}^n} \bar{f} H(A, V) f \, d\lambda \\ &= (H(A, V)f, f).\end{aligned}$$

Now we can apply Proposition 9.48 to obtain the desired result. □

Finally, we still mention the gauge invariance of the spectrum of $H(A, V)$.

Proposition 10.12. *Let $A, A' \in C^2(\mathbb{R}^n, \mathbb{R}^n)$ be such that $dA = dA'$. Suppose that $V \in L^2_{\text{loc}}(\mathbb{R}^n)$ and $V \geq 0$. Then $\sigma(H(A, V)) = \sigma(H(A', V))$.*

Proof. By the Poincaré lemma we have $A' = A + dg$, where $g \in C^1(\mathbb{R}^n)$.

Let $X_j = (-i\frac{\partial}{\partial x_j} - A_j)$ and $X_j' = (-i\frac{\partial}{\partial x_j} - A_j')$ for $j = 1, \ldots, n$. Then

$$X_j' = e^{ig} \, X_j \, e^{-ig}.$$

Hence

$$H(A', V) = e^{ig} \, H(A, V) \, e^{-ig}.$$

Therefore the operators $H(A', V)$ and $H(A, V)$ are unitarily equivalent and hence by Lemma 9.31 have the same spectrum. □

In our first application of methods from complex analysis, we will concentrate on Schrödinger operators with magnetic field on $\mathbb{R}^2$ of the form

$$-\triangle_A + B,$$

where the 2-form dA is given by $dA = B dx \wedge dy$. So in this case the Schrödinger operator

$$-\triangle_A + B = H(A, B)$$

is completely determined by the vector field A.

Let φ be a subharmonic $\mathcal{C}^2$-function. We consider the inhomogeneous $\overline{\partial}$-equation $\overline{\partial}u = f$ for $f \in L^2(\mathbb{C}, e^{-\varphi})$. The canonical solution operator to $\overline{\partial}$ gives a solution with minimal $L^2(\mathbb{C}, e^{-\varphi})$-norm. We substitute $v = u\,e^{-\varphi/2}$ and $g = f\,e^{-\varphi/2}$, and the equation becomes

$$\overline{D}v = g\,,$$

where

$$\overline{D} = e^{-\varphi/2}\,\frac{\partial}{\partial \overline{z}}\,e^{\varphi/2}. \tag{10.25}$$

Then u is the minimal solution to the $\overline{\partial}$-equation in $L^2(\mathbb{C}, e^{-\varphi})$ if and only if v is the solution to $\overline{D}v = g$ that is minimal in $L^2(\mathbb{C})$.

The formal adjoint of $\overline{D}$ is $\overline{D}^* = -e^{\varphi/2}\frac{\partial}{\partial z}e^{-\varphi/2}$. We define

$$\operatorname{dom}(\overline{D}) = \{f \in L^2(\mathbb{C}) : \overline{D}f \in L^2(\mathbb{C})\}$$

and likewise for $\overline{D}^*$. Then $\overline{D}$ and $\overline{D}^*$ are closed unbounded linear operators from $L^2(\mathbb{C})$ to itself. Further, we define

$$\operatorname{dom}(\overline{D}\,\overline{D}^*) = \{u \in \operatorname{dom}(\overline{D}^*) : \overline{D}^* u \in \operatorname{dom}(\overline{D})\},$$

and we define $\overline{D}\,\overline{D}^*$ as $\overline{D} \circ \overline{D}^*$ on this domain. Any function of the form $e^{\varphi/2}g$ with $g \in \mathcal{C}_0^2(\mathbb{C})$ belongs to $\operatorname{dom}(\overline{D}\,\overline{D}^*)$, and hence $\operatorname{dom}(\overline{D}\,\overline{D}^*)$ is dense in $L^2(\mathbb{C})$. Since

$$\overline{D} = \frac{\partial}{\partial \overline{z}} + \frac{1}{2}\frac{\partial \varphi}{\partial \overline{z}} \quad \text{and} \quad \overline{D}^* = -\frac{\partial}{\partial z} + \frac{1}{2}\frac{\partial \varphi}{\partial z},$$

we see that

$$\mathcal{S} = \overline{D}\,\overline{D}^* = -\frac{\partial^2}{\partial z \partial \overline{z}} - \frac{1}{2}\frac{\partial \varphi}{\partial \overline{z}}\frac{\partial}{\partial z} + \frac{1}{2}\frac{\partial \varphi}{\partial z}\frac{\partial}{\partial \overline{z}} + \frac{1}{4}\left|\frac{\partial \varphi}{\partial z}\right|^2 + \frac{1}{2}\frac{\partial^2 \varphi}{\partial z \partial \overline{z}}. \tag{10.26}$$

It follows that

$$4\mathcal{S} = -\left(\frac{\partial^2}{\partial x^2} + \frac{\partial^2}{\partial y^2}\right) + \frac{i}{2}\frac{\partial \varphi}{\partial x}\frac{\partial}{\partial y} - \frac{i}{2}\frac{\partial \varphi}{\partial y}\frac{\partial}{\partial x} + \frac{1}{4}\left(\left(\frac{\partial \varphi}{\partial x}\right)^2 + \left(\frac{\partial \varphi}{\partial y}\right)^2\right) + \frac{1}{2}\triangle\varphi,$$

which implies that $\mathcal{S}$ is a Schrödinger operator with magnetic field, namely

$$\mathcal{S} = \frac{1}{4}(-\triangle_A + B), \tag{10.27}$$

where the 1-form $A = A_1\,dx + A_2\,dy$ is related to the weight φ by

$$A_1 = -\partial_y\varphi/2, \quad A_2 = \partial_x\varphi/2, \tag{10.28}$$

$$\triangle_A = \left(\frac{\partial}{\partial x} - iA_1\right)^2 + \left(\frac{\partial}{\partial y} - iA_2\right)^2, \tag{10.29}$$

and the magnetic field $B dx \wedge dy$ satisfies

$$B(x,y) = \frac{1}{2} \triangle \varphi(x,y).$$

Both operators $\overline{D}\,\overline{D}^*$ and $\overline{D}^*\,\overline{D}$ are nonnegative self-adjoint operators; see Lemmas 4.28 and 4.29. By Proposition 10.10 we now know that the operator $\mathcal{S}$ with its domain described above is the uniquely determined self-adjoint extension of the Schrödinger operator with magnetic field $\frac{1}{4}(-\triangle_A + B)$.

Since $4\overline{D}\,\overline{D}^* = -\triangle_A + \frac{1}{2}\triangle\varphi$, it follows that $((-\triangle_A + \frac{1}{2}\triangle\varphi)f, f) \geq 0$ for $f \in \mathcal{C}_0^2(\mathbb{C})$. Similarly, we show that $4\overline{D}^*\,\overline{D} = -\triangle_A - \frac{1}{2}\triangle\varphi$, which implies

$$-2\triangle_A \geq -\triangle_A + \frac{1}{2}\triangle\varphi \geq -\triangle_A\,.$$

It follows that

$$\overline{D}^*\,\overline{D} = e^{-\varphi/2}\,\overline{\partial}_\varphi^*\,\overline{\partial}\,e^{\varphi/2} \tag{10.30}$$

and that

$$\overline{D}\,\overline{D}^* = e^{-\varphi/2}\,\overline{\partial}\,\overline{\partial}_\varphi^*\,e^{\varphi/2}, \tag{10.31}$$

where $\overline{\partial}_\varphi^* = -\frac{\partial}{\partial z} + \frac{\partial\varphi}{\partial z}$. For $n = 1$, we have

$$\Box_\varphi = \overline{\partial}\,\overline{\partial}_\varphi^*,$$

which means that

$$\overline{D}\,\overline{D}^* = e^{-\varphi/2}\,\Box_\varphi\,e^{\varphi/2}. \tag{10.32}$$

Theorem 10.13. *Let φ be a subharmonic $\mathcal{C}^2$-function on $\mathbb{C}$ such that*

$$\triangle\varphi(z) > \epsilon > 0.$$

Then the Schrödinger operator with magnetic field

$$\mathcal{S} = \overline{D}\,\overline{D}^* = \frac{1}{4}\left(-\triangle_A + \frac{1}{2}\triangle\varphi\right) \tag{10.33}$$

has a bounded inverse on $L^2(\mathbb{C})$,

$$(\overline{D}\,\overline{D}^*)^{-1} = e^{-\varphi/2}\,N_\varphi\,e^{\varphi/2}, \tag{10.34}$$

where $N_\varphi = \Box_\varphi^{-1}$.

Proof. We know that $4\overline{D}^*\,\overline{D} = -\triangle_A - \frac{1}{2}\triangle\varphi$, which implies that

$$-\triangle_A \geq \frac{1}{2}\triangle\varphi \geq 0.$$

Hence, for $u \in \mathcal{C}_0^\infty(\mathbb{C})$, we have

$$((-\triangle_A + B)u, u) = (-\triangle_A\, u, u) + (Bu, u) \geq (Bu, u), \tag{10.35}$$

and by a similar density argument as in Proposition 6.2 we obtain from (10.35) together with the assumption on φ that

$$\|\overline{D}^* u\| \geq C\|u\| \tag{10.36}$$

for all $u \in \mathrm{dom}(\overline{D}^*)$.

Now we take $\mathcal{V} = \mathrm{dom}(\overline{D}^*)$ endowed with the topology given by the inner product $(\overline{D}^* u, \overline{D}^* v)$, the sesquilinear form $a(u,v) = (\overline{D}^* u, \overline{D}^* v)$, and $H = L^2(\mathbb{C})$ and apply Proposition 4.36 to get the desired result. □

10.4 Witten Laplacians

For several complex variables, the situation is more complicated.

Let $\varphi : \mathbb{C}^n \longrightarrow \mathbb{R}$ be a $\mathcal{C}^2$-weight function. We consider the $\overline{\partial}$-complex

$$L^2(\mathbb{C}^n, e^{-\varphi}) \stackrel{\overline{\partial}}{\longrightarrow} L^2_{(0,1)}(\mathbb{C}^n, e^{-\varphi}) \stackrel{\overline{\partial}}{\longrightarrow} L^2_{(0,2)}(\mathbb{C}^n, e^{-\varphi}).$$

For $v \in L^2(\mathbb{C}^n)$, let

$$\overline{D}_1 v = \sum_{k=1}^n \left(\frac{\partial v}{\partial \overline{z}_k} + \frac{1}{2}\frac{\partial \varphi}{\partial \overline{z}_k} v \right) d\overline{z}_k,$$

and for $g = \sum_{j=1}^n g_j\, d\overline{z}_j \in L^2_{(0,1)}(\mathbb{C}^n)$, let

$$\overline{D}_1^* g = \sum_{j=1}^n \left(\frac{1}{2}\frac{\partial \varphi}{\partial z_j} g_j - \frac{\partial g_j}{\partial z_j} \right),$$

where the derivatives are taken in the sense of distributions.

It is easy to see that $\overline{\partial} u = f$ for $u \in L^2(\mathbb{C}^n, e^{-\varphi})$ and $f \in L^2_{(0,1)}(\mathbb{C}^n, e^{-\varphi})$ if and only if $\overline{D}_1 v = g$, where $v = u\, e^{-\varphi/2}$ and $g = f\, e^{-\varphi/2}$. It is also clear that the necessary condition $\overline{\partial} f = 0$ for solvability holds if and only if $\overline{D}_2 g = 0$. Here

$$\overline{D}_2 g = \sum_{j,k=1}^{n} \left(\frac{\partial g_j}{\partial \overline{z}_k} + \frac{1}{2} \frac{\partial \varphi}{\partial \overline{z}_k} g_j \right) d\overline{z}_k \wedge d\overline{z}_j,$$

and

$$\overline{D}_2^* h = \sum_{j,k=1}^{n} \left(\frac{1}{2} \frac{\partial \varphi}{\partial z_k} h_{kj} - \frac{\partial h_{kj}}{\partial z_k} \right) d\overline{z}_j$$

for a suitable $(0,2)$-form $h = \sum'_{|J|=2} h_J \, d\overline{z}_J$.

We consider the corresponding $\overline{D}$-complex

$$L^2(\mathbb{C}^n) \underset{\overline{D}_1^*}{\overset{\overline{D}_1}{\rightleftarrows}} L^2_{(0,1)}(\mathbb{C}^n) \underset{\overline{D}_2^*}{\overset{\overline{D}_2}{\rightleftarrows}} L^2_{(0,2)}(\mathbb{C}^n).$$

The so-called Witten Laplacians (see [50]) $\Delta_\varphi^{(0,0)}$ and $\Delta_\varphi^{(0,1)}$ are defined by

$$\begin{aligned} \Delta_\varphi^{(0,0)} &= \overline{D}_1^* \overline{D}_1, \\ \Delta_\varphi^{(0,1)} &= \overline{D}_1 \overline{D}_1^* + \overline{D}_2^* \overline{D}_2. \end{aligned} \tag{10.37}$$

A computation shows that

$$\overline{D}_1^* \overline{D}_1 v = \sum_{j=1}^{n} \left(\frac{1}{2} \frac{\partial \varphi}{\partial z_j} \frac{\partial v}{\partial \overline{z}_j} + \frac{1}{4} \frac{\partial \varphi}{\partial z_j} \frac{\partial \varphi}{\partial \overline{z}_j} v - \frac{1}{2} \frac{\partial \varphi}{\partial \overline{z}_j} \frac{\partial v}{\partial z_j} - \frac{1}{2} \frac{\partial^2 \varphi}{\partial z_j \partial \overline{z}_j} v - \frac{\partial^2 v}{\partial z_j \partial \overline{z}_j} \right)$$

and that

$$\begin{aligned} (\overline{D}_1 \overline{D}_1^* + \overline{D}_2^* \overline{D}_2) g = \sum_{k=1}^{n} \Bigg[\sum_{j=1}^{n} \Bigg(& \frac{1}{2} \frac{\partial \varphi}{\partial z_j} \frac{\partial g_k}{\partial \overline{z}_j} + \frac{1}{4} \frac{\partial \varphi}{\partial z_j} \frac{\partial \varphi}{\partial \overline{z}_j} g_k \\ & - \frac{1}{2} \frac{\partial \varphi}{\partial \overline{z}_j} \frac{\partial g_k}{\partial z_j} - \frac{1}{2} \frac{\partial^2 \varphi}{\partial z_j \partial \overline{z}_j} g_k - \frac{\partial^2 g_k}{\partial z_j \partial \overline{z}_j} + \frac{\partial^2 \varphi}{\partial z_j \partial \overline{z}_k} g_j \Bigg) \Bigg] d\overline{z}_k. \end{aligned} \tag{10.38}$$

More generally, we set $Z_k = \frac{\partial}{\partial \overline{z}_k} + \frac{1}{2} \frac{\partial \varphi}{\partial \overline{z}_k}$ and $Z_k^* = -\frac{\partial}{\partial z_k} + \frac{1}{2} \frac{\partial \varphi}{\partial z_k}$, and we consider $(0,q)$-forms $h = \sum'_{|J|=q} h_J \, d\overline{z}_J$, where $\sum'$ means that we sum up only increasing multi-indices $J = (j_1, \dots, j_q)$, and where $d\overline{z}_J = d\overline{z}_{j_1} \wedge \cdots \wedge d\overline{z}_{j_q}$. We define

$$\overline{D}_{q+1} h = \sum_{k=1}^{n} \sum_{|J|=q}{}' Z_k(h_J) \, d\overline{z}_k \wedge d\overline{z}_J$$

and

$$\overline{D}_q^* h = \sum_{k=1}^{n} \sum_{|J|=q}{}' Z_k^*(h_J) \, d\overline{z}_k \rfloor d\overline{z}_J,$$

where $d\overline{z}_k \rfloor d\overline{z}_J$ denotes the contraction or interior multiplication by $d\overline{z}_k$, i. e., we have

$$\langle \alpha, d\overline{z}_k \rfloor d\overline{z}_J \rangle = \langle d\overline{z}_k \wedge \alpha, d\overline{z}_J \rangle$$

for each $(0, q-1)$-form α.

The complex Witten Laplacian on $(0,q)$-forms is then given by

$$\Delta_\varphi^{(0,q)} = \overline{D}_q \overline{D}_q^* + \overline{D}_{q+1}^* \overline{D}_{q+1} \tag{10.39}$$

for $q = 1, \dots, n-1$.

The general $\overline{D}$-complex has the form

$$L^2_{(0,q-1)}(\mathbb{C}^n) \underset{\overline{D}_q^*}{\overset{\overline{D}_q}{\rightleftarrows}} L^2_{(0,q)}(\mathbb{C}^n) \underset{\overline{D}_{q+1}^*}{\overset{\overline{D}_{q+1}}{\rightleftarrows}} L^2_{(0,q+1)}(\mathbb{C}^n).$$

It follows that

$$\overline{D}_{q+1} \Delta_\varphi^{(0,q)} = \Delta_\varphi^{(0,q+1)} \overline{D}_{q+1} \quad \text{and} \quad \overline{D}_{q+1}^* \Delta_\varphi^{(0,q+1)} = \Delta_\varphi^{(0,q)} \overline{D}_{q+1}^*.$$

We remark that

$$\overline{D}_q^* h = \sum_{k=1}^n {\sum_{|J|=q}}' Z_k^*(h_J)\, d\overline{z}_k \rfloor d\overline{z}_J = {\sum_{|K|=q-1}}' \sum_{k=1}^n Z_k^*(h_{kK})\, d\overline{z}_K.$$

In particular, for a function $v \in L^2(\mathbb{C}^n)$, we get

$$\Delta_\varphi^{(0,0)} v = \overline{D}_1^* \overline{D}_1 v = \sum_{j=1}^n Z_j^* Z_j(v),$$

and for a $(0,1)$-form $g = \sum_{\ell=1}^n g_\ell\, d\overline{z}_\ell \in L^2_{(0,1)}(\mathbb{C}^n)$, we obtain

$$\begin{aligned}
\Delta_\varphi^{(0,1)} g &= (\overline{D}_1 \overline{D}_1^* + \overline{D}_2^* \overline{D}_2) g \\
&= \sum_{j,k,\ell=1}^n \{Z_j(Z_k^*(g_\ell))\, d\overline{z}_j \wedge (d\overline{z}_k \rfloor d\overline{z}_\ell) + Z_k^*(Z_j(g_\ell))\, d\overline{z}_k \rfloor (d\overline{z}_j \wedge d\overline{z}_\ell)\} \\
&= \sum_{j,k,\ell=1}^n \{Z_k^*(Z_j(g_\ell))\, (d\overline{z}_j \wedge (d\overline{z}_k \rfloor d\overline{z}_\ell) + d\overline{z}_k \rfloor (d\overline{z}_j \wedge d\overline{z}_\ell)) \\
&\quad + [Z_j, Z_k^*](g_\ell)\, d\overline{z}_j \wedge (d\overline{z}_k \rfloor d\overline{z}_\ell)\} \\
&= \sum_{j,\ell=1}^n Z_j^* Z_j(g_\ell)\, d\overline{z}_\ell + \sum_{j,k,\ell=1}^n \frac{\partial^2 \varphi}{\partial \overline{z}_j \partial z_k} g_\ell\, \delta_{k\ell} d\overline{z}_j \\
&= (\Delta_\varphi^{(0,0)} \otimes I) g + M_\varphi g,
\end{aligned}$$

where we used that for (0, 1)-forms α, a, b, we have

$$\alpha \rfloor (a \wedge b) = (\alpha \rfloor a) \wedge b - a \wedge (\alpha \rfloor b),$$

which implies that

$$\begin{aligned} &d\bar{z}_j \wedge (d\bar{z}_k \rfloor d\bar{z}_\ell) + d\bar{z}_k \rfloor (d\bar{z}_j \wedge d\bar{z}_\ell) \\ &\quad = d\bar{z}_j \wedge (d\bar{z}_k \rfloor d\bar{z}_\ell) + (d\bar{z}_k \rfloor d\bar{z}_j) \wedge d\bar{z}_\ell - d\bar{z}_j \wedge (d\bar{z}_k \rfloor d\bar{z}_\ell) \\ &\quad = (d\bar{z}_k \rfloor d\bar{z}_j) \wedge d\bar{z}_\ell = \delta_{k\ell}\, d\bar{z}_\ell, \end{aligned}$$

and where we set

$$M_\varphi g = \sum_{j=1}^{n} \left(\sum_{k=1}^{n} \frac{\partial^2 \varphi}{\partial z_k \partial \bar{z}_j} g_k \right) d\bar{z}_j$$

and

$$(\Delta_\varphi^{(0,0)} \otimes I)\, g = \sum_{k=1}^{n} \Delta_\varphi^{(0,0)}\, g_k\, d\bar{z}_k.$$

By Proposition 6.5 we now obtain the following:

Theorem 10.14. *Let $\varphi : \mathbb{C}^n \longrightarrow \mathbb{R}$ be a $\mathcal{C}^2$-plurisubharmonic function and suppose that the lowest eigenvalue μ_φ of the Levi matrix M_φ of φ satisfies* (6.5). *Then the operator $\Delta_\varphi^{(0,1)}$ has a bounded inverse on $L^2_{(0,1)}(\mathbb{C}^n)$,*

$$(\bar{D}_1 \bar{D}_1^* + \bar{D}_2^* \bar{D}_2)^{-1} = (\Delta_\varphi^{(0,1)})^{-1} = e^{-\varphi/2}\, N_\varphi\, e^{\varphi/2}, \tag{10.40}$$

where $N_\varphi = \Box_\varphi^{-1}$.

10.5 Dirac and Pauli operators

There is an interesting connection to Dirac and Pauli operators: recall (10.29) and define the Dirac operator $\mathcal{D}$ by

$$\mathcal{D} = \left(-i\frac{\partial}{\partial x} - A_1 \right) \sigma_1 + \left(-i\frac{\partial}{\partial y} - A_2 \right) \sigma_2 = \mathcal{A}_1 \sigma_1 + \mathcal{A}_2 \sigma_2, \tag{10.41}$$

where

$$\sigma_1 = \begin{pmatrix} 0 & 1 \\ 1 & 0 \end{pmatrix}, \quad \sigma_2 = \begin{pmatrix} 0 & -i \\ i & 0 \end{pmatrix}.$$

Hence we can write

$$\mathcal{D} = \begin{pmatrix} 0 & \mathcal{A}_1 - i\mathcal{A}_2 \\ \mathcal{A}_1 + i\mathcal{A}_2 & 0 \end{pmatrix}.$$

We remark that $i(\mathcal{A}_2\mathcal{A}_1 - \mathcal{A}_1\mathcal{A}_2) = B$, and hence it turns out that the square of $\mathcal{D}$ is diagonal with the Pauli operators $P_\pm$ on the diagonal:

$$\begin{aligned} \mathcal{D}^2 &= \begin{pmatrix} \mathcal{A}_1^2 - i(\mathcal{A}_2\mathcal{A}_1 - \mathcal{A}_1\mathcal{A}_2) + \mathcal{A}_2^2 & 0 \\ 0 & \mathcal{A}_1^2 + i(\mathcal{A}_2\mathcal{A}_1 - \mathcal{A}_1\mathcal{A}_2) + \mathcal{A}_2^2 \end{pmatrix} \\ &= \begin{pmatrix} P_- & 0 \\ 0 & P_+ \end{pmatrix}, \end{aligned}$$

where

$$P_\pm = \left(-i\frac{\partial}{\partial x} - A_1\right)^2 + \left(-i\frac{\partial}{\partial y} - A_2\right)^2 \pm B = - \triangle_A \pm B.$$

By Lemmas 4.28 and 4.29 the Pauli operators $P_\pm$ are nonnegative self-adjoint operators.

It follows that

$$4\mathcal{S} = P_+$$

is the Schrödinger operator with magnetic field and that

$$4\,\Delta_\varphi^{(0,0)} = P_-. \tag{10.42}$$

In addition, we obtain that $\mathcal{D}^2$ is self-adjoint and likewise $\mathcal{D}$ by the spectral theorem.

Finally, we consider decoupled weights $\varphi(z_1,\dots,z_n) = \sum_{j=1}^n \varphi_j(z_j)$. In this case the operator $\Delta_\varphi^{(0,1)}$ acts diagonally on $(0,1)$-forms, each component E_k of the diagonal being

$$E_k = \frac{1}{4}P_+^{(k)} + \frac{1}{4}\sum_{j\neq k} P_-^{(j)}, \tag{10.43}$$

where

$$P_\pm^{(\ell)} = \left(-i\frac{\partial}{\partial x_\ell} - A_1^{(\ell)}\right)^2 + \left(-i\frac{\partial}{\partial y_\ell} - A_2^{(\ell)}\right)^2 \pm B^{(\ell)} \tag{10.44}$$

with $z_\ell = x_\ell + iy_\ell$, $A_1^{(\ell)} = -\frac{1}{2}\frac{\partial\varphi_\ell}{\partial y_\ell}$, $A_2^{(\ell)} = \frac{1}{2}\frac{\partial\varphi_\ell}{\partial x_\ell}$, and $B^{(\ell)} = \frac{1}{2}\triangle\varphi_\ell$, $\ell = 1,\dots,n$. This follows from (10.38) for a decoupled weight and from (10.42).

10.6 Notes

The regularity result on second-order elliptic partial differential operators is taken from [27]. Basic facts about Schrödinger operators with magnetic fields can be found in [21,

51, 52, 54]. Kato's inequality, sometimes also called the diamagnetic inequality, is an important tool for basic questions about Schrödinger operators; see, for instance, [51]. The relationship to the $\overline{\partial}$-equation appears in [8, 40, 46]. In the following, we will describe the spectral analysis of these operators in more detail. The weighted $\overline{\partial}$-complex in several variables leads to the Witten Laplacian. The Witten Laplacians were introduced by E. Witten on a compact manifold in connection with a Morse function, which is a function with nondegenerate critical points. The main idea of E. Witten was that the dimension of the kernels of these Laplacians, which are related by the Hodge theory to the Betti numbers, can be estimated from above by the dimension of the eigenspace associated with the small eigenvalues of these operators (see [50, 55, 56, 61, 95, 96]). The Dirac and Pauli operators are discussed in [21, 54, 92].

11 Compactness

In this chapter, we study the compactness of the $\overline{\partial}$-Neumann operator. We give a general characterization using a description of precompact subsets in L^2-spaces. Therefore we begin this part with a general characterization of precompact subsets in L^2-spaces, followed by basic facts about Sobolev spaces, such as Gårding's inequality and the Rellich–Kondrachov lemma. We point out that our approach includes the case of $L^2(\mathbb{C}^n, e^{-\varphi})$, where the classical Rellich–Kondrachov lemma cannot be used. In our discussion of Schrödinger operators with magnetic field, we follow the approach of Iwatsuka [59] and give a general characterization of compact resolvents and a necessary condition for compact resolvents in terms of the magnetic field. A sufficient condition for the compactness of the $\overline{\partial}$-Neumann operator is given in terms of the behavior at infinity of the eigenvalues of the Levi matrix of the weight function. We also show that properties P and $\tilde{P}$ imply the compactness of the $\overline{\partial}$-Neumann operator for smoothly bounded pseudoconvex domains. In addition, we characterize the compactness in the Segal–Bargmann space.

11.1 Precompact sets in L^2-spaces

A set A is precompact (i. e., $\overline{A}$ is compact) in a Banach space X if and only if for every positive number ϵ, there is a finite subset N_ϵ of points of X such that $A \subset \bigcup_{y\in N_\epsilon} B_\epsilon(y)$. Such a set N_ϵ is called a finite ϵ-net for A.

We recall the Arzelà–Ascoli theorem: Let Ω be a bounded domain in $\mathbb{R}^n$. A subset K of $C(\overline{\Omega})$ is precompact in $C(\overline{\Omega})$ if the following two conditions hold:

(i) There exists a constant M such that $|\phi(x)| \leq M$ for all $\phi \in K$ and $x \in \Omega$ (boundedness);

(ii) For every $\epsilon > 0$, there exists $\delta > 0$ such that if $\phi \in K,\ x, y \in \Omega$, and $|x - y| < \delta$, then $|\phi(x) - \phi(y)| < \epsilon$ (equicontinuity).

Let $(\chi_\epsilon)_\epsilon$ be an approximation to the identity (see Section 5.1). Recall that $u * \chi_\epsilon \in C^\infty(\mathbb{R}^n)$ if $u \in L^1_{\text{loc}}(\mathbb{R}^n)$ (Lemma 5.3).

In a similar way, we prove the following result: If Ω is a domain in $\mathbb{R}^n$ and $u \in L^2(\Omega)$, then $u * \chi_\epsilon \in L^2(\Omega)$, and

$$\|u * \chi_\epsilon\|_2 \leq \|u\|_2, \quad \lim_{\epsilon\to 0+} \|u * \chi_\epsilon - u\|_2 = 0.$$

Let $\Omega \subseteq \mathbb{R}^n$ be a domain, and let u bea complex-valued function on Ω. Let

$$\tilde{u}(x) = \begin{cases} u(x), & x \in \Omega, \\ 0, & x \in \mathbb{R}^n \setminus \Omega. \end{cases}$$

https://doi.org/10.1515/9783111182926-011

Theorem 11.1. *A bounded subset $\mathcal{A}$ of $L^2(\Omega)$ is precompact in $L^2(\Omega)$ if and only if for every $\epsilon > 0$, there exist $\delta > 0$ and a subset $\omega \Subset \Omega$ such that for all $u \in \mathcal{A}$ and $h \in \mathbb{R}^n$ with $|h| < \delta$, we have the following inequalities:*

$$\int_{\Omega} \left|\tilde{u}(x+h) - \tilde{u}(x)\right|^2 d\lambda(x) < \epsilon^2, \qquad \int_{\Omega\setminus\overline{\omega}} \left|u(x)\right|^2 d\lambda(x) < \epsilon^2. \tag{11.1}$$

Proof. Let $\tau_h u(x) = u(x+h)$ denote the translate of u by h. First, assume that $\mathcal{A}$ is precompact. Since $\mathcal{A}$ has a finite $\epsilon/6$-net, and since $C_0(\Omega)$, the space of continuous functions with compact support in Ω, is dense in $L^2(\Omega)$, there exists a finite set $S \subset C_0(\Omega)$ such that for each $u \in \mathcal{A}$, there exists $\phi \in S$ satisfying $\|u - \phi\|_2 < \epsilon/3$. Let ω be the union of the supports of finitely many functions in S. Then $\omega \Subset \Omega$, and the second inequality follows immediately. To prove the first inequality, choose a closed ball B_r of radius r centered at the origin and containing ω. Note that $(\tau_h\phi - \phi)(x) = \phi(x+h) - \phi(x)$ is uniformly continuous and vanishes outside B_{r+1}, provided that $|h| < 1$. Hence

$$\lim_{|h|\to 0} \int_{\mathbb{R}^n} \left|\tau_h\phi(x) - \phi(x)\right|^2 d\lambda(x) = 0,$$

the convergence being uniform for $\phi \in S$. For $|h|$ sufficiently small, we have $\|\tau_h\phi - \phi\|_2 < \epsilon/3$. If $\phi \in S$ satisfies $\|u - \phi\|_2 < \epsilon/3$, then also $\|\tau_h\tilde{u} - \tau_h\phi\|_2 < \epsilon/3$. Hence for $|h|$ sufficiently small (independent of $u \in \mathcal{A}$), we have

$$\|\tau_h\tilde{u} - \tilde{u}\|_2 \le \|\tau_h\tilde{u} - \tau_h\phi\|_2 + \|\tau_h\phi - \phi\|_2 + \|\phi - \tilde{u}\|_2 < \epsilon,$$

and the first inequality follows.

It is sufficient to prove the converse for the particular case $\Omega = \mathbb{R}^n$, as it follows for general Ω from its application in this case to the set $\tilde{\mathcal{A}} = \{\tilde{u} : u \in \mathcal{A}\}$.

Let $\epsilon > 0$ be given and choose $\omega \subset\subset \mathbb{R}^n$ such that for all $u \in \mathcal{A}$,

$$\int_{\mathbb{R}^n\setminus\overline{\omega}} \left|u(x)\right|^2 d\lambda(x) < \frac{\epsilon}{3}.$$

For any $\eta > 0$, the function $u * \chi_\eta \in C^\infty(\mathbb{R}^n)$, and in particular it belongs to $C(\overline{\omega})$. If $\phi \in C_0(\mathbb{R}^n)$, then by Hölder's inequality

$$\begin{aligned}\left|\chi_\eta * \phi(x) - \phi(x)\right|^2 &= \left|\int_{\mathbb{R}^n} \chi_\eta(y)(\phi(x-y) - \phi(x))\, d\lambda(y)\right|^2 \\ &\le \int_{B_\eta} \chi_\eta(y)\left|\tau_{-y}\phi(x) - \phi(x)\right|^2 d\lambda(y).\end{aligned}$$

Hence

$$\|\chi_\eta * \phi - \phi\|_2 \le \sup_{h \in B_\eta} \|\tau_h \phi - \phi\|_2.$$

If $u \in L^2(\mathbb{R}^n)$, then let $(\phi_j)_j$ be a sequence in $C_0(\mathbb{R}^n)$ converging to u in L^2 norm. Then $(\chi_\eta * \phi_j)_j$ is a sequence converging to $\chi_\eta * u$ in $L^2(\mathbb{R}^n)$. Since also $\tau_h \phi_j \to \tau_h u$ in $L^2(\mathbb{R}^n)$, we have

$$\|\chi_\eta * u - u\|_2 \le \sup_{h \in B_\eta} \|\tau_h u - u\|_2.$$

From the first inequality in our assumption we derive that $\lim_{|h| \to 0} \|\tau_h u - u\|_2 = 0$ uniformly for $u \in \mathcal{A}$. Hence $\lim_{\eta \to 0} \|\chi_\eta * u - u\|_2 = 0$ uniformly for $u \in \mathcal{A}$. Fix $\eta > 0$ such that

$$\int_{\overline{\omega}} |\chi_\eta * u(x) - u(x)|^2 \, d\lambda(x) < \frac{\epsilon}{6}$$

for all $u \in \mathcal{A}$.

We show that $\{\chi_\eta * u : u \in \mathcal{A}\}$ satisfies the conditions of the Arzelà–Ascoli theorem on $\overline{\omega}$ and hence is precompact in $C(\overline{\omega})$. We have

$$|\chi_\eta * u(x)| \le \left(\sup_{y \in \mathbb{R}^n} \chi_\eta^2(y)\right)^{1/2} \|u\|_2,$$

which is bounded uniformly for $x \in \mathbb{R}^n$ and $u \in \mathcal{A}$, since $\mathcal{A}$ is bounded in $L^2(\mathbb{R}^n)$ and η is fixed. Similarly,

$$|\chi_\eta * u(x+h) - \chi_\eta * u(x)| \le \left(\sup_{y \in \mathbb{R}^n} \chi_\eta^2(y)\right)^{1/2} \|\tau_h u - u\|_2,$$

and so $\lim_{|h| \to 0} \chi_\eta * u(x+h) = \chi_\eta * u(x)$ uniformly for $x \in \mathbb{R}^n$ and $u \in \mathcal{A}$. Thus $\{\chi_\eta * u : u \in \mathcal{A}\}$ is precompact in $C(\overline{\omega})$, and there exists a finite set $\{\psi_1, \dots, \psi_m\}$ of functions in $C(\overline{\omega})$ such that if $u \in \mathcal{A}$, then for some j, $1 \le j \le m$, and all $x \in \overline{\omega}$, we have

$$|\psi_j(x) - \chi_\eta * u(x)| < \sqrt{\frac{\epsilon}{6|\overline{\omega}|}}.$$

This, together with the inequality $(|a| + |b|)^2 \le 2(|a|^2 + |b|^2)$, implies that

$$\begin{aligned}
\int_{\mathbb{R}^n} |u(x) - \tilde{\psi}_j(x)|^2 \, d\lambda(x) &= \int_{\mathbb{R}^n \setminus \overline{\omega}} |u(x)|^2 \, d\lambda(x) + \int_{\overline{\omega}} |u(x) - \psi_j(x)|^2 \, dx \\
&< \frac{\epsilon}{3} + 2 \int_{\overline{\omega}} \left(|u(x) - \chi_\eta * u(x)|^2 + |\chi_\eta * u(x) - \psi_j(x)|^2\right) d\lambda(x) \\
&< \frac{\epsilon}{3} + 2\left(\frac{\epsilon}{6} + \frac{\epsilon}{6|\overline{\omega}|} |\overline{\omega}|\right) = \epsilon.
\end{aligned}$$

Hence $\mathcal{A}$ has a finite $\epsilon^{1/2}$-net in $L^2(\mathbb{R}^n)$ and is therefore precompact in $L^2(\mathbb{R}^n)$. □

Remark 11.2. (a) With the same proof, we get:

A bounded subset $\mathcal{A}$ of $L^2(\Omega)$ is precompact in $L^2(\Omega)$ if and only if the following two conditions are satisfied:

(i) for every $\epsilon > 0$ and for each $\omega \Subset \Omega$, there exists $\delta > 0$ such that for all $u \in \mathcal{A}$ and $h \in \mathbb{R}^n$ with $|h| < \delta$, we have

$$\int_\omega |\tilde{u}(x+h) - \tilde{u}(x)|^2 \, d\lambda(x) < \epsilon^2; \tag{11.2}$$

(ii) for every $\epsilon > 0$, there exists $\omega \Subset \Omega$ such that for every $u \in \mathcal{A}$,

$$\int_{\Omega\setminus\overline{\omega}} |u(x)|^2 \, d\lambda(x) < \epsilon^2. \tag{11.3}$$

(b) An analogous result holds in weighted spaces $L^2(\mathbb{C}^n, e^{-\varphi})$.

11.2 Sobolev spaces and Gårding's inequality

In the following, we will prove a simple version of Gårding's inequality (coercive estimate), which will be used to investigate the compactness of the $\overline{\partial}$-Neumann operator. For a comprehensive treatment of Gårding's inequality, see, for instance, [31] or [16].

Definition 11.3. If Ω is a bounded open set in $\mathbb{R}^n$, then we define the Sobolev space $H^k(\Omega)$ for a nonnegative integer k to be the completion of $\mathcal{C}^\infty(\overline{\Omega})$ with respect to the norm

$$\left[\sum_{|\alpha|\le k} \int_\Omega |\partial^\alpha f|^2 \, d\lambda\right]^{1/2}, \tag{11.4}$$

where $\alpha = (\alpha_1, \dots, \alpha_n)$ is a multi-index, $|\alpha| = \sum_{j=1}^n \alpha_j$, and

$$\partial^\alpha f = \frac{\partial^{|\alpha|} f}{\partial x_1^{\alpha_1} \dots \partial x_n^{\alpha_n}}.$$

By $H_0^k(\Omega)$ we denote the closure of $\mathcal{C}_0^\infty(\Omega)$ with respect to the norm (11.4).

If Ω is a domain with a $\mathcal{C}^1$-boundary, then $H^k(\Omega)$ coincides with

$$W^k(\Omega) = \{f \in L^2(\Omega) : \partial^\alpha f \in L^2(\Omega), \ |\alpha| \le k\},$$

where the derivatives are taken in the sense of distributions (Proposition 5.10).

Theorem 11.4. *Let Ω be a bounded open set in $\mathbb{R}^n$ with C^1-boundary. Let D be a Dirichlet form of order 1 given by*

$$D(u,v) = \sum_{j,k=1}^{n} (\partial_j u, b_{jk}\partial_k v) + \sum_{k=1}^{n} (\partial_k u, b_k v) + \sum_{k=1}^{n} (u, b'_k \partial_k v) + (u, bv) \tag{11.5}$$

for $u, v \in H^1(\Omega)$, where $\partial_j = \frac{\partial}{\partial x_j}$, b_{jk}, b_k, b'_k, b are C^∞-coefficients, and the b_{jk} are real-valued. Suppose that there exists a constant $C_0 > 0$ such that

$$\Re \sum_{j,k=1}^{n} b_{jk}(x)\xi_j\xi_k \geq C_0|\xi|^2, \quad \xi \in \mathbb{R}^n, \quad x \in \overline{\Omega}; \tag{11.6}$$

then we say that D is strongly elliptic on Ω.

Then there exist constants $C > 0$ and $M \geq 0$ such that

$$\Re D(u,u) \geq C\|u\|^2_{W^1(\Omega)} - M\|u\|^2_{L^2(\Omega)}, \quad u \in H^1(\Omega); \tag{11.7}$$

then we say that D is coercive over $H^1(\Omega)$ (Gårding's inequality).

Proof. We first set $a_{jk} = \frac{1}{2}(b_{jk} + b_{kj})$. Since the b_{jk} are real, strong ellipticity means that for some constant $C_0 > 0$,

$$\sum_{j,k=1}^{n} a_{jk}\xi_j\xi_k = \sum_{j,k=1}^{n} b_{jk}\xi_j\xi_k \geq C_0|\xi|^2$$

for all $\xi \in \mathbb{R}^n$. Thus (a_{jk}) is symmetric ($a_{jk} = a_{kj}$), so if $\xi \in \mathbb{C}^n$ is any complex vector, then

$$\Re \sum_{j,k=1}^{n} b_{jk}\xi_j\overline{\xi}_k = \sum_{j,k=1}^{n} a_{jk}\xi_j\overline{\xi}_k \geq C_0|\xi|^2.$$

Setting $\xi = \nabla u$ for $u \in H^1(\Omega)$, we obtain

$$\Re \sum_{j,k=1}^{n} b_{jk}(\partial_j u)(\partial_k \overline{u}) \geq C_0 \sum_{k=1}^{n} |\partial_k u|^2,$$

so integration over Ω yields

$$\Re \sum_{j,k=1}^{n} (\partial_j u, b_{jk}\partial_k v) \geq C_0 \sum_{k=1}^{n} \|\partial_k u\|^2_{L^2(\Omega)} = C_0\big(\|u\|^2_{W^1(\Omega)} - \|u\|^2_{L^2(\Omega)}\big).$$

Also, for some $C_1 > 0$ (independent of u), we have

$$\big|(\partial_k u, b_k u)\big| \leq \|u\|_{W^1(\Omega)}\|b_k u\|_{L^2(\Omega)} \leq C_1\|u\|_{W^1(\Omega)}\|u\|_{L^2(\Omega)},$$

$$\begin{aligned}|(u, b_k' \partial_k u)| &\le \|u\|_{L^2(\Omega)} \|b_k' \partial_k u\|_{L^2(\Omega)} \le C_1 \|u\|_{W^1(\Omega)} \|u\|_{L^2(\Omega)},\\ |(u, bu)| &\le C_1 \|u\|_{L^2(\Omega)}^2 \le C_1 \|u\|_{W^1(\Omega)} \|u\|_{L^2(\Omega)}.\end{aligned}$$

Setting $C_2 = (2n+1)C_1$, we have

$$\Re D(u,u) \ge C_0 \big(\|u\|_{W^1(\Omega)}^2 - \|u\|_{L^2(\Omega)}^2\big) - C_2 \|u\|_{W^1(\Omega)} \|u\|_{L^2(\Omega)}.$$

By the inequality $cd \le \frac{1}{2}(c^2 + d^2)$ for $c, d > 0$ we have

$$C_2 \|u\|_{W^1(\Omega)} \|u\|_{L^2(\Omega)} \le \frac{C_0}{2} \|u\|_{W^1(\Omega)}^2 + \frac{C_2^2}{2C_0} \|u\|_{L^2(\Omega)}^2,$$

so

$$\Re D(u,u) \ge \frac{C_0}{2} \|u\|_{W^1(\Omega)}^2 - \frac{2C_0^2 + C_2^2}{2C_0} \|u\|_{L^2(\Omega)}^2, \tag{11.8}$$

which proves Gårding's inequality. □

Our next aim is to prove the classical Rellich–Kondrachov lemma, which states that the embedding of $W^1(\Omega)$ into $L^2(\Omega)$ is compact, provided that Ω is a bounded domain with C^1-boundary.

Lemma 11.5 (Rellich–Kondrachov). *Let Ω be a bounded domain with C^1-boundary. Then the embedding $j : W^1(\Omega) \longrightarrow L^2(\Omega)$ is compact.*

Proof. We have to show that the unit ball in $W^1(\Omega)$ is precompact in $L^2(\Omega)$. For this purpose, we apply Proposition 5.11 and consider the extension of elements of the unit ball in $W^1(\Omega)$ to $\mathbb{R}^n$. Let $\mathcal{F}$ denote the set of all these extensions. Then, by Proposition 5.11(iii), $\mathcal{F}$ is a bounded set in $L^2(\mathbb{R}^n)$. By Lemma 5.3 we know that for each $\epsilon > 0$, there exists $N > 0$ such that

$$\|\chi_{1/k} * f - f\|_{L^2(\mathbb{R}^n)} \le \epsilon \tag{11.9}$$

for all $f \in \mathcal{F}$ and $k > N$.

By Hölder's inequality we have

$$\|\chi_{1/k} * f\|_{L^\infty(\mathbb{R}^n)} \le C_k \|f\|_{L^2(\mathbb{R}^n)} \tag{11.10}$$

for all $f \in \mathcal{F}$, where $C_k = \|\chi_{1/k}\|_{L^2(\mathbb{R}^n)}$.

Hence we can now verify the second condition in Theorem 11.1: Given $\epsilon > 0$, there exists $\omega \Subset \Omega$ such that

$$\|f\|_{L^2(\Omega \setminus \omega)} < \epsilon$$

for each f in the unit ball of $W^1(\Omega)$: indeed, we consider the extensions to $\mathbb{R}^n$ and write

$$\|f\|_{L^2(\Omega\setminus\omega)} \le \|f - \chi_{1/k} * f\|_{L^2(\mathbb{R}^n)} + \|\chi_{1/k} * f\|_{L^2(\Omega\setminus\omega)}$$

using Proposition 5.11(iii) and (11.9), and in view of (11.10), we have to choose ω such that $|\Omega \setminus \omega|$ is small enough.

We are left to verify the first condition of Remark 11.2: Let $\omega \Subset \Omega$ and $\epsilon > 0$, and consider first a function $u \in C^\infty(\overline{\Omega})$. Let $h \in \mathbb{R}^n$ be such that $|h| < \operatorname{dist}(\omega, b\Omega)$, and set

$$v(t) := u(x + th), \quad t \in [0, 1].$$

Then $v'(t) = h \cdot \nabla u(x + th)$, and

$$u(x+h) - u(x) = v(1) - v(0) = \int_0^1 v'(t)\,dt = \int_0^1 h \cdot \nabla u(x+th)\,dt. \tag{11.11}$$

Hence we obtain

$$|u(x+h) - u(x)|^2 \le |h|^2 \int_0^1 |\nabla u(x+th)|^2\,dt \tag{11.12}$$

and

$$\begin{aligned}\int_\omega |u(x+h) - u(x)|^2\,d\lambda(x) &\le |h|^2 \int_\omega \int_0^1 |\nabla u(x+th)|^2\,dt\,d\lambda(x)\\ &= |h|^2 \int_0^1 \int_\omega |\nabla u(x+th)|^2\,d\lambda(x)\,dt\\ &= |h|^2 \int_0^1 \int_{\omega+th} |\nabla u(x)|^2\,d\lambda(x)\,dt.\end{aligned}$$

If $|h| < \operatorname{dist}(\omega, b\Omega)$, then there exists $\omega' \Subset \Omega$ such that $\omega + th \subset \omega'$ for each $t \in [0, 1]$. Therefore we get the estimate

$$\|\tau_h u - u\|^2_{L^2(\omega)} \le |h|^2 \int_{\omega'} |\nabla u(x)|^2\,d\lambda(x). \tag{11.13}$$

If u belongs to the unit ball in $W^1(\Omega)$, then we approximate u by functions in $C^\infty(\overline{\Omega})$ (Proposition 5.10), apply (11.13), and pass to the limit, getting

$$\|\tau_h u - u\|^2_{L^2(\omega)} \le |h|^2 \int_{\omega'} |\nabla u(x)|^2\,d\lambda(x) \le |h|^2,$$

which shows that the first condition of Remark 11.2 holds. □

Remark 11.6.

(a) If $\Omega \subset \mathbb{R}^n$, $n \geq 2$, is a bounded domain with C^1-boundary, it even follows that

$$W^1(\Omega) \subset L^q(\Omega), \quad q \in [1, 2n/(n-2))$$

and that the embedding is also compact (see, for instance, [12]).

(b) It follows immediately from Lemma 11.5 that the embedding of

$$\iota : H_0^1(\Omega) \longrightarrow L^2(\Omega)$$

is compact.

By Theorem 2.5 the adjoint

$$\iota^* : L^2(\Omega) \longrightarrow (H_0^1(\Omega))' =: H_0^{-1}(\Omega)$$

is also compact.

In the following, we will still describe $H_0^{-1}(\Omega)$ as a certain space of distributions, which will be helpful later on. Recall that the dual-norm is given by

$$\|f\|_{H_0^{-1}(\Omega)} = \sup\{|f(u)| : u \in H_0^1(\Omega),\ \|u\|_{H_0^1(\Omega)} \leq 1\}.$$

Proposition 11.7. *Let $f \in H_0^{-1}(\Omega)$. Then there exist functions $f_0, f_1, \ldots, f_n$ in $L^2(\Omega)$ such that*

$$f(v) = \int_\Omega \Big(f_0 v + \sum_{j=1}^n f_j v_{x_j} \Big) d\lambda \,, \tag{11.14}$$

$$\|f\|_{H_0^{-1}(\Omega)} = \inf\left\{ \Big(\int_\Omega \sum_{j=0}^n |f_j^2| \, d\lambda \Big)^{1/2} : f \text{ satisfies (11.14)} \right\}. \tag{11.15}$$

We write

$$f = f_0 - \sum_{j=1}^n \frac{\partial f_j}{\partial x_j}$$

whenever (11.14) holds.

Proof. For $u, v \in H_0^1(\Omega)$, the inner product is given by

$$(u, v) = \int_\Omega \nabla u \cdot \nabla v \, d\lambda + \int_\Omega uv \, d\lambda.$$

If $f \in H_0^{-1}(\Omega)$, then the Riesz representation (Theorem 1.6) implies that there exists a unique function $u \in H_0^1(\Omega)$ such that

$$f(v) = (u, v) \quad \forall v \in H_0^1(\Omega),$$

and hence

$$f(v) = \int_\Omega \nabla u \cdot \nabla v \, d\lambda + \int_\Omega uv \, d\lambda, \tag{11.16}$$

which gives (11.14), where $f_0 = u$ and $f_j = u_{x_j}$, $j = 1, \dots, n$. By the Cauchy–Schwarz inequality we obtain

$$\|f\|_{H_0^{-1}(\Omega)} \leq \left(\int_\Omega \sum_{j=0}^{n} |f_j|^2 \, d\lambda \right)^{1/2},$$

and setting $v = u/\|u\|_{H_0^1(\Omega)}$ in (11.16), we deduce

$$\|f\|_{H_0^{-1}(\Omega)} = \left(\int_\Omega \sum_{j=0}^{n} |f_j|^2 \, d\lambda \right)^{1/2},$$

which gives (11.15). □

11.3 Magnetic Schrödinger operators with compact resolvent

In this section, we consider the Schrödinger operators with magnetic field $H(A, V)$ as in Proposition 10.8 and suppose that the electric potential V satisfies $V(x) \geq 0$ for all $x \in \mathbb{R}^n$. Recall that $H(A, V)$ has a compact resolvent if and only if the essential spectrum $\sigma_e(H(A, V))$ is empty (Proposition 9.39). We follow Iwatsuka [59] for a general characterization of magnetic Schrödinger operators with compact resolvent and a necessary condition in terms of A and V; compare also with Propositions 9.46 and 9.48 about the bottom of the spectrum and of the essential spectrum of a self-adjoint semibounded operator.

Proposition 11.8. *Let $s \in \mathbb{R}$. The following conditions are equivalent:*
(a) $\inf \sigma_e(H(A, V)) \geq s$.
(b) $\lim_{R\to\infty} e_{A,V}(\Omega_R) \geq s$, *where* $\Omega_R = \{x \in \mathbb{R}^n : |x| > R\}$, *and*

$$e_{A,V}(\Omega_R) = \inf\left\{ \frac{(H(A, V)\phi, \phi)}{(\phi, \phi)} : \phi \in C_0^\infty(\Omega_R), \phi \not\equiv 0 \right\}.$$

(c) *There exists a continuous function* $\Lambda : \mathbb{R}^n \longrightarrow \mathbb{R}$ *such that*

$$\liminf_{|x|\to\infty} \Lambda(x) \geq s \quad \text{and}$$

$$(H(A,V)\phi,\phi) \geq \int_{\mathbb{R}^n} \Lambda(x)|\phi(x)|^2\, d\lambda(x)$$

for all $\phi \in C_0^\infty(\mathbb{R}^n)$.

Proof. (a)⇒(b): Since $e_{A,V}(\Omega_R)$ is increasing in R, the limit $\lim_{R\to\infty} e_{A,V}(\Omega_R)$ exists. Suppose that (a) holds but (b) does not. Then $s > 0$, and there is $s' < s$ such that $e_{A,V}(\Omega_R) < s'$ for all $R > 0$. We can find a sequence $(\phi_k)_k \subset C_0^\infty(\mathbb{R}^n)$ such that $\|\phi_k\| = 1$ and $(H(A,V)\phi_k,\phi_k) < s'$, where $\operatorname{supp}\phi_k \subset \{x \in \mathbb{R}^n : a_k < |x| < a_{k+1}\}$ for some increasing sequence $(a_k)_k$ tending to infinity. Let E be the uniquely determined resolution of the identity on the Borel subsets of $\mathbb{R}$ such that

$$(H(A,V)\phi_k,\phi_k) = \int_{-\infty}^{\infty} t\, d\|E(t)\phi_k\|^2.$$

By (a) we have that $E(t) = E((-\infty,t])$ is compact for $t < s$. Hence $E(t)\phi_k$ converges strongly to 0 as $k \to \infty$, since ϕ_k converges weakly to 0 in $L^2(\mathbb{R}^n)$ (see Proposition 4.17). Now we get for $0 \leq t < s$ that

$$\begin{aligned}(H(A,V)\phi_k,\phi_k) &= \int_0^\infty \mu\, d\|E(\mu)\phi_k\|^2 \\ &\geq \int_t^\infty t\, d\|E(\mu)\phi_k\|^2 \\ &= t(\|\phi_k\|^2 - \|E(t)\phi_k\|)^2 \to t \quad \text{as } k \to \infty.\end{aligned}$$

Hence $\liminf_{k\to\infty}(H(A,V)\phi_k,\phi_k) \geq t$ for all $t \in [0,s)$, contradicting

$$\limsup_{k\to\infty}(H(A,V)\phi_k,\phi_k) \leq s'$$

for our choice of the sequence $(\phi_k)_k$.

(b)⇒(c): Let ϕ_0,ϕ_1 be real-valued smooth functions on $\mathbb{R}$ such that $\operatorname{supp}\phi_0 \subset (-\infty,2)$, $\operatorname{supp}\phi_1 \subset (1,\infty)$, and $\phi_0(r)^2 + \phi_1(r)^2 = 1$ for all $r \in \mathbb{R}$. We define the sequence $(\zeta_k)_{k\geq 0}$ of real-valued functions in $C_0^\infty(\mathbb{R}^n)$ by $\zeta_0(x) = \phi_0(|x|)$ for $x \in \mathbb{R}^n$,

$$\zeta_1(x) = \begin{cases}\phi_1(|x|) & \text{if } |x| \leq 2, \\ \phi_0(|x|/2) & \text{if } |x| \geq 2,\end{cases} \tag{11.17}$$

and $\zeta_k(x) = \zeta_1(x/2^{k-1})$ for $x \in \mathbb{R}^n$ and $k \geq 2$. Then we have $\operatorname{supp}\zeta_0 \subset \{x : |x| < 2\}$, $\operatorname{supp}\zeta_k \subset \{x : 2^{k-1} < |x| < 2^{k+1}\}$ for $k \geq 1$, and $\sum_{k=0}^\infty \zeta_k^2 = 1$. By Proposition 10.8 we know that

$$(H(A,V)\phi,\psi)=\sum_{j=1}^{n}(X_j\phi,X_j\psi)+(V^{1/2}\phi,V^{1/2}\psi)$$

for $\phi,\psi\in C_0^\infty(\mathbb{R}^n)$, where $X_j=(-i\frac{\partial}{\partial x_j}-A_j)$, and as $(H(A,V)\phi,\phi)$ is always real, we can write

$$(H(A,V)\phi,\phi)=\left(H(A,V)\phi,\sum_{k=0}^{\infty}\zeta_k^2\phi\right)=\Re(\sum_{k=0}^{\infty}(H(A,V)\phi,\zeta_k^2\phi).$$

Now we compute

$$\begin{aligned}\Re((X_j(\phi)(X_j(\zeta_k^2\phi))^{-})&=\zeta_k^2|X_j(\phi)|^2+2\zeta_k\frac{\partial\zeta_k}{\partial x_j}\Re(X_j(i|\phi|^2))\\&=|X_j(\zeta_k^2\phi)|^2-\left|\frac{\partial\zeta_k}{\partial x_j}\phi\right|^2.\end{aligned}$$

Hence we get that

$$(H(A,V)\phi,\phi)=\sum_{k=0}^{\infty}(H(A,V)(\zeta_k\phi),\zeta_k\phi)-\sum_{k=0}^{\infty}\|(\nabla\zeta_k)\phi\|^2.$$

Let $e_0=e_{A,V}(\{x:|x|<2\})$ and $e_k=e_{A,V}(\Omega_{2^{k-1}})$ for $k\geq 1$. Define

$$\Lambda(x)=\sum_{k=0}^{\infty}e_k\zeta_k(x)^2-\sum_{k=0}^{\infty}|\nabla\zeta_k(x)|^2.$$

Then

$$(H(A,V)\phi,\phi)\geq\sum_{k=0}^{\infty}e_k\|\zeta_k\phi\|^2-\sum_{k=0}^{\infty}\|(\nabla\zeta_k)\phi\|^2=\int_{\mathbb{R}^n}\Lambda(x)|\phi(x)|^2\,d\lambda(x).$$

Now let $\epsilon>0$. Using (b), we can find some $k_0\geq 2$ such that $e_k\geq s-\epsilon$ for all $k\geq k_0$. If $2^k\leq|x|<2^{k+1}$ for $k\geq k_0$, then by the definitions of $(\zeta_k)_k$ and Λ we get that

$$\begin{aligned}\Lambda(x)&=e_k\zeta_k(x)^2+e_{k+1}\zeta_{k+1}(x)^2-(|\nabla\zeta_k(x)|^2+|\nabla\zeta_{k+1}(x)|^2)\\&\geq(s-\epsilon)-\frac{5}{2^{2k}}\sup_{x\in\mathbb{R}^n}|\nabla\zeta_1(x)|^2.\end{aligned}$$

Letting $|x|\to\infty$, we get (c).

(c)⇒(a): Let $\tau\in\sigma_e(H(A,V))$. By Lemma 9.36 there exists a sequence $(u_k)_k$ in $\mathrm{dom}(H(A,V))$ such that $\|u_k\|=1$ for each $k\in\mathbb{N}$, u_k converges weakly to 0, and

$$\lim_{k\to\infty}\|(H(A,V)-\tau I)u_k\|=0.$$

We denote $\tilde{\Lambda}(x) = \min(\Lambda(x), s)$ and have by (c) that $\tilde{\Lambda}(x) \to s$ as $|x| \to \infty$ and

$$(H(A,V)\phi,\phi) \geq \int_{\mathbb{R}^n} \tilde{\Lambda}(x)\,|\phi(x)|^2\, d\lambda(x).$$

Then $H(A,V) \geq s + K$, where K denotes the multiplication operator by $(\tilde{\Lambda}(x) - s)^{1/2}$. We endow $\operatorname{dom}(H(A,V))$ with the norm $(\|u\|^2 + \|H(A,V)u\|^2)^{1/2}$ and consider K as an operator defined on this normed space. We claim that the image of the unit ball of this space under K is a precompact set in $L^2(\mathbb{R}^n)$. For this, we can apply Theorem 11.1 and show that the first condition in Theorem 11.1 is satisfied by using the method in the proof of Lemma 11.5. The second condition in Theorem 11.1 is satisfied since for each $\epsilon > 0$, there exists $\omega \Subset \mathbb{R}^n$ such that

$$\int_{\mathbb{R}^n \setminus \omega} (\tilde{\Lambda}(x) - s)|\phi(x)|^2\, d\lambda(x) < \epsilon,$$

because $(\tilde{\Lambda}(x) - s) \to 0$ as $|x| \to \infty$. We have that $(u_k)_k$ is a bounded sequence in $L^2(\mathbb{R}^n)$ and

$$\|H(A,V)u_k\| = \|(H(A,V) - \tau I)u_k + \tau u_k\| \leq \|(H(A,V) - \tau I)u_k\| + \|\tau u_k\|,$$

and hence also $(H(A,V)u_k)_k$ is bounded. On the other side, we now know that $Ku_k \to 0$ strongly as $k \to \infty$, and hence

$$(H(A,V)u_k, u_k) \geq s + (Ku_k, u_k),$$

where the left-hand side converges to τ, and the right-hand side converges to s as $k \to \infty$. So $\tau \geq s$ for any $\tau \in \sigma_e(H(A,V))$, and we get (a). □

Now we are able to prove a general characterization of magnetic Schrödinger operators with compact resolvent.

Theorem 11.9. *The following conditions are equivalent:*
(a) *$H(A,V)$ has a compact resolvent.*
(b) *$e_{A,V}(\Omega_R) \to \infty$ as $R \to \infty$, where $\Omega_R = \{x \in \mathbb{R}^n : |x| > R\}$.*
(c) *$e_{A,V}(Q_x) \to \infty$ as $|x| \to \infty$, where $Q_x = \{y \in \mathbb{R}^n : |x - y| < 1\}$.*
(d) *There exists a continuous function $\Lambda : \mathbb{R}^n \longrightarrow \mathbb{R}$ such that*

$$\Lambda(x) \to \infty \quad \text{as } |x| \to \infty \quad \text{and}$$
$$(H(A,V)\phi,\phi) \geq \int_{\mathbb{R}^n} \Lambda(x)|\phi(x)|^2\, d\lambda(x)$$

for all $\phi \in C_0^\infty(\mathbb{R}^n)$.

Proof. Since $H(A,V)$ has compact resolvent if and only if $\sigma_e(H(A,V)) = \emptyset$, the equivalence of (a) and (b) follows from the equivalence of (a) and (b) of Proposition 11.8.

Since $e_{A,V}(\Omega) \ge e_{A,V}(\Omega')$ if $\Omega \subset \Omega'$, (b) implies (c).

It is also easy to see that (d) implies (a): since $\Lambda(x) \to \infty$ as $|x| \to \infty$, we have $\liminf_{|x|\to\infty} \Lambda(x) \ge s$ for any $s \in \mathbb{R}$. Hence we can apply (c) of Proposition 11.8 and get (a) of Proposition 11.8 for any $s \in \mathbb{R}$. This implies $\sigma_e(H(A,V)) = \emptyset$, and we get (a).

It remains to prove that (c) implies (d): for this purpose, let ϕ_0 be a real-valued $C^\infty(\mathbb{R}^n)$ such that $\phi_0(x) = 1$ if $|x| \le 1/2$ and $\operatorname{supp}\phi_0 \subset Q_{(0,\dots,0)}$. Now let

$$L = \{(l_1,\dots,l_n) : l_j = k_j/\sqrt{n};\ k_j \in \mathbb{Z},\ j = 1,\dots,n\}$$

and define $\Phi(x) = \sum_{l\in L} \phi_0(x-l)^2$. Then we have $\Phi(x) \ge 1$ for all $x \in \mathbb{R}^n$, since for all $x \in \mathbb{R}^n$, there exists a point $l \in L$ such that $|x-l| \le 1/2$. Set

$$\zeta_l(x) = \phi_0(x-l)/\sqrt{\Phi(x)}.$$

Then we have $\zeta_l \in C_0^\infty(\mathbb{R}^n)$ and $\zeta_l(x) = \zeta_{(0,\dots,0)}(x-l)$ for $l \in L$, $\sum_{l\in L}\zeta_l(x)^2 = 1$, and $\operatorname{supp}\zeta_l \subset Q_l$.

In a similar way as in the proof that (b) implies (c) of Proposition 11.8, we get now

$$(H(A,V)\phi,\phi) \ge \int_{\mathbb{R}^n} \Lambda(x)|\phi(x)|^2\, d\lambda(x)$$

for all $\phi \in C_0^\infty(\mathbb{R}^n)$, where

$$\Lambda(x) = \sum_{l\in L} e_l\zeta_l(x)^2 - \sum_{l\in L}|\nabla\zeta_l(x)|^2$$

with $e_l = e_{A,V}(Q_l)$. Since, by (c), $e_l \to \infty$ as $|l| \to \infty$, we obtain $\sum_{l\in L} e_l\zeta_l(x)^2 \to \infty$ as $|x| \to \infty$. Since $M(x+l) = M(x)$ for $l \in L$, $M(x) \equiv \sum_{l\in L}|\nabla\zeta_l(x)|^2$ is bounded on $\mathbb{R}^n$, so we get $\Lambda(x) \to \infty$ as $|x| \to \infty$ and thus arrive at (d). □

The next aim is to find a necessary condition for magnetic Schrödinger operators with compact resolvent in terms of the magnetic field and electric potential. For this purpose, the gauge invariance of the spectrum of $H(A,V)$ (Proposition 10.12) will be of importance.

A vector potential A is considered as a 1-form $A = \sum_j A_j dx_j$. The 2-form

$$B = dA = \sum_{j<k}\left(\frac{\partial A_k}{\partial x_j} - \frac{\partial A_j}{\partial x_k}\right) dx_j \wedge dx_k = \sum_{j<k} B_{jk}\, dx_j \wedge dx_k$$

is the magnetic field. Since $d^2 = 0$, we have that B is closed, i. e.,

$$\frac{\partial B_{jk}}{\partial x_i} + \frac{\partial B_{ki}}{\partial x_j} + \frac{\partial B_{ij}}{\partial x_k} = 0. \tag{11.18}$$

We need a Poincaré lemma with L^2-estimates.

Lemma 11.10. *Let B be a C^∞ skew-symmetric matrix-valued function on $\mathbb{R}^n$ satisfying* (11.18)*, and let Ω be a bounded convex open set in $\mathbb{R}^n$. Then there exists a constant C, dependent only on n and* $\operatorname{diam}(\Omega) \equiv \sup_{x,y\in\Omega} |x-y|$*, and a C^∞ 1-form b on $\mathbb{R}^n$ such that $db = B$ and*

$$\int_\Omega \sum_j |b_j|^2 \, d\lambda \le C \int_\Omega \sum_{j<k} |B_{jk}|^2 \, d\lambda. \tag{11.19}$$

Proof. For $x, y \in \mathbb{R}^n$ and $j = 1, \dots, n$, define

$$\tilde{b}_j(x,y) = \sum_{k=1}^n (x_k - y_k) \int_0^1 B_{kj}(y + t(x-y)) t \, dt.$$

Then $\tilde{b}_j \in C^\infty(\mathbb{R}^n \times \mathbb{R}^n)$. We claim that $\frac{\partial \tilde{b}_j}{\partial x_m} - \frac{\partial \tilde{b}_m}{\partial x_j} = B_{mj}$ for $j, m = 1, \dots, n$ as functions in the first variable.

We get

$$\frac{\partial \tilde{b}_j}{\partial x_m} = \int_0^1 B_{mj}(y + t(x-y)) t \, dt + \sum_{k=1}^n (x_k - y_k) \int_0^1 \frac{\partial B_{kj}}{\partial x_m}(y + t(x-y)) t^2 \, dt$$

and

$$\frac{\partial \tilde{b}_m}{\partial x_j} = \int_0^1 B_{jm}(y + t(x-y)) t \, dt + \sum_{k=1}^n (x_k - y_k) \int_0^1 \frac{\partial B_{km}}{\partial x_j}(y + t(x-y)) t^2 \, dt,$$

and hence

$$\frac{\partial \tilde{b}_j}{\partial x_m} - \frac{\partial \tilde{b}_m}{\partial x_j} = 2 \int_0^1 B_{mj}(y + t(x-y)) t \, dt + \sum_{k=1}^n (x_k - y_k) \int_0^1 \frac{\partial B_{mj}}{\partial x_k}(y + t(x-y)) t^2 \, dt,$$

where we used (11.18). By the chain rule we obtain

$$\int_0^1 \frac{\partial B_{mj}}{\partial t}(y + t(x-y)) t^2 \, dt = \sum_{k=1}^n (x_k - y_k) \int_0^1 \frac{\partial B_{mj}}{\partial x_k}(y + t(x-y)) t^2 \, dt,$$

and integration by parts gives

$$\int_0^1 \frac{\partial B_{mj}}{\partial t}(y+t(x-y))t^2\,dt = B_{mj}(x) - 2\int_0^1 B_{mj}(y+t(x-y))t\,dt,$$

so $\frac{\partial \tilde{b}_j}{\partial x_m} - \frac{\partial \tilde{b}_m}{\partial x_j} = B_{mj}$. Using the Schwarz inequality, we have

$$\begin{aligned}|\tilde{b}_j(x,y)|^2 &\le \left(\sum_{k=1}^n |x_k - y_k|^2\right)\left(\sum_{k=1}^n \int_0^1 |B_{jk}(y+t(x-y))|^2\,dt \int_0^1 t^2 dt\right)\\ &= \frac{1}{3}|x-y|^2 \sum_{k=1}^n \int_0^1 |B_{jk}(y+t(x-y))|^2\,dt.\end{aligned}$$

Summing up, we get

$$\sum_{j=1}^n |\tilde{b}_j(x,y)|^2 \le |x-y|^2 \sum_{j<k} \int_0^1 |B_{jk}(y+t(x-y))|^2\,dt. \tag{11.20}$$

Now define

$$b_j(x) = \frac{1}{|\Omega|}\int_\Omega \tilde{b}_j(x,y)\,d\lambda(y)$$

and $b = \sum_{j=1}^n b_j dx_j$. Then we have $b_j \in C^\infty(\mathbb{R}^n)$ and $db = B$. Again, by the Schwarz inequality, we have

$$\sum_{j=1}^n |b_j(x)|^2 = \frac{1}{|\Omega|^2}\sum_{j=1}^n \left|\int_\Omega \tilde{b}_j(x,y)\,d\lambda(y)\right|^2 \le \frac{1}{|\Omega|}\int_\Omega \sum_{j=1}^n |\tilde{b}_j(x,y)|^2\,d\lambda(y).$$

Hence by (11.20)

$$\int_\Omega \sum_{j=1}^n |b_j(x)|^2\,d\lambda(x) \le \frac{\operatorname{diam}(\Omega)^2}{|\Omega|}\int_0^1 I(t)\,dt,$$

where

$$I(t) = \int_{\Omega\times\Omega} |B(y+t(x-y))|^2\,d\lambda(x)d\lambda(y),$$

and $|B|^2 = \sum_{j<k}|B_{jk}|^2$. If $1/2 \le t \le 1$ and $z = tx + (1-t)y$ for y fixed, then we get

$$I(t) = \int_\Omega \left(\int_{t\Omega+(1-t)y} |B(z)|^2 t^{-n}\,d\lambda(z)\right)d\lambda(y)$$

$$\leq 2^n \int_{\Omega\times\Omega} |B(z)|^2 \, d\lambda(z)\, d\lambda(y)$$
$$= 2^n |\Omega| \int_{\Omega} |B(z)|^2 \, d\lambda(z),$$

where we have used that Ω is convex. For $0 \leq t \leq 1/2$, we have the same estimate $I(t) \leq 2^n|\Omega| \int_\Omega |B(z)|^2 \, d\lambda(z)$, since $I(t) = I(1-t)$. Finally, we get (11.19) with $C = 2^n \operatorname{diam}(\Omega)^2$. □

To find a necessary condition for magnetic Schrödinger operators with compact resolvent, we first prove the following:

Lemma 11.11. *Let $A = \sum_{j=1}^n A_j dx_j$ be a smooth 1-form on $\mathbb{R}^n$, and let V be a nonnegative continuous function on $\mathbb{R}^n$. Suppose that $H(A,V)$ has compact resolvent. Then*

$$\inf_{b\in G_A(Q_x)} \int_{Q_x} \left(\sum_{j=1}^n |b_j(y)|^2 + V(y) \right) d\lambda(y) \to \infty \tag{11.21}$$

as $|x| \to \infty$, where $G_A(Q_x) = \{b \text{ smooth 1-form on } Q_x : db = dA\}$ and $Q_x = \{y : |y-x| < 1\}$.

Proof. Let $X_j = -i\frac{\partial}{\partial x_j} - b_j$ for $j = 1, \dots, n$. For real-valued smooth functions f, we have

$$|X_j f(y)|^2 = \left| \frac{\partial f}{\partial x_j}(y) \right|^2 + |b_j(y) f(y)|^2.$$

Let ϕ be a real-valued function in $C_0^\infty(Q_0)$ such that $\phi \not\equiv 0$ and $|\phi(y)| \leq 1$ for all $y \in Q_0$, and let $\phi_x(y) = \phi(y-x)$. Then, as in the proof of Proposition 10.7, we have

$$(H(b,V)\phi_x, \phi_x) = \int_{Q_x} \left(|\nabla\phi_x(y)|^2 + \left(\sum_{j=1}^n |b_j(y)|^2 + V(y) \right) \phi_x(y)^2 \right) d\lambda(y).$$

Since $\sum_{j=1}^n |b_j(y)|^2 + V(y) \geq (\sum_{j=1}^n |b_j(y)|^2 + V(y))\phi_x(y)^2$, we have

$$\begin{aligned}
\int_{Q_x} \left(\sum_{j=1}^n |b_j(y)|^2 + V(y) \right) d\lambda(y) &\geq (H(b,V)\phi_x, \phi_x) - \|\nabla\phi\|^2 \\
&\geq e_{b,V}(Q_x)\|\phi\|^2 - \|\nabla\phi\|^2 \\
&= e_{A,V}(Q_x)\|\phi\|^2 - \|\nabla\phi\|^2
\end{aligned}$$

by Proposition 10.12. Since $e_{A,V}(Q_x) \to \infty$ as $|x| \to \infty$ (Theorem 11.9(c)). we get the desired result. □

Finally, we can describe the necessary condition for a compact resolvent in the following way.

Theorem 11.12. *Let $A = \sum_{j=1}^{n} A_j dx_j$ be a smooth 1-form on $\mathbb{R}^n$, and let V be a nonnegative continuous function on $\mathbb{R}^n$. Suppose that $H(A, V)$ has compact resolvent. Then*

$$\int_{Q_x} \Big(\sum_{j<k} |B_{jk}(y)|^2 + V(y) \Big)\, d\lambda \to \infty \quad \text{as } |x| \to \infty, \tag{11.22}$$

where $B = dA$. In particular, if $H(A, 0)$ has compact resolvent, then

$$\int_{Q_x} \sum_{j<k} |B_{jk}(y)|^2 \, d\lambda(y) \to \infty \quad \text{as } |x| \to \infty.$$

Proof. For each x, there exists a C^∞ 1-form b_x such that

$$\int_{Q_x} \sum_{j=1}^{n} |(b_x)_j(y)|^2 \, d\lambda(y) \le C \int_{Q_x} \sum_{j<k} |B_{jk}(y)|^2 \, d\lambda(y),$$

where $B = db_x$, and C is independent of x since $\operatorname{diam}(Q_x) = 2$ (Lemma 11.10). This implies

$$\inf_{b \in G_A(Q_x)} \int_{Q_x} \Big(\sum_{j=1}^{n} |b_j(y)|^2 \Big)\, d\lambda(y) \le C \int_{Q_x} \sum_{j<k} |B_{jk}(y)|^2 \, d\lambda(y),$$

and we can apply Lemma 11.11 to get the desired result. □

11.4 Compactness in weighted spaces

Now we will apply the description of compact subsets in L^2-spaces to derive a general characterization of compactness of the $\overline{\partial}$-Neumann operator and a sufficient condition for compactness in terms of the weight function. In a certain sense, we adopt the real method using Sobolev spaces to the situation in $\mathbb{C}^n$ for the complex $\Box$-operator.

Definition 11.13. Let

$$\mathcal{W}^{Q_\varphi} = \{ u \in L^2_{(0,1)}(\mathbb{C}^n, e^{-\varphi}) : u \in \operatorname{dom}(\overline{\partial}) \cap \operatorname{dom}(\overline{\partial}_\varphi^*) \}$$

with norm

$$\|u\|_{Q_\varphi} = (\|\overline{\partial} u\|_\varphi^2 + \|\overline{\partial}_\varphi^* u\|_\varphi^2)^{1/2}. \tag{11.23}$$

Remark. $\mathcal{W}^{Q_\varphi}$ coincides with the form domain $\operatorname{dom}(\overline{\partial}) \cap \operatorname{dom}(\overline{\partial}_\varphi^*)$ of Q_φ (see Proposition 6.5).

Proposition 11.14. *Suppose that the weight function φ is plurisubharmonic and that the lowest eigenvalue μ_φ of the Levi matrix M_φ satisfies*

$$\lim_{|z|\to\infty} \mu_\varphi(z) = +\infty. \tag{11.24}$$

Then the embedding

$$j_\varphi : \mathcal{W}^{Q_\varphi} \hookrightarrow L^2_{(0,1)}(\mathbb{C}^n, e^{-\varphi}) \tag{11.25}$$

is compact.

Proof. For $u \in \mathcal{W}^{Q_\varphi}$, by Proposition 6.3 we have

$$\|\overline{\partial} u\|_\varphi^2 + \|\overline{\partial}_\varphi^* u\|_\varphi^2 \geq (M_\varphi u, u)_\varphi.$$

This implies

$$\|\overline{\partial} u\|_\varphi^2 + \|\overline{\partial}_\varphi^* u\|_\varphi^2 \geq \int_{\mathbb{C}^n} \mu_\varphi(z)\, |u(z)|^2\, e^{-\varphi(z)}\, d\lambda(z). \tag{11.26}$$

We show that the unit ball in $\mathcal{W}^{Q_\varphi}$ is relatively compact in $L^2_{(0,1)}(\mathbb{C}^n, e^{-\varphi})$. For this purpose, we use the characterization of compact subsets in L^2-spaces from Theorem 11.1.

We start with the first condition in Theorem 11.1. Let $u = \sum_{j=1}^n u_j\, dz_j$ be a $(0,1)$-form with coefficients in C_0^∞. For each u_j and for $t \in \mathbb{R}$ and $h = (h_1, \dots, h_n) \in \mathbb{C}^n$, let

$$v_j(t) := u_j(z + th).$$

Note that

$$|v_j'(t)| \leq |h| \left[\sum_{k=1}^n \left(\left| \frac{\partial u_j}{\partial x_k}(z + th) \right|^2 + \left| \frac{\partial u_j}{\partial y_k}(z + th) \right|^2 \right) \right]^{1/2},$$

where $z_k = x_k + iy_k$ for $k = 1, \dots, n$. Since

$$u_j(z + h) - u_j(z) = v_j(1) - v_j(0) = \int_0^1 v_j'(t)\, dt,$$

we can now estimate, for $|h| < R$,

$$\begin{aligned}
&\int_{\mathbb{B}_R} |\tau_h u_j(z) - u_j(z)|^2 e^{-\varphi(z)}\, d\lambda(z) \\
&\quad = \int_{\mathbb{B}_R} |\tau_h(\chi_R u_j)(z) - \chi_R u_j(z)|^2 e^{-\varphi(z)}\, d\lambda(z) \\
&\quad \leq |h|^2 \int_{\mathbb{B}_R} \left[\int_0^1 \sum_{k=1}^n \left(\left| \frac{\partial(\chi_R u_j)}{\partial x_k}(z + th) \right|^2 + \left| \frac{\partial(\chi_R u_j)}{\partial y_k}(z + th) \right|^2 \right) dt \right] e^{-\varphi(z)}\, d\lambda(z)
\end{aligned}$$

$$\leq C_{\varphi,R}\, |h|^2 \int_{\mathbb{B}_{3R}} \sum_{k=1}^{n} \left(\left| \frac{\partial(\chi_R u_j)}{\partial x_k}(z) \right|^2 + \left| \frac{\partial(\chi_R u_j)}{\partial y_k}(z) \right|^2 \right) e^{-\varphi(z)}\, d\lambda(z)$$

for $j = 1, \dots, n$, where χ_R is a $\mathcal{C}^\infty$ cut-off function that is identically 1 on $\mathbb{B}_{2R}$ and zero outside $\mathbb{B}_{3R}$. It is clear that the corresponding Dirichlet form of $\Box_\varphi$ satisfies the assumptions of Theorem 11.4 in $\mathbb{B}_{3R}$ (by (6.11) the second-order terms reduce to the Laplacian), so by Gårding's inequality for $\mathbb{B}_{3R}$,

$$\begin{aligned} \|\chi_R u\|^2_{W^1,\varphi} &\leq C'_{\varphi,R}(\|\overline{\partial}(\chi_R u)\|^2_\varphi + \|\overline{\partial}^*_\varphi(\chi_R u)\|^2_\varphi + \|\chi_R u\|^2_\varphi) \\ &\leq C''_{\varphi,R}(\|\overline{\partial} u\|^2_\varphi + \|\overline{\partial}^*_\varphi u\|^2_\varphi + \|u\|^2_\varphi). \end{aligned}$$

We can control the last integral by the norm $\|u\|^2_{Q_\varphi}$. Since we started from the unit ball in $\mathcal{W}^{Q_\varphi}$, we get that the first condition in Theorem 11.1 is satisfied.

The second condition is satisfied for the unit ball of $\mathcal{W}^{Q_\varphi}$ since we have

$$\int_{\mathbb{C}^n \setminus \mathbb{B}_R} |u(z)|^2 e^{-\varphi(z)}\, d\lambda(z) \leq \int_{\mathbb{C}^n \setminus \mathbb{B}_R} \frac{\mu_\varphi(z)|u(z)|^2}{\inf\{\mu_\varphi(z) : |z| \geq R\}} e^{-\varphi(z)} d\lambda(z).$$

So formula (11.26), together with assumption (11.24), shows that

$$\int_{\mathbb{C}^n \setminus \mathbb{B}_R} |u(z)|^2 e^{-\varphi(z)}\, d\lambda(z) \leq \frac{\|u\|^2_{Q_\varphi}}{\inf\{\mu_\varphi(z) : |z| \geq R\}} < \epsilon \tag{11.27}$$

if R is big enough.

We remark that

$$L^2_{(0,1)}(\mathbb{C}^n, e^{-\varphi}) \cong L^2(\mathbb{C}^n, e^{-\varphi}) \oplus \cdots \oplus L^2(\mathbb{C}^n, e^{-\varphi})$$

with n factors and that the precompactness of a subset in $L^2_{(0,1)}(\mathbb{C}^n, e^{-\varphi})$ is equivalent to being contained in a product of precompact subsets of $L^2(\mathbb{C}^n, e^{-\varphi})$. □

We are now able to give a concise proof of the main result in [46] or [36]; see [42], where the original proof uses a number of more involved methods from real analysis.

Proposition 11.15. *Let φ be a plurisubharmonic $\mathcal{C}^2$-weight function. If the lowest eigenvalue $\mu_\varphi(z)$ of the Levi matrix M_φ satisfies* (11.24), *then N_φ is compact.*

Proof. By Proposition 11.14 the embedding $\mathcal{W}^{Q_\varphi} \hookrightarrow L^2_{(0,1)}(\mathbb{C}^n, e^{-\varphi})$ is compact. The inverse N_φ of $\Box_\varphi$ is continuous as an operator from $L^2_{(0,1)}(\mathbb{C}^n, e^{-\varphi})$ into $\mathcal{W}^{Q_\varphi}$, which follows from Proposition 6.5. Therefore we have that N_φ is compact as an operator from $L^2_{(0,1)}(\mathbb{C}^n, e^{-\varphi})$ into itself (see Remark 6.6). □

Proposition 11.16. *Let φ be a plurisubharmonic $\mathcal{C}^2$-weight function. Let $1 \le q \le n$ and suppose that the sum s_q of any q (equivalently: the smallest q) eigenvalues of M_φ satisfies*

$$\lim_{|z|\to\infty} s_q(z) = +\infty. \tag{11.28}$$

Then $N_{\varphi,q} : L^2_{(0,q)}(\mathbb{C}^n, e^{-\varphi}) \longrightarrow L^2_{(0,q)}(\mathbb{C}^n, e^{-\varphi})$ is compact.

Proof. For $(0,q)$-forms, use (6.11) to show that Garding's inequality can be applied and Proposition 6.7 to show that

$$\|\overline{\partial} u\|_\varphi^2 + \|\overline{\partial}_\varphi^* u\|_\varphi^2 \ge \int_{\mathbb{C}^n} s_q(z)\,|u(z)|^2\, e^{-\varphi(z)}\, d\lambda(z). \tag{11.29}$$

Now we can continue as in the proof of Proposition 11.14. □

Example. We consider the plurisubharmonic weight function

$$\varphi(z,w) = |z|^2|w|^2 + |w|^4$$

on $\mathbb{C}^2$. The Levi matrix of φ has the form

$$\begin{pmatrix} |w|^2 & \overline{z}w \\ \overline{w}z & |z|^2 + 4|w|^2 \end{pmatrix},$$

and the eigenvalues are

$$\mu_{\varphi,1}(z,w) = \frac{1}{2}(5|w|^2 + |z|^2 - \sqrt{9|w|^4 + 10|z|^2|w|^2 + |z|^4})$$

and

$$\mu_{\varphi,2}(z,w) = \frac{1}{2}(5|w|^2 + |z|^2 + \sqrt{9|w|^4 + 10|z|^2|w|^2 + |z|^4}).$$

It follows that (11.24) fails, but

$$s_2(z,w) = \frac{1}{4} \triangle \varphi(z,w) = |z|^2 + 5|w|^2,$$

and hence (11.28) is satisfied.

We point out that since $\mu_{\varphi,1}(z,0) = 0$ for all $z \in \mathbb{C}$, we have

$$\liminf_{|(z,w)|\to\infty} \mu_{\varphi,1}(z,w) = 0,$$

so condition (6.5) is not satisfied, and it is not even clear that N_φ is bounded.

Notice that

$$N_\varphi : L^2_{(0,1)}(\mathbb{C}^n, e^{-\varphi}) \longrightarrow L^2_{(0,1)}(\mathbb{C}^n, e^{-\varphi})$$

can be written in the form

$$N_\varphi = j_\varphi \circ j_\varphi^*,$$

where

$$j_\varphi^* : L^2_{(0,1)}(\mathbb{C}^n, e^{-\varphi}) \longrightarrow \mathcal{W}^{Q_\varphi}$$

is the adjoint operator to j_φ.

This means that N_φ is compact if and only if j_φ is compact, and summarizing the above results, we get the following:

Theorem 11.17. *Let $1 \leq q \leq n$, and let $\varphi : \mathbb{C}^n \longrightarrow \mathbb{R}^+$ be a plurisubharmonic $\mathcal{C}^2$-weight function. Suppose that $s_q(z) > \epsilon > 0$.*

Then the $\overline{\partial}$-Neumann operator

$$N_{\varphi,q} : L^2_{(0,q)}(\mathbb{C}^n, e^{-\varphi}) \longrightarrow L^2_{(0,q)}(\mathbb{C}^n, e^{-\varphi})$$

is compact if and only if for each $\epsilon > 0$, there exists $R > 0$ such that

$$\int_{\mathbb{C}^n \setminus \mathbb{B}_R} |u(z)|^2 e^{-\varphi(z)} \, d\lambda(z) \leq \epsilon(\|\overline{\partial} u\|_\varphi^2 + \|\overline{\partial}_\varphi^* u\|_\varphi^2) \tag{11.30}$$

for each $u \in \mathrm{dom}(\overline{\partial}) \cap \mathrm{dom}(\overline{\partial}_\varphi^*)$.

To derive a necessary condition for compactness of $N_{\varphi,q}$, we will apply $\Box_\varphi$ to $(0,q)$-forms with coefficients (for the standard basis) belonging to the weighted space of entire functions $A^2(\mathbb{C}^n, e^{-\varphi})$; see Theorem 8.11. The point is that $\Box_\varphi$ restricts to a multiplication operator on this space.

Theorem 11.18. *Let $\varphi : \mathbb{C}^n \to \mathbb{R}$ be a plurisubharmonic $\mathcal{C}^2$-function and suppose that the corresponding weighted space $A^2(\mathbb{C}^n, e^{-\varphi})$ of entire functions is infinite dimensional. If there is $1 \leq q \leq n$ such that the $\overline{\partial}$-Neumann operator $N_{\varphi,q}$ is compact, then*

$$\limsup_{z \to \infty} \operatorname{tr}(M_\varphi(z)) = +\infty. \tag{11.31}$$

Proof. Let g be a smooth $(0,q)$-form, $g = \sum'_{|J|=q} g_J \, d\overline{z}_J$. If the coefficients g_J of g are holomorphic, then $\Box_\varphi$ is given by

$$\square_\varphi g = \bar\partial\bar\partial_\varphi^* g = \bar\partial\left(-\sum_{|K|=q-1}{}' \sum_{k=1}^{n} \left(\frac{\partial}{\partial z_k} - \frac{\partial\varphi}{\partial z_k} \right) g_{kK}\, d\bar z_K \right)$$
$$= -\sum_{j=1}^{n} \sum_{|K|=q-1}{}' \sum_{k=1}^{n} \left(\frac{\partial^2}{\partial\bar z_j \partial z_k} - \frac{\partial\varphi}{\partial z_k}\frac{\partial}{\partial\bar z_j} - \frac{\partial^2\varphi}{\partial\bar z_j\partial z_k} \right) g_{kK}\, d\bar z_j \wedge d\bar z_K,$$

which reduces to

$$\sum_{|K|=q-1}{}' \sum_{j,k=1}^{n} \frac{\partial^2\varphi}{\partial\bar z_j\partial z_k} g_{kK}\, d\bar z_j \wedge d\bar z_K. \tag{11.32}$$

Now assume that (11.31) is wrong, so that there exists $C \geq 0$ such that

$$\operatorname{tr}(M_\varphi(z)) \leq C \tag{11.33}$$

for all $z \in \mathbb{C}^n$. Since φ is assumed to be plurisubharmonic, M_φ is nonnegative, and hence (11.33) is equivalent to the boundedness of all entries of M_φ. Indeed, $\operatorname{tr}(M_\varphi^* M_\varphi) = \operatorname{tr}(M_\varphi^2) \leq \operatorname{tr}(M_\varphi)^2$, and the Hilbert–Schmidt norm $A \mapsto \operatorname{tr}(A^*A)^{1/2}$ on the space of complex $n \times n$ matrices is equivalent to the max norm (i. e., the maximum of the entries). It now follows from (11.32) that all $(0, q)$-forms g with coefficients in $A^2(\mathbb{C}^n, e^{-\varphi})$ belong to $\operatorname{dom} \square_\varphi$ and that

$$\square_\varphi g = \sum_{|K|=q-1}{}' \sum_{j,k=1}^{n} \frac{\partial^2\varphi}{\partial\bar z_j\partial z_k} g_{kK}\, d\bar z_j \wedge d\bar z_K.$$

Hence, if all such g belong to a bounded set of $L^2_{(0,q)}(\mathbb{C}^n, e^{-\varphi})$, then the image under $\square_\varphi$ of this bounded set will be again bounded in $L^2_{(0,q)}(\mathbb{C}^n, e^{-\varphi})$. In addition, we have

$$N_{\varphi,q}\left(\sum_{|K|=q-1}{}' \sum_{j,k=1}^{n} \frac{\partial^2\varphi}{\partial\bar z_j\partial z_k} g_{kK}\, d\bar z_j \wedge d\bar z_K \right) = g,$$

so that the identity on the (infinite-dimensional) space of $(0, q)$-forms with coefficients in $A^2(\mathbb{C}^n, e^{-\varphi})$ is the composition of the compact operator $N_{\varphi,q}$ and a bounded operator, and hence it is compact. This contradicts $\dim(A^2(\mathbb{C}^n, e^{-\varphi})) = \infty$. □

Next, we adapt a method from [91, Proposition 4.5] and use the last result to show that compactness percolates up the weighted $\bar\partial$-complex.

Proposition 11.19. *Let φ be a plurisubharmonic $\mathcal{C}^2$-function. Suppose that $1 \leq q \leq n-1$ and that (11.28) holds. If $N_{\varphi,q}$ is compact, then $N_{\varphi,q+1}$ is also compact.*

Proof. For $u = \sum'_{|J|=q+1} u_J\, d\bar z_J \in \operatorname{dom}(\bar\partial) \cap \operatorname{dom}(\bar\partial_\varphi^*)$, we define the $(0, q)$-forms

$$v_k := \sum_{|K|=q}{}' u_{kK}\, d\bar z_K, \quad k = 1, \dots, n.$$

We claim that $v_k \in \operatorname{dom}(\overline{\partial}) \cap \operatorname{dom}(\overline{\partial}^*_\varphi)$. The components of $\overline{\partial} v_k$ are linear combinations of terms $\partial u_J / \partial \overline{z}_j$, and their weighted L^2-norms are controlled by $\|\overline{\partial} u\|^2_\varphi + \|\overline{\partial}^*_\varphi u\|^2_\varphi$, since by (6.11) we have

$$\|\overline{\partial} u\|^2_\varphi + \|\overline{\partial}^*_\varphi u\|^2_\varphi \geq \sum_{|J|=q+1}{}' \sum_{j=1}^n \left\| \frac{\partial u_J}{\partial \overline{z}_j} \right\|^2_\varphi .$$

This shows that $v_k \in \operatorname{dom}(\overline{\partial})$. To prove that $v_k \in \operatorname{dom}(\overline{\partial}^*_\varphi)$, we first observe that for $\alpha = \sum'_{|K|=q} \alpha_K \, d\overline{z}_K \in L^2_{(0,q)}(\mathbb{C}^n, e^{-\varphi})$ and for k fixed, we have

$$\begin{aligned} (d\overline{z}_k \wedge \alpha, u)_\varphi &= \Big(\sum_{|K|=q}{}' \alpha_K (d\overline{z}_k \wedge d\overline{z}_K), \sum_{|J|=q+1}{}' u_J \, d\overline{z}_J \Big)_\varphi \\ &= (\alpha, v_k)_\varphi, \end{aligned}$$

where we used the inner products in $L^2_{(0,q+1)}(\mathbb{C}^n, e^{-\varphi})$ and $L^2_{(0,q)}(\mathbb{C}^n, e^{-\varphi})$, respectively. Now for $\beta \in \operatorname{dom}(\overline{\partial})$, we get

$$\begin{aligned} (\overline{\partial}\beta, v_k)_\varphi &= (d\overline{z}_k \wedge \overline{\partial}\beta, u)_\varphi = -(\overline{\partial}(d\overline{z}_k \wedge \beta), u)_\varphi \\ &= -(d\overline{z}_k \wedge \beta, \overline{\partial}^*_\varphi u)_\varphi = -(\beta, \gamma_k)_\varphi, \end{aligned}$$

where $\gamma_k = \sum'_{|L|=q-1} (\overline{\partial}^*_\varphi u)_{kL} \, d\overline{z}_L$. The last equality follows as before, and we get that $v_k \in \operatorname{dom}(\overline{\partial}^*_\varphi)$ and

$$\overline{\partial}^*_\varphi v_k = -\gamma_k . \tag{11.34}$$

Since $N_{\varphi,q}$ is compact, we can apply Theorem 11.17: for each $\epsilon > 0$, there exists $R > 0$ such that

$$\int_{\mathbb{C}^n \setminus \mathbb{B}_R} |v_k(z)|^2 e^{-\varphi(z)} \, d\lambda(z) \leq \epsilon (\|\overline{\partial} v_k\|^2_\varphi + \|\overline{\partial}^*_\varphi v_k\|^2_\varphi) \leq C\epsilon (\|\overline{\partial} u\|^2_\varphi + \|\overline{\partial}^*_\varphi u\|^2_\varphi)$$

for $k = 1, \ldots, n$, where the last inequality follows from (6.11),

$$\|\overline{\partial} v_k\|^2_\varphi = \sum_{|K|=q}{}' \sum_{j=1}^n \left\| \frac{\partial u_{kK}}{\partial \overline{z}_j} \right\|^2_\varphi \leq C' (\|\overline{\partial} u\|^2_\varphi + \|\overline{\partial}^*_\varphi u\|^2_\varphi),$$

and

$$\|\overline{\partial}^*_\varphi v_k\|^2_\varphi = \|\gamma_k\|^2_\varphi = \sum_{|L|=q-1}{}' \|(\overline{\partial}^*_\varphi u)_{kL}\|^2_\varphi \leq \|\overline{\partial}^*_\varphi u\|^2_\varphi .$$

From the definition of v_k it follows that

$$\int_{\mathbb{C}^n\setminus\mathbb{B}_R} |u(z)|^2 e^{-\varphi(z)}\,d\lambda(z) = \frac{1}{q+1}\sum_{k=1}^n \int_{\mathbb{C}^n\setminus\mathbb{B}_R} |v_k(z)|^2 e^{-\varphi(z)}\,d\lambda(z)$$
$$\leq \frac{\tilde{C}\epsilon}{q+1}(\|\bar\partial u\|_\varphi^2 + \|\bar\partial_\varphi^* u\|_\varphi^2).$$

Since u was an arbitrary $(0,q)$-form in $\operatorname{dom}(\bar\partial) \cap \operatorname{dom}(\bar\partial_\varphi^*)$ we can again apply Theorem 11.17 to get the desired conclusion. □

Using the last result together with the necessary condition for Schrödinger operators with magnetic field (Theorem 11.12), we can now find a stronger necessary condition for compactness of $N_{\varphi,q}$.

Let $u = u\,d\bar z_1 \wedge \cdots \wedge d\bar z_n$ be a smooth $(0,n)$-form belonging to the domain of $\Box_{\varphi,n}$. For $1 \leq j \leq n$, denote by K_j the increasing multiindex $K_j := (1,\dots,j-1,j+1,\dots,n)$ of length $n-1$. Then

$$\bar\partial_\varphi^* u = \sum_{j=1}^n (-1)^{j+1}\left(\frac{\partial\varphi}{\partial z_j}u - \frac{\partial u}{\partial z_j}\right) d\bar z_{K_j}.$$

Hence

$$\bar\partial\bar\partial_\varphi^* u = \left[\sum_{j=1}^n \frac{\partial}{\partial \bar z_j}\left(\frac{\partial\varphi}{\partial z_j}u - \frac{\partial u}{\partial z_j}\right)\right] d\bar z_1 \wedge \cdots \wedge d\bar z_n$$
$$= \left[\sum_{j=1}^n \left(\frac{\partial^2\varphi}{\partial z_j \partial \bar z_j}u + \frac{\partial\varphi}{\partial z_j}\frac{\partial u}{\partial \bar z_j} - \frac{\partial^2 u}{\partial z_j \partial \bar z_j}\right)\right] d\bar z_1 \wedge \cdots \wedge d\bar z_n.$$

Conjugation with the unitary operator $L^2(\mathbb{C}^n, e^{-\varphi}) \to L^2(\mathbb{C}^n)$ of multiplication by $e^{-\varphi/2}$ gives

$$e^{-\varphi/2}\Box_{\varphi,n}\,e^{\varphi/2}f = \sum_{j=1}^n\left(-\frac{\partial^2 f}{\partial z_j\partial\bar z_j} - \frac{1}{2}\frac{\partial\varphi}{\partial\bar z_j}\frac{\partial f}{\partial z_j} + \frac{1}{2}\frac{\partial\varphi}{\partial z_j}\frac{\partial f}{\partial\bar z_j} + \frac{1}{4}\frac{\partial\varphi}{\partial\bar z_j}\frac{\partial\varphi}{\partial z_j}f + \frac{1}{2}\frac{\partial^2\varphi}{\partial z_j\partial\bar z_j}f\right), \tag{11.35}$$

where $f \in L^2(\mathbb{C}^n)$, and we just wrote down the coefficient of the corresponding $(0,n)$-form. This operator can be expressed by real variables in the form

$$e^{-\varphi/2}\Box_{\varphi,n}\,e^{\varphi/2}f = \frac{1}{4}(-\Delta_A + V)f \tag{11.36}$$

with

$$\Delta_A = \sum_{j=1}^n\left[\left(-\frac{\partial}{\partial x_j} - \frac{i}{2}\frac{\partial\varphi}{\partial y_j}\right)^2 + \left(-\frac{\partial}{\partial y_j} + \frac{i}{2}\frac{\partial\varphi}{\partial x_j}\right)^2\right]$$

and $V = 2\operatorname{tr}(M_\varphi)$. It follows that $-\Delta_A + V$ is a Schrödinger operator on $L^2(\mathbb{R}^{2n})$ with magnetic vector potential

$$A = \frac{1}{2}\left(-\frac{\partial\varphi}{\partial y_1}, \frac{\partial\varphi}{\partial x_1}, \dots, -\frac{\partial\varphi}{\partial y_n}, \frac{\partial\varphi}{\partial x_n}\right),$$

where $z_j = x_j + iy_j$, $j = 1, \dots, n$, and nonnegative electric potential V in the case where φ is plurisubharmonic. Note that A corresponds to the real 1-form $-\frac{i}{2}(\partial - \bar\partial)\varphi$, and the associated magnetic field is $dA = i\partial\bar\partial\varphi$. By essential self-adjointness, $\Box_{\varphi,n}$ agrees with the unique self-adjoint extension of the Schrödinger operator $H(A, V)$; see Proposition 10.10. We can now apply Theorem 11.12.

Theorem 11.20. *Let $\varphi: \mathbb{C}^n \to \mathbb{R}$ be smooth and plurisubharmonic. Then $\Box_{\varphi,n}$ has a compact resolvent on $L^2_{0,n}(\mathbb{C}^n, e^{-\varphi})$ if and only if the magnetic Schrödinger operator $H(A, V) = -\Delta_A + V$ has a compact resolvent on $L^2(\mathbb{R}^{2n})$. In particular, if there is $1 \le q \le n$ such that $\Box_{\varphi,q}$ has compact resolvent, then*

$$\lim_{|z|\to\infty} \int_{Q_z} \operatorname{tr}(M_\varphi)^2\, d\lambda = +\infty. \tag{11.37}$$

Proof. It remains to prove the last statement. If $\Box_{\varphi,q}$ has a compact resolvent, then so does $\Box_{\varphi,n}$ (see Proposition 11.19), and if the magnetic Schrödinger operator has a compact resolvent, then

$$\int_{Q_z} (|M_\varphi|^2 + 2\operatorname{tr}(M_\varphi))d\lambda \to \infty \tag{11.38}$$

as $|z| \to \infty$ by Theorem 11.12, where $|B|^2 = \sum_{j,k=1}^n |B_{j,k}|^2$, since the magnetic field is $i\partial\bar\partial\varphi$. Because $M_\varphi \ge 0$, the above is equivalent to $\int_{Q_z} (\operatorname{tr}(M_\varphi)^2 + \operatorname{tr}(M_\varphi))\, d\lambda \to \infty$ as $|z| \to \infty$. By Hölder's inequality,

$$\int_{Q_z} \operatorname{tr}(M_\varphi)\, d\lambda \le |Q_z|^{1/2}\left(\int_{Q_z} \operatorname{tr}(M_\varphi)^2 d\lambda\right)^{1/2},$$

and we see that (11.38) implies $f(\int_{Q_z} \operatorname{tr}(M_\varphi)^2\, d\lambda) \to \infty$ as $|z| \to \infty$, where $f: [0, \infty) \to [0, \infty)$ is $f(t) = t + t^{1/2}$. Because f is a monotone bijection, this is equivalent to $\int_{Q_z} \operatorname{tr}(M_\varphi)^2\, d\lambda \to \infty$ as $|z| \to \infty$. □

For a further study of compactness, we define weighted Sobolev spaces and prove, under suitable conditions, the Rellich–Kondrachov lemma for these weighted Sobolev spaces. We will also have to consider their dual spaces, which already appeared in [10] and [62].

Definition 11.21. For $k \in \mathbb{N}$, let

$$W^k(\mathbb{C}^n, e^{-\varphi}) := \{f \in L^2(\mathbb{C}^n, e^{-\varphi}) : D^\alpha f \in L^2(\mathbb{C}^n, e^{-\varphi}) \text{ for } |\alpha| \le k\},$$

where $D^\alpha = \frac{\partial^{|\alpha|}}{\partial^{\alpha_1} x_1 \dots \partial^{\alpha_{2n}} y_n}$ for $(z_1, \dots, z_n) = (x_1, y_1, \dots, x_n, y_n)$ with norm

$$\|f\|_{k,\varphi}^2 = \sum_{|\alpha| \le k} \|D^\alpha f\|_\varphi^2.$$

We will also need weighted Sobolev spaces with negative exponent. It turns out that for our purposes, it is more reasonable to consider the dual spaces of the following spaces.

Definition 11.22. Let

$$X_j = \frac{\partial}{\partial x_j} - \frac{\partial \varphi}{\partial x_j} \quad \text{and} \quad Y_j = \frac{\partial}{\partial y_j} - \frac{\partial \varphi}{\partial y_j}$$

for $j = 1, \dots, n$ and define

$$W^1(\mathbb{C}^n, e^{-\varphi}, \nabla\varphi) = \{f \in L^2(\mathbb{C}^n, e^{-\varphi}) \ : X_j f,\ Y_j f \in L^2(\mathbb{C}^n, e^{-\varphi}), j = 1, \dots, n\}$$

with norm

$$\|f\|_{\varphi,\nabla\varphi}^2 = \|f\|_\varphi^2 + \sum_{j=1}^n (\|X_j f\|_\varphi^2 + \|Y_j f\|_\varphi^2).$$

In the next step, we will analyze the dual space of $W^1(\mathbb{C}^n, e^{-\varphi}, \nabla\varphi)$; compare with Proposition 11.7.

By the mapping $f \mapsto (f, X_j f, Y_j f)$ the space $W^1(\mathbb{C}^n, e^{-\varphi}, \nabla\varphi)$ can be identified with a product of $L^2(\mathbb{C}^n, e^{-\varphi})$, and hence each continuous linear functional L on $W^1(\mathbb{C}^n, e^{-\varphi}, \nabla\varphi)$ is represented (in a nonunique way) by

$$L(f) = \int_{\mathbb{C}^n} f g_0 e^{-\varphi}\, d\lambda + \sum_{j=1}^n \int_{\mathbb{C}^n} ((X_j f) g_j + (Y_j f) h_j) e^{-\varphi}\, d\lambda$$

for some $g_j, h_j \in L^2(\mathbb{C}^n, e^{-\varphi})$. For $f \in \mathcal{C}_0^\infty(\mathbb{C}^n)$, it follows that

$$L(f) = \int_{\mathbb{C}^n} f g_0 e^{-\varphi}\, d\lambda - \sum_{j=1}^n \int_{\mathbb{C}^n} f\left(\frac{\partial g_j}{\partial x_j} + \frac{\partial h_j}{\partial y_j}\right) e^{-\varphi}\, d\lambda.$$

Since $\mathcal{C}_0^\infty(\mathbb{C}^n)$ is dense in $W^1(\mathbb{C}^n, e^{-\varphi}, \nabla\varphi)$, we have shown the following:

Lemma 11.23. *Each element $u \in (W^1(\mathbb{C}^n, e^{-\varphi}, \nabla\varphi))' = W^{-1}(\mathbb{C}^n, e^{-\varphi}, \nabla\varphi)$ can be represented in a non unique way by*

$$u = g_0 + \sum_{j=1}^n \left(\frac{\partial g_j}{\partial x_j} + \frac{\partial h_j}{\partial y_j}\right),$$

where $g_j, h_j \in L^2(\mathbb{C}^n, e^{-\varphi})$.

The dual norm $\|u\|_{-1,\varphi,\nabla\varphi} := \sup\{|u(f)| : \|f\|_{\varphi,\nabla\varphi} \le 1\}$ *can be expressed in the form*

$$\|u\|^2_{-1,\varphi,\nabla\varphi} = \inf\left\{ \|g_0\|^2 + \sum_{j=1}^n (\|g_j\|^2 + \|h_j\|^2) \right\},$$

where the infimum is taken over all families (g_j, h_j) *in* $L^2(\mathbb{C}^n, e^{-\varphi})$ *representing the functional u.*

In particular, each function in $L^2(\mathbb{C}^n, e^{-\varphi})$ can be identified with an element of $W^{-1}(\mathbb{C}^n, e^{-\varphi}, \nabla\varphi)$.

Proposition 11.24. *Suppose that* φ *is a* $\mathcal{C}^2$*-function satisfying*

$$\lim_{|z|\to\infty} (\theta|\nabla\varphi(z)|^2 + \triangle\varphi(z)) = +\infty$$

for some $\theta \in (0,1)$, *where*

$$|\nabla\varphi(z)|^2 = \sum_{k=1}^n \left(\left|\frac{\partial\varphi}{\partial x_k}\right|^2 + \left|\frac{\partial\varphi}{\partial y_k}\right|^2 \right).$$

Then the embedding of $W^1(\mathbb{C}^n, e^{-\varphi}, \nabla\varphi)$ *into* $L^2(\mathbb{C}^n, e^{-\varphi})$ *is compact.*

Proof. We adapt methods from Proposition 11.14 (see also [10, 61] and [62]) and use the general result that an operator between Hilbert spaces is compact if and only if the image of a weakly convergent sequence is strongly convergent (Proposition 4.17).

For the vector fields X_j from Definition 11.22 and their formal adjoints X_j^* in the weighted space $L^2(\mathbb{C}^n, e^{-\varphi})$, we have $X_j^* = -\frac{\partial}{\partial x_j}$,

$$(X_j + X_j^*)f = -\frac{\partial\varphi}{\partial x_j} f \quad \text{and} \quad [X_j, X_j^*]f = -\frac{\partial^2\varphi}{\partial x_j^2} f$$

for $f \in \mathcal{C}_0^\infty(\mathbb{C}^n)$, and

$$([X_j, X_j^*]f, f)_\varphi = \|X_j^* f\|^2_\varphi - \|X_j f\|^2_\varphi, \tag{11.39}$$

$$\|(X_j + X_j^*)f\|^2_\varphi \le (1 + 1/\epsilon)\|X_j f\|^2_\varphi + (1+\epsilon)\|X_j^* f\|^2_\varphi \tag{11.40}$$

for each $\epsilon > 0$. Similar relations hold for the vector fields Y_j. Now we set

$$\Psi(z) = |\nabla\varphi(z)|^2 + (1+\epsilon) \triangle \varphi(z).$$

By (11.39) and (11.40) it follows that

$$(\Psi f, f)_\varphi \leq (2 + \epsilon + 1/\epsilon) \sum_{j=1}^{n} (\|X_j f\|_\varphi^2 + \|Y_j f\|_\varphi^2).$$

Since $\mathcal{C}_0^\infty(\mathbb{C}^n)$ is dense in $W^1(\mathbb{C}^n, e^{-\varphi}, \nabla\varphi)$ by definition, this inequality holds for all $f \in W^1(\mathbb{C}^n, e^{-\varphi}, \nabla\varphi)$.

If $(f_k)_k$ is a sequence in $W^1(\mathbb{C}^n, e^{-\varphi}, \nabla\varphi)$ converging weakly to 0, then $(f_k)_k$ is a bounded sequence in $W^1(\mathbb{C}^n, e^{-\varphi}, \nabla\varphi)$, and our assumption implies that

$$\Psi(z) = |\nabla\varphi(z)|^2 + (1 + \epsilon) \triangle \varphi(z)$$

is positive in a neighborhood of ∞. So we obtain

$$\begin{aligned} \int_{\mathbb{C}^n} |f_k(z)|^2 e^{-\varphi(z)}\, d\lambda(z) &\leq \int_{|z|<R} |f_k(z)|^2 e^{-\varphi(z)}\, d\lambda(z) \\ &\quad + \int_{|z|\geq R} \frac{\Psi(z)|f_k(z)|^2}{\inf\{\Psi(z) : |z| \geq R\}} e^{-\varphi(z)}\, d\lambda(z) \\ &\leq C_{\varphi,R} \|f_k\|_{L^2(B(0,R))}^2 + \frac{C_\epsilon \|f_k\|_{\varphi,\nabla\varphi}^2}{\inf\{\Psi(z) : |z| \geq R\}}. \end{aligned}$$

Hence the assumption and the fact that the embedding

$$W^1(B(0,R)) \hookrightarrow L^2(B(0,R))$$

is compact (Rellich–Kondrachov Lemma 11.5) show that a subsequence of $(f_k)_k$ tends to 0 in $L^2(\mathbb{C}^n, e^{-\varphi})$. □

Remark 11.25. It follows that the adjoint to the above embedding of $L^2(\mathbb{C}^n, e^{-\varphi})$ into $(W^1(\mathbb{C}^n, e^{-\varphi}, \nabla\varphi))' = W^{-1}(\mathbb{C}^n, e^{-\varphi}, \nabla\varphi)$ (in the sense of Lemma 11.23) is also compact.

Remark 11.26. Note that we do not need the plurisubharmonicity of the weight function in Proposition 11.24. If the weight function is plurisubharmonic, then we can drop θ in the assumptions of Proposition 11.24.

The following proposition reformulates the compactness condition for the case of a bounded pseudoconvex domain in $\mathbb{C}^n$; see [9, 91]. The difference to the compactness estimate for a bounded pseudoconvex domain is that here we have to assume a condition on the weight function implying a corresponding Rellich–Kondrachov lemma.

Proposition 11.27. *Suppose that the weight function φ satisfies* (6.5) *and*

$$\lim_{|z|\to\infty} (\theta|\nabla\varphi(z)|^2 + \triangle\varphi(z)) = +\infty$$

for some $\theta \in (0,1)$. *Then the following statements are equivalent.*

(1) *The* $\overline{\partial}$*-Neumann operator* $N_{1,\varphi}$ *is a compact operator from* $L^2_{(0,1)}(\mathbb{C}^n, e^{-\varphi})$ *into itself.*

(2) *The embedding of the space* $\mathrm{dom}(\overline{\partial}) \cap \mathrm{dom}(\overline{\partial}^*_\varphi)$*, provided with the graph norm* $u \mapsto (\|u\|^2_\varphi + \|\overline{\partial}u\|^2_\varphi + \|\overline{\partial}^*_\varphi u\|^2_\varphi)^{1/2}$*, into* $L^2_{(0,1)}(\mathbb{C}^n, e^{-\varphi})$ *is compact.*

(3) *For every positive* ϵ*, there exists a constant* C_ϵ *such that*

$$\|u\|^2_\varphi \le \epsilon(\|\overline{\partial}u\|^2_\varphi + \|\overline{\partial}^*_\varphi u\|^2_\varphi) + C_\epsilon \|u\|^2_{-1,\varphi,\nabla\varphi}$$

for all $u \in \mathrm{dom}(\overline{\partial}) \cap \mathrm{dom}(\overline{\partial}^*_\varphi)$.

(4) *For every positive* ϵ*, there exists* $R > 0$ *such that*

$$\int_{\mathbb{C}^n \setminus \mathbb{B}_R} |u(z)|^2 e^{-\varphi(z)}\, d\lambda(z) \le \epsilon(\|\overline{\partial}u\|^2_\varphi + \|\overline{\partial}^*_\varphi u\|^2_\varphi)$$

for all $u \in \mathrm{dom}(\overline{\partial}) \cap \mathrm{dom}(\overline{\partial}^*_\varphi)$.

(5) *The operators*

$$\overline{\partial}^*_\varphi N_{1,\varphi} : L^2_{(0,1)}(\mathbb{C}^n, e^{-\varphi}) \cap \ker(\overline{\partial}) \longrightarrow L^2(\mathbb{C}^n, e^{-\varphi})$$

and

$$\overline{\partial}^*_\varphi N_{2,\varphi} : L^2_{(0,2)}(\mathbb{C}^n, e^{-\varphi}) \cap \ker(\overline{\partial}) \longrightarrow L^2_{(0,1)}(\mathbb{C}^n, e^{-\varphi})$$

are both compact.

Proof. (1) and (4) are equivalent by Theorem 11.17. Next, we show that (1) and (5) are equivalent: suppose that $N_{1,\varphi}$ is compact. For $f \in L^2_{(0,1)}(\mathbb{C}^n, e^{-\varphi})$, it follows that

$$\|\overline{\partial}^*_\varphi N_{1,\varphi} f\|^2_\varphi \le \langle f, N_{1,\varphi} f\rangle_\varphi \le \epsilon \|f\|^2_\varphi + C_\epsilon \|N_{1,\varphi} f\|^2_\varphi.$$

Hence, by Proposition 2.1, $\overline{\partial}^*_\varphi N_{1,\varphi}$ is compact. Applying the formula

$$N_{1,\varphi} = (\overline{\partial}^*_\varphi N_{1,\varphi})^* (\overline{\partial}^*_\varphi N_{1,\varphi}) + (\overline{\partial}^*_\varphi N_{2,\varphi})(\overline{\partial}^*_\varphi N_{2,\varphi})^*$$

(see Proposition 4.64), we get that also $\overline{\partial}^*_\varphi N_{2,\varphi}$ is compact. The converse follows easily from the same formula.

Now we show (5) $\Longrightarrow$ (3) $\Longrightarrow$ (2) $\Longrightarrow$ (1). We follow the lines of [91], where the case of a bounded pseudoconvex domain is handled.

Assume (5): if (3) does not hold, then there exist $\epsilon_0 > 0$ and a sequence $(u_n)_n$ in $\mathrm{dom}(\overline{\partial}) \cap \mathrm{dom}(\overline{\partial}^*_\varphi)$ such that $\|u_n\|_\varphi = 1$ and

$$\|u_n\|^2_\varphi \ge \epsilon_0(\|\overline{\partial}u_n\|^2_\varphi + \|\overline{\partial}^*_\varphi u_n\|^2_\varphi) + n\|u_n\|^2_{-1,\varphi,\nabla\varphi}$$

for each $n \geq 1$, which implies that $u_n \to 0$ in $W^{-1}_{(0,1)}(\mathbb{C}^n, e^{-\varphi}, \nabla\varphi)$. Since u_n can be written in the form

$$u_n = (\overline{\partial}^*_\varphi N_{1,\varphi})^* \,\overline{\partial}^*_\varphi u_n + (\overline{\partial}^*_\varphi N_{2,\varphi})\,\overline{\partial} u_n,$$

(5) implies that there exists a subsequence of $(u_n)_n$ converging in $L^2_{(0,1)}(\mathbb{C}^n, e^{-\varphi})$ and the limit must be 0, which contradicts $\|u_n\|_\varphi = 1$.

To show that (3) implies (2), we consider a bounded sequence in $\mathrm{dom}(\overline{\partial}) \cap \mathrm{dom}(\overline{\partial}^*_\varphi)$. By Proposition 6.3 this sequence is also bounded in $L^2_{(0,1)}(\mathbb{C}^n, e^{-\varphi})$.

Now Proposition 11.24 implies that it has a subsequence converging in $W^{-1}_{(0,1)}(\mathbb{C}^n, e^{-\varphi}, \nabla\varphi)$. Finally, use (3) to show that this subsequence is a Cauchy sequence in $L^2_{(0,1)}(\mathbb{C}^n, e^{-\varphi})$, and therefore (2) holds.

Assume (2): by Proposition 6.3 and the basic facts about $N_{1,\varphi}$ it follows that

$$N_{1,\varphi} : L^2_{(0,1)}(\mathbb{C}^n, e^{-\varphi}) \longrightarrow \mathrm{dom}\,(\overline{\partial}) \cap \mathrm{dom}(\overline{\partial}^*_\varphi)$$

is continuous in the graph topology, and hence

$$N_{1,\varphi} : L^2_{(0,1)}(\mathbb{C}^n, e^{-\varphi}) \longrightarrow \mathrm{dom}\,(\overline{\partial}) \cap \mathrm{dom}(\overline{\partial}^*_\varphi) \hookrightarrow L^2_{(0,1)}(\mathbb{C}^n, e^{-\varphi})$$

is compact. □

Remark 11.28. If

$$\lim_{|z|\to\infty} \mu_\varphi(z) = +\infty,$$

then the condition of the Rellich–Kondrachov Lemma 11.24 is satisfied.

This follows from the fact that for the trace $\mathrm{tr}(M_\varphi)$ of the Levi matrix, we have

$$\mathrm{tr}(M_\varphi) = \frac{1}{4} \triangle \varphi,$$

and since for any invertible $(n \times n)$ matrix T,

$$\mathrm{tr}(M_\varphi) = \mathrm{tr}(T M_\varphi T^{-1}),$$

it follows that $\mathrm{tr}(M_\varphi)$ equals the sum of all eigenvalues of M_φ. Hence our assumption on the lowest eigenvalue μ_φ of the Levi matrix implies that the assumption of Proposition 11.24 is satisfied.

Remark 11.29. We mention that for the weight $\varphi(z) = |z|^2$, the $\overline{\partial}$-Neumann operator fails to be compact (see Chapter 15), but the condition

$$\lim_{|z|\to\infty} (\theta|\nabla\varphi(z)|^2 + \triangle\varphi(z)) = +\infty$$

of the Rellich–Kondrachov lemma is satisfied.

Remark 11.30. Let $A^2_{(0,1)}(\mathbb{C}^n, e^{-\varphi})$ denote the space of $(0,1)$-forms with holomorphic coefficients belonging to $L^2(\mathbb{C}^n, e^{-\varphi})$.

We point out that assuming (11.24) implies directly – without use of Sobolev spaces – that the embedding of the space

$$A^2_{(0,1)}(\mathbb{C}^n, e^{-\varphi}) \cap \operatorname{dom}(\overline{\partial}^*_\varphi)$$

provided with the graph norm $u \mapsto (\|u\|^2_\varphi + \|\overline{\partial}^*_\varphi u\|^2_\varphi)^{1/2}$ into $A^2_{(0,1)}(\mathbb{C}^n, e^{-\varphi})$ is compact. Compare Proposition 11.27(2).

For this purpose, let $u \in A^2_{(0,1)}(\mathbb{C}^n, e^{-\varphi}) \cap \operatorname{dom}(\overline{\partial}^*_\varphi)$. Then we obtain from the proof of Proposition 6.3 that

$$\|\overline{\partial}^*_\varphi u\|^2_\varphi = \int_{\mathbb{C}^n} \sum_{j,k=1}^n \frac{\partial^2 \varphi}{\partial z_j \partial \overline{z}_k} u_j \overline{u}_k \, e^{-\varphi} \, d\lambda.$$

For $u = \sum_{j=1}^n u_j \, d\overline{z}_j$, let us identify $u(z)$ with the vector $(u_1(z), \dots, u_n(z)) \in \mathbb{C}^n$. Then denoting by $\langle \cdot, \cdot \rangle$ the standard inner product in $\mathbb{C}^n$, we have

$$\langle u(z), u(z) \rangle = \sum_{j=1}^n |u_j(z)|^2 \quad \text{and} \quad \langle M_\varphi u(z), u(z) \rangle = \sum_{j,k=1}^n \frac{\partial^2 \varphi(z)}{\partial z_j \partial \overline{z}_k} u_j(z) \overline{u_k(z)}.$$

Note that the lowest eigenvalue μ_φ of the Levi matrix M_φ can be expressed as

$$\mu_\varphi(z) = \inf_{u(z) \neq 0} \frac{\langle M_\varphi u(z), u(z) \rangle}{\langle u(z), u(z) \rangle}.$$

So we get

$$\begin{aligned}
\int_{\mathbb{C}^n} \langle u, u \rangle e^{-\varphi} \, d\lambda &\leq \int_{\mathbb{B}_R} \langle u, u \rangle e^{-\varphi} \, d\lambda + \Big[\inf_{\mathbb{C}^n \setminus \mathbb{B}_R} \mu_\varphi(z) \Big]^{-1} \int_{\mathbb{C}^n \setminus \mathbb{B}_R} \mu_\varphi(z) \, \langle u, u \rangle e^{-\varphi} \, d\lambda \\
&\leq \int_{\mathbb{B}_R} \langle u, u \rangle e^{-\varphi} \, d\lambda + \Big[\inf_{\mathbb{C}^n \setminus \mathbb{B}_R} \mu_\varphi(z) \Big]^{-1} \int_{\mathbb{C}^n} \langle M_\varphi u, u \rangle e^{-\varphi} \, d\lambda.
\end{aligned}$$

For a given $\epsilon > 0$, choose R so large that

$$\Big[\inf_{\mathbb{C}^n \setminus \mathbb{B}_R} \mu_\varphi(z) \Big]^{-1} < \epsilon.$$

For simplicity, we can also use polycylinders $\mathbb{D}_r$ of the form

$$\mathbb{D}_r := \{(z_1, \dots, z_n) : |z_j| < r, j = 1, \dots, n\}$$

instead of the balls $\mathbb{B}_R$. Now we use the fact that for Bergman spaces of holomorphic functions, the embedding of $A^2(\mathbb{D}_s)$ into $A^2(\mathbb{D}_r)$ is compact for $r < s$. To show this, we consider the orthonormal basis

$$\phi_\alpha^s(z) = \left(\frac{(\alpha_1+1)\dots(\alpha_n+1)}{\pi^n} \right)^{1/2} \frac{z_1^{\alpha_1}\dots z_n^{\alpha_n}}{s^{|\alpha|+n}}$$

in $A^2(\mathbb{D}_s)$ and observe that the squares of the norms of ϕ_α^s in $A^2(\mathbb{D}_r)$ are

$$\left(\frac{r}{s}\right)^{2|\alpha|+2n}.$$

As

$$\sum_\alpha \left(\frac{r}{s}\right)^{2|\alpha|+2n} < \infty,$$

we get from Proposition 2.10 that the embedding of $A^2(\mathbb{D}_s)$ into $A^2(\mathbb{D}_r)$ is even a Hilbert–Schmidt operator, and we can conclude the proof of our statement as in Proposition 11.24.

Inspired by a result on Schrödinger operators with magnetic field of Iwatsuka [59] we prove another characterization of compactness, which will be used later.

Proposition 11.31. *Let $\varphi : \mathbb{C}^n \longrightarrow \mathbb{R}^+$ be a plurisubharmonic $\mathcal{C}^2$-weight function. The $\overline{\partial}$-Neumann operator*

$$N_\varphi : L^2_{(0,1)}(\mathbb{C}^n, e^{-\varphi}) \longrightarrow L^2_{(0,1)}(\mathbb{C}^n, e^{-\varphi})$$

is compact if and only if there is a smooth function $\Lambda : \mathbb{C}^n \longrightarrow \mathbb{R}$ such that $\Lambda(z) \to \infty$ as $|z| \to \infty$ and

$$\|\overline{\partial}u\|_\varphi^2 + \|\overline{\partial}_\varphi^* u\|_\varphi^2 \geq \int_{\mathbb{C}^n} \Lambda\, |u|^2 e^{-\varphi}\, d\lambda \tag{11.41}$$

for each $u \in \mathcal{W}^{Q_\varphi}$.

Proof. Suppose (11.41) holds. Then for each $\epsilon > 0$, there exists $R > 0$ such that $\Lambda \geq 1/\epsilon$ on $\mathbb{C}^n \setminus \mathbb{B}_R$. This implies

$$\|\overline{\partial}u\|_\varphi^2 + \|\overline{\partial}_\varphi^* u\|_\varphi^2 \geq \int_{\mathbb{C}^n} \Lambda\, |u|^2 e^{-\varphi}\, d\lambda \geq \frac{1}{\epsilon} \int_{\mathbb{C}^n \setminus \mathbb{B}_R} |u|^2 e^{-\varphi}\, d\lambda,$$

which means that (11.30) holds.

We indicate that the condition of Theorem 11.17 can be written in the following form: for each $\epsilon > 0$, there exists $R(\epsilon) > 0$ such that

$$\|u\|_{L^2_{(0,1)}(\mathbb{C}^n\setminus\mathbb{B}_{R(\epsilon)},\varphi)} \leq \epsilon\|u\|_{Q_\varphi}.$$

Hence for all $u \in \mathcal{W}^{Q_\varphi}$ and $j \in \mathbb{N}$, we have

$$2^j \int\limits_{\mathbb{C}^n\setminus\mathbb{B}_{R(1/2^j)}} |u|^2 e^{-\varphi}\, d\lambda \leq \frac{1}{2^j}\|u\|^2_{Q_\varphi},$$

and hence

$$\begin{aligned}\int\limits_{\mathbb{C}^n} |u|^2 e^{-\varphi}\, d\lambda &\leq \int\limits_{\mathbb{B}_{R(1/2)}} 1\cdot|u|^2 e^{-\varphi}\, d\lambda + \int\limits_{\mathbb{B}_{R(1/4)}\setminus\mathbb{B}_{R(1/2)}} 2\cdot|u|^2 e^{-\varphi}\, d\lambda \\ &\quad + \int\limits_{\mathbb{B}_{R(1/8)}\setminus\mathbb{B}_{R(1/4)}} 4\cdot|u|^2 e^{-\varphi}\, d\lambda + \cdots \\ &\leq (C+1)\|u\|^2_{Q_\varphi}.\end{aligned}$$

Now it is easy to define a smooth function Λ tending to ∞ as $|z|$ tends to ∞ such that (11.41) holds. □

11.5 Compactness in the Segal–Bargmann space

In this chapter, we use Montel's theorem (see, for instance, [43]) to characterize compact subsets of the Segal–Bargmann space $A^2(\mathbb{C}^n, e^{-|z|^2})$. Later (Chapter 16), we will use this characterization to describe properties of certain unbounded operators on the Segal–Bargmann space, which are connected to the uncertainty principle.

Theorem 11.32. *A bounded closed subset $\mathcal{A}$ of $A^2(\mathbb{C}^n, e^{-|z|^2})$ is compact if and only if for every $\epsilon > 0$, there exists a number $R > 0$ such that for every $f \in \mathcal{A}$,*

$$\int\limits_{\mathbb{C}^n\setminus\mathbb{B}_R} |f(z)|^2 e^{-|z|^2}\, d\lambda(z) < \epsilon^2. \tag{11.42}$$

Proof. For $f \in A^2(\mathbb{C}^n, e^{-|z|^2})$ and $R > 0$, let

$$\|f\|_R = \Big(\int\limits_{\mathbb{B}_R} |f(z)|^2 e^{-|z|^2}\, d\lambda(z)\Big)^{1/2}, \quad \|f\|_{(R)} = \Big(\int\limits_{\mathbb{C}^n\setminus\mathbb{B}_R} |f(z)|^2 e^{-|z|^2}\, d\lambda(z)\Big)^{1/2}. \tag{11.43}$$

Then

$$\|f\|^2_{(R)} = \|f\|^2 - \|f\|^2_R \quad \text{and} \quad \big|\,\|f\|^2 - \|f\|^2_R\,\big| \to 0 \tag{11.44}$$

as $R \to \infty$. Assume that $\mathcal{A}$ is compact and that (11.42) fails. Then there is $\epsilon > 0$ such that there is no $R > 0$ with

$$\|f\|^2 - \|f\|_R^2 < \epsilon^2$$

for all $f \in \mathcal{A}$. Hence, for each $R_1 > 0$, there is $f_1 \in \mathcal{A}$ with $\|f_1\|^2 - \|f_1\|_{R_1}^2 > \epsilon^2$. We construct a sequence in $\mathcal{A}$ that has no Cauchy subsequence. We have

$$\|f_1\|_{(R_1)} = \left(\|f_1\|^2 - \|f_1\|_{R_1}^2\right)^{1/2} > \epsilon$$

and can find $R_2 > R_1$ such that

$$\|f_1\|_{(R_2)} = \left(\|f_1\|^2 - \|f_1\|_{R_2}^2\right)^{1/2} < \frac{\epsilon}{2}.$$

For R_2, we can find $f_2 \in \mathcal{A}$ with $\|f_2\|_{(R_2)} > \epsilon$. Hence

$$\|f_2 - f_1\| \geq \|f_2 - f_1\|_{(R_2)} \geq \left| \|f_2\|_{(R_2)} - \|f_1\|_{(R_2)} \right| > \frac{\epsilon}{2}.$$

We continue in the same way to find f_3 and notice that $\|f_3 - f_2\| > \epsilon/2$, and since $\|f_1\|_{(R_3)} \leq \|f_1\|_{(R_2)}$, we have also $\|f_3 - f_1\| > \epsilon/2$. Now it is clear that there is no subsequence of $(f_k)_k$ that is a Cauchy sequence, and we proved that (11.42) is necessary for compactness.

To show the sufficiency, we recall (1.23). For the bounded set $\mathcal{A}$, we have $\|f\| \leq C_A$ for each $f \in \mathcal{A}$, where $C_A > 0$ is a constant. By (1.23), for each compact subset $K \subset \mathbb{C}^n$, there is a constant $M_K > 0$ such that

$$\sup_{z \in K} |f(z)| \leq M_K$$

for all $f \in \mathcal{A}$. This means that $\mathcal{A}$ is uniformly bounded in the compact open topology, and Montel's theorem implies that each sequence $(f_k)_k$ in $\mathcal{A}$ has a convergent subsequence, which is again denoted by $(f_k)_k$; it is convergent in the compact open topology, and it is also a Cauchy sequence in this topology, which means that for each compact ball $\mathbb{B}_R$ and for each $\epsilon > 0$, there exists $N > 0$ such that

$$\sup_{z \in \mathbb{B}_R} |f_k(z) - f_m(z)| < \epsilon \tag{11.45}$$

for all $k, m > N$. It remains to show that $(f_k)_k$ is a Cauchy sequence in $A^2(\mathbb{C}^n, e^{-|z|^2})$. For this purpose, we use assumption (11.42) and estimate

$$\begin{aligned}
\|f_k - f_m\|^2 &\leq \int_{\mathbb{B}_R} |f_k(z) - f_m(z)|^2 e^{-|z|^2}\, d\lambda(z) + \int_{\mathbb{C}^n \setminus \mathbb{B}_R} |f_k(z) - f_m(z)|^2 e^{-|z|^2}\, d\lambda(z) \\
&\leq \sup_{z \in \mathbb{B}_R} |f_k(z) - f_m(z)|^2 \int_{\mathbb{C}^n} e^{-|z|^2}\, d\lambda(z) + \epsilon \\
&\leq C\epsilon^2 + \epsilon^2.
\end{aligned}$$

□

11.6 Bounded pseudoconvex domains

Finally, we investigate the compactness of the $\overline{\partial}$-Neumann operator on a bounded pseudoconvex domain. For this purpose we consider the following potential theoretic concepts.

Definition 11.33. Let $\Omega \Subset \mathbb{C}^n$ be a smoothly bounded pseudoconvex domain. It satisfies property (P) if for each $M > 0$, there exist a neighborhood U of $\partial\Omega$ and a plurisubharmonic function $\varphi_M \in C^2(U)$ with $0 \leq \varphi_M \leq 1$ on U such that

$$\sum_{j,k=1}^{n} \frac{\partial^2 \varphi_M}{\partial z_j \partial \overline{z}_k}(p) t_j \overline{t}_k \geq M\|t\|^2$$

for all $p \in U$ and $t \in \mathbb{C}^n$.

Definition 11.34. Ω satisfies property ($\tilde{\mathrm{P}}$) if the following holds: there is a constant C such that for all $M > 0$, there exists a C^2-function φ_M in a neighborhood U (depending on M) of $b\Omega$ such that

(i)

$$\left| \sum_{j=1}^{n} \frac{\partial \varphi_M}{\partial z_j}(z) t_j \right|^2 \leq C \sum_{j,k=1}^{n} \frac{\partial^2 \varphi_M}{\partial z_j \partial \overline{z}_k}(z) t_j \overline{t}_k$$

and

(ii)

$$\sum_{j=1}^{n} \frac{\partial^2 \varphi_M}{\partial z_j \partial \overline{z}_k}(z) t_j \overline{t}_k \geq M\|t\|^2$$

for all $z \in U$ and $t \in \mathbb{C}^n$.

Remark 11.35. Catlin [13] showed that property (P) implies the compactness of the $\overline{\partial}$-operator N on $L^2_{(0,1)}(\Omega)$, and McNeal [75] showed that property ($\tilde{\mathrm{P}}$) also implies the compactness of the $\overline{\partial}$-operator N on $L^2_{(0,1)}(\Omega)$. It is not difficult to show that property (P) implies property ($\tilde{\mathrm{P}}$): If (φ_M) is the family of functions from the definition of property (P), then (e^{φ_M}) will work for ($\tilde{\mathrm{P}}$); see also [91].

We can now use a similar approach as for weighted spaces to prove Catlin's result. We use again Theorem 11.1.

To show that the unit ball in $\mathrm{dom}(\overline{\partial}) \cap \mathrm{dom}(\overline{\partial}^*)$ in the graph norm $f \mapsto (\|\overline{\partial} f\|^2 + \|\overline{\partial}^* f\|^2)^{\frac{1}{2}}$ satisfies condition (i) of Theorem 11.1, we first remark that the compactly supported forms are not dense in $\mathrm{dom}(\overline{\partial}) \cap \mathrm{dom}(\overline{\partial}^*)$, but the forms in $\mathrm{dom}(\overline{\partial}^*)$ with coefficients in $C^\infty(\overline{\Omega})$ are dense (Proposition 5.14). So if $\omega \Subset \Omega$, the we choose $\omega \Subset \omega_1 \Subset \omega_2 \Subset \Omega$ and a cut-off function ψ such that $\psi(z) = 1$ for $z \in \omega_1$ and $\psi(z) = 0$ for $z \in \Omega \setminus \omega_2$.

For $u \in \operatorname{dom}(\bar{\partial}) \cap \operatorname{dom}(\bar{\partial}^*)$, we define $\tilde{u} = \psi u$ and remark that the domain of $\bar{\partial}^*$ is preserved under multiplication by a function in $\mathcal{C}^1(\overline{\Omega})$ (Remark 5.15), and therefore $\tilde{u}$ has compactly supported coefficients and belongs to $\operatorname{dom}(\bar{\partial}) \cap \operatorname{dom}(\bar{\partial}^*)$. The graph norm of $\tilde{u}$ is bounded by a constant C depending only on $\omega, \omega_1, \omega_2, \Omega$ if u belongs to the unit ball in the graph norm. By construction we have

$$\|\tau_h u - u\|_{L^2(\omega)} = \|\tau_h \tilde{u} - \tilde{u}\|_{L^2(\omega)}$$

if $|h|$ is small enough, and hence we can use Gårding's inequality for $\omega \Subset \Omega$ to show that condition (i) holds.

To verify condition (ii), we use property (P) and the following version of the Kohn–Morrey formula:

$$\int_\Omega \sum_{j,k=1}^n \frac{\partial^2 \varphi_M}{\partial z_j \partial \bar{z}_k} u_j \bar{u}_k \, e^{-\varphi_M} \, d\lambda \leq \|\bar{\partial} u\|_{\varphi_M}^2 + \|\bar{\partial}_{\varphi_M}^* u\|_{\varphi_M}^2, \tag{11.46}$$

here we used that Ω is pseudoconvex, which means that the boundary terms in the Kohn–Morrey formula can be neglected; this follows easily from Theorem 7.2. Now we point out that the weighted $\bar{\partial}$-complex is equivalent to the unweighted one and that the expression $\sum_{j=1}^n \frac{\partial \varphi_M}{\partial z_j} u_j$ appearing in $\bar{\partial}_{\varphi_M}^* u$ can be controlled by the complex Hessian $\sum_{j,k=1}^n \frac{\partial^2 \varphi_M}{\partial z_j \partial \bar{z}_k} u_j \bar{u}_k$, which follows from the fact that property (P) implies property $(\tilde{\mathrm{P}})$. Of course, we also use the boundedness of the weight φ_M on $\Omega \Subset \mathbb{C}^n$. In this way, the same reasoning as in the weighted case shows that property (P) implies condition (11.3). Therefore condition (P) gives that the unit ball of $\operatorname{dom}(\bar{\partial}) \cap \operatorname{dom}(\bar{\partial}^*)$ in the graph norm $f \mapsto (\|\bar{\partial} f\|^2 + \|\bar{\partial}^* f\|^2)^{\frac{1}{2}}$ is relatively compact in $L^2_{(0,1)}(\Omega)$ and hence that the $\bar{\partial}$-Neumann operator is compact.

Now let

$$j : \operatorname{dom}(\bar{\partial}) \cap \operatorname{dom}(\bar{\partial}^*) \hookrightarrow L^2_{(0,1)}(\Omega)$$

denote the embedding. By (4.42) we know that

$$N = j \circ j^*.$$

Hence N is compact if and only if j is compact, where $\operatorname{dom}(\bar{\partial}) \cap \operatorname{dom}(\bar{\partial}^*)$ is endowed with the graph norm $f \mapsto (\|\bar{\partial} f\|^2 + \|\bar{\partial}^* f\|^2)^{\frac{1}{2}}$.

Theorem 11.36. *Let $\Omega \Subset \mathbb{C}^n$ be a smoothly bounded pseudoconvex domain. The $\bar{\partial}$-Neumann operator N is compact if and only if for each $\epsilon > 0$, there exists $\omega \Subset \Omega$ such that*

$$\int_{\Omega \setminus \omega} |u(z)|^2 \, d\lambda(z) \leq \epsilon (\|\bar{\partial} u\|^2 + \|\bar{\partial}^* u\|^2)$$

for each $u \in \operatorname{dom}(\overline{\partial}) \cap \operatorname{dom}(\overline{\partial}^*)$.

This follows from the above remarks about the embedding j and the fact that the two conditions (11.1) are also necessary for a bounded set in L^2 to be relatively compact.

Remark 11.37.

(a) Let $1 \le q \le n$. We say that Ω satisfies property (P_q) if the following holds: for every positive number M, there exist a neighborhood U of $b\Omega$ and a C^2-function φ_M on U such that $0 \le \varphi_M(z) \le 1$, $z \in U$, and such that for any $z \in U$, the sum of any q (equivalently, the smallest q) eigenvalues of the Hermitian form $(\frac{\partial^2 \varphi_M}{\partial z_j \partial \overline{z}_k}(z))_{j,k}$ is at least M. Using the methods from above, we can show that property (P_q) implies the compactness of the operator N_q; compare with Proposition 11.16 and [91]. For this purpose, we consider the following version of the Kohn–Morrey formula:

$$\int_\Omega \sum_{|K|=q-1}' \sum_{j,k=1}^{n} \frac{\partial^2 \varphi_M}{\partial z_j \partial \overline{z}_k} u_{jK} \overline{u}_{kK}\, e^{-\varphi_M}\, d\lambda \le \|\overline{\partial} u\|^2_{\varphi_M} + \|\overline{\partial}^*_{\varphi_M} u\|^2_{\varphi_M}, \tag{11.47}$$

where $u \in \operatorname{dom}(\overline{\partial}) \cap \operatorname{dom}(\overline{\partial}^*)$ is a $(0, q)$-form, and follow the lines of Proposition 11.16 (see also [91, Theorem 4.8]).

(b) In view of Remark 11.6(a), it is interesting to compare the Sobolev embedding

$$W^1(\Omega) \subset L^q(\Omega), \quad q \in [1, 2n/(n-2)),$$

where the derivatives are taken with respect to the real variables $x_j = \Re z_j$ and $y_j = \Im z_j$ for $j = 1, \dots, n$, with the embedding of the space

$$\mathcal{W}(\Omega) := \{u \in L^2_{(0,1)}(\Omega) : u \in \operatorname{dom}(\overline{\partial}) \cap \operatorname{dom}(\overline{\partial}^*)\},$$

endowed with graph norm, into $L^2_{(0,1)}(\Omega)$. We have the following result: If $\Omega \subset\subset \mathbb{C}^n$ is a smoothly bounded pseudoconvex domain and

$$\|u\|_q \le C(\|\overline{\partial} u\|^2 + \|\overline{\partial}^* u\|^2)^{1/2} \tag{11.48}$$

for some $q > 2$ and all $u \in \operatorname{dom}(\overline{\partial}) \cap \operatorname{dom}(\overline{\partial}^*)$, then the $\overline{\partial}$-Neumann operator

$$N : L^2_{(0,1)}(\Omega) \longrightarrow L^2_{(0,1)}(\Omega)$$

is compact.

To show this, we have to check that the unit ball in $\mathcal{W}(\Omega)$ is precompact in $L^2_{(0,1)}(\Omega)$. By Theorem 11.36 we have to show that for each $\epsilon > 0$, there exists $\omega \Subset \Omega$ such that

$$\int_{\Omega \setminus \omega} |u(z)|^2\, d\lambda(z) < \epsilon^2$$

for all u in the unit ball of $\mathcal{W}(\Omega)$.

By (11.48) and Hölder's inequality we have

$$\Big(\int\limits_{\Omega\setminus\omega}|u(z)|^2\,d\lambda(z)\Big)^{\frac{1}{2}} \le \Big(\int\limits_{\Omega\setminus\omega}|u(z)|^q\,d\lambda(z)\Big)^{\frac{1}{q}}\cdot|\Omega\setminus\omega|^{\frac{1}{2}-\frac{1}{q}}$$
$$\le C\,|\Omega\setminus\omega|^{\frac{1}{2}-\frac{1}{q}}.$$

Now we can choose $\omega \subset\subset \Omega$ such that the last term is $<\epsilon$.

In a similar way as for Proposition 11.27, we obtain compactness estimates for the $\overline{\partial}$-Neumann operator on a smoothly bounded domain. Here we use the standard Sobolev spaces $W^1(\Omega)$ and the classical Rellich–Kondrachov lemma without weights.

Proposition 11.38. *Let Ω be a smoothly bounded pseudoconvex domain. Then the following statements are equivalent.*

(1) *The $\overline{\partial}$-Neumann operator N_1 is a compact operator from $L^2_{(0,1)}(\Omega)$ into itself.*

(2) *The embedding of the space* $\mathrm{dom}(\overline{\partial})\cap\mathrm{dom}(\overline{\partial}^*)$, *provided with the graph norm $u\mapsto(\|u\|^2+\|\overline{\partial}u\|^2+\|\overline{\partial}^*u\|^2)^{1/2}$, into $L^2_{(0,1)}(\Omega)$ is compact.*

(3) *For every positive ϵ, there exists a constant C_ϵ such that*

$$\|u\|^2 \le \epsilon(\|\overline{\partial}u\|^2+\|\overline{\partial}^*u\|^2)+C_\epsilon\|u\|^2_{-1}$$

for all $u\in\mathrm{dom}(\overline{\partial})\cap\mathrm{dom}(\overline{\partial}^*)$.

(4) *For every positive ϵ, there exists $\omega\Subset\Omega$ such that*

$$\int\limits_{\Omega\setminus\omega}|u(z)|^2\,d\lambda(z)\le\epsilon(\|\overline{\partial}u\|^2+\|\overline{\partial}^*u\|^2)$$

for all $u\in\mathrm{dom}(\overline{\partial})\cap\mathrm{dom}(\overline{\partial}^*)$.

(5) *The operators*

$$\overline{\partial}^*N_1 : L^2_{(0,1)}(\Omega)\cap\ker(\overline{\partial})\longrightarrow L^2(\Omega)\quad\text{and}$$
$$\overline{\partial}^*N_2 : L^2_{(0,2)}(\Omega)\cap\ker(\overline{\partial})\longrightarrow L^2_{(0,1)}(\Omega)$$

are both compact.

11.7 Notes

In this chapter, we tried to point out some parallel aspects in the classical theory of Sobolev spaces and in the properties of the $\overline{\partial}$-Neumann operator. The treatment of compact subsets in L^2-spaces stems from [1], and basic properties of Sobolev spaces and the Rellich–Kondrachov lemma are taken from [12] and [27]. We refer the reader to these books for more general results on higher derivatives and L^p-spaces. The sufficient condition for the compactness of the $\overline{\partial}$-Neumann operator in weighted L^2-spaces was originally discovered from the theory of Schrödinger operators with magnetic field and from the Witten Laplacian (see [46]); here we presented the complex analytic approach from [42] and [36]. We also discuss necessary conditions for compactness, where we use a corresponding result on Schrödinger operators with magnetic fields [59] and methods; from complex analysis; see [6], [7], and [22], where there are many other results concerning the question of compactness. A different proof of the compactness criterium in the Segal–Bargmann space can be found in [25]. A thorough treatment of the compactness of the $\overline{\partial}$-Neumann operator for bounded pseudoconvex domains can be found in the book of Straube [91], with all its important implications concerning regularity of the $\overline{\partial}$-Neumann operator in Sobolev spaces and far-reaching connections to potential theory. A characterization of the compactness of the $\overline{\partial}$-Neumann operator in terms of geometric properties of pseudoconvex domains or in terms of the weight functions is still open. Fu and Straube [33] solved the problem for bounded convex domains. It turned out that in this case, the compactness of the $\overline{\partial}$-Neumann operator on the level of holomorphic $(0,q)$-forms already implies the compactness on the whole $L^2_{(0,q)}(\Omega)$. In counterexamples to compactness, the obstruction often already occurs on the level of holomorphic $(0,q)$-forms. We will draw attention to this phenomenon in the next chapters.

12 The $\overline{\partial}$-Neumann operator and the Bergman projection

In this chapter, we investigate the connection between the $\overline{\partial}$-Neumann operator and commutators of the Bergman projection with multiplication operators. In [14], it is shown that the compactness of the $\overline{\partial}$-Neumann operator N on $L^2_{(0,1)}(\Omega)$ implies the compactness of the commutator $[P, M]$, where P is the Bergman projection, and M is a pseudodifferential operator of order 0. Here we show that the compactness of the $\overline{\partial}$-Neumann operator N restricted to $(0, 1)$-forms with holomorphic coefficients is equivalent to the compactness of the commutator $[P, M]$ defined on the whole $L^2(\Omega)$. In addition, we derive a formula for the $\overline{\partial}$-Neumann operator restricted to $(0, 1)$-forms with holomorphic coefficients expressed by commutators of the Bergman projection and the multiplication operators by z and $\bar{z}$. To show the equivalences to the compactness mentioned above, we will need a complex version of the Stone–Weierstraß theorem.

12.1 The Stone–Weierstraß theorem

In the sequel, we will need a generalization of the well-known theorem of Weierstraß that any real-valued continuous function on a compact interval $[a, b]$ is the uniform limit of a sequence of polynomials on $[a, b]$. We will consider a compact subset K in $\mathbb{C}^n$ and will show that any complex-valued continuous function on K can be uniformly approximated on K by polynomials in the variables $z = (z_1, \dots, z_n)$ and $\overline{z} = (\overline{z}_1, \dots, \overline{z}_n)$. We prove a more general version and will need some appropriate notions and several lemmas.

A subset $\mathcal{A}$ of the space $\mathcal{C}(K)$ of continuous real-valued or complex-valued functions (endowed with the sup-norm) is said to separate points if for every $z, w \in K$ such that $z \neq w$, there exists $f \in \mathcal{A}$ such that $f(z) \neq f(w)$. We call $\mathcal{A}$ an algebra if it is a real (complex) vector subspace of $\mathcal{C}(K)$ such that $fg \in \mathcal{A}$ (pointwise multiplication) for all $f, g \in \mathcal{A}$. If $\mathcal{A}$ is a subset of the space $\mathcal{C}(K)$ of real-valued continuous functions, then $\mathcal{A}$ is called a lattice if $\max(f, g)$ and $\min(f, g)$ are in $\mathcal{A}$ for all $f, g \in \mathcal{A}$. Since the algebra and lattice operations are continuous in sup-norm of $\mathcal{C}(K)$, we easily sees that if $\mathcal{A}$ is an algebra or a lattice, then is its the closure $\overline{\mathcal{A}}$ in the sup-norm.

Lemma 12.1. *For each $\epsilon > 0$, there is a polynomial p on $\mathbb{R}$ such that $p(0) = 0$ and*

$$\big| |x| - p(x) \big| < \epsilon \quad \textit{for } x \in [-1, 1].$$

Proof. For $|t| < 1$, the function $(1 - t)^{1/2}$ has the Taylor series expansion

$$(1 - t)^{1/2} = 1 + \sum_{k=1}^{\infty} \frac{(-1)}{2} \frac{1}{2} \cdots \frac{2k - 3}{2} \frac{t^k}{k!}$$

https://doi.org/10.1515/9783111182926-012

$$= 1 - \sum_{k=1}^{\infty} c_k t^k$$

with $c_k > 0$. It is also absolutely convergent for $t = -1$. The monotone convergence theorem applied to the counting measure on $\mathbb{N}$ implies that

$$\sum_{k=1}^{\infty} c_k = \lim_{t\to 1} \sum_{k=1}^{\infty} c_k t^k = 1 - \lim_{t\to 1}(1-t)^{1/2} = 1,$$

from which we get that the series $1 - \sum_{k=1}^{\infty} c_k t^k$ converges absolutely and uniformly on $[-1, 1]$ to $(1-t)^{1/2}$.

Given $\epsilon > 0$, we can find a suitable partial sum $q(t)$ such that

$$\left|(1-t)^{1/2} - q(t)\right| < \epsilon/2, \quad t \in [-1, 1].$$

We set $t = 1 - x^2$ and $r(x) = q(1 - x^2)$ and obtain a polynomial r such that

$$\big| |x| - r(x) \big| < \epsilon/2, \quad x \in [-1, 1].$$

In particular, $|r(0)| < \epsilon/2$, so setting $p(x) = r(x) - r(0)$, we get the desired polynomial. □

First, we consider the space of real-valued continuous functions.

Lemma 12.2. *Let $\mathcal{A}$ be a closed subalgebra of the space $C(K)$ of real-valued continuous functions. Then $|f| \in \mathcal{A}$ whenever $f \in \mathcal{A}$, and $\mathcal{A}$ is a lattice.*

Proof. We write $\|f\| = \sup_{z\in K} |f(z)|$ for $f \in C(K)$. If $f \in \mathcal{A}$ and $f \neq 0$, then $h := f/\|f\|$ maps K into $[-1, 1]$. If p is as in Lemma 12.1, then we have

$$\big\| |h| - p \circ h \big\| < \epsilon.$$

Since $p(0) = 0$, p has no constant term, and therefore $p \circ h \in \mathcal{A}$, as $\mathcal{A}$ is an algebra. Since $\mathcal{A}$ is closed, it follows that $|h| \in \mathcal{A}$ and $|f| = \|f\| \, |h| \in \mathcal{A}$. This proves the first statement. The second follows from

$$\max(f, g) = \frac{1}{2}(f + g + |f - g|), \quad \min(f, g) = \frac{1}{2}(f + g - |f - g|).$$

□

Lemma 12.3. *Let $\mathcal{A}$ be a closed lattice in the space $C(K)$ of real-valued continuous functions. Let $f \in C(K)$ and suppose that for all $z, w \in K$, there exists $g_{zw} \in \mathcal{A}$ such that $f(z) = g_{zw}(z)$ and $f(w) = g_{zw}(w)$. Then $f \in \mathcal{A}$.*

Proof. For $\epsilon > 0$ and $z, w \in K$, let

$$U_{zw} = \{\zeta \in K : f(\zeta) < g_{zw}(\zeta) + \epsilon\}, \quad V_{zw} = \{\zeta \in K : f(\zeta) > g_{zw}(\zeta) - \epsilon\}.$$

These sets are open and contain z and w. Fix $w \in K$. Then $\{U_{zw}\}_{z\in K}$ covers K, so there is a finite subcover $\{U_{z_j w}\}_{j=1}^n$. Let $g_w = \max(g_{z_1 w}, \ldots, g_{z_n w})$. Then $f < g_w + \epsilon$ on K, and $f > g_w - \epsilon$ on $V_w = \bigcap_{j=1}^n V_{z_j w}$, which is open and contains w. So $\{V_w\}_{w\in K}$ is an open covering of K, which has a finite subcover $\{V_{w_k}\}_{k=1}^m$. Let $g = \min(g_{w_1}, \ldots, g_{w_m})$. Then $\|f - g\| < \epsilon$, and since $\mathcal{A}$ is a closed lattice, $g \in \mathcal{A}$ and $f \in \mathcal{A}$.

Before we prove the first version of the Stone–Weierstraß theorem, we remark that $\mathbb{R}^2$ as an algebra under coordinatewise addition and multiplication has only the following subalgebras: $\mathbb{R}^2$, $\{(0,0)\}$, and the linear spans of $(1,0), (0,1)$, and $(1,1)$, because a nonzero subalgebra $\mathcal{A} \subset \mathbb{R}^2$ has the property that with $(0,0) \neq (a,b) \in \mathcal{A}$, the pair $(a^2, b^2) \in \mathcal{A}$, so if $a \neq 0, b \neq 0$, and $a \neq b$, then (a,b) and (a^2,b^2) are linearly independent, so $\mathcal{A} = \mathbb{R}^2$. The other cases – $a \neq 0 = b, a = 0 \neq b$, and $a = b \neq 0$ – give the other three subalgebras.

With this in mind, we can now prove the first version of the Stone–Weierstraß theorem.

Theorem 12.4. *If $\mathcal{A}$ is a closed subalgebra of the space $C(K)$ of real-valued continuous functions that separates points, then either $\mathcal{A} = C(K)$, or $\mathcal{A} = \{f \in C(K) : f(z_0) = 0\}$ for some $z_0 \in K$. The first alternative holds if and only if $\mathcal{A}$ contains the constant functions.*

Proof. For $z, w \in K$ such that $z \neq w$, let $\mathcal{A}_{zw} = \{(f(z), f(w)) : f \in \mathcal{A}\}$. Then since $f \mapsto (f(z), f(w))$ is an algebra homomorphism, $\mathcal{A}_{zw}$ is a subalgebra of $\mathbb{R}^2$ in the sense of the remark from above. If $\mathcal{A}_{zw} = \mathbb{R}^2$ for all $z, w \in K$, then Lemmas 12.1 and 12.2 imply that $\mathcal{A} = C(K)$. Otherwise, there exist $z, w \in K$ for which $\mathcal{A}_{zw}$ is a proper subalgebra of $\mathbb{R}^2$. It cannot be $\{(0,0)\}$ or the linear span of $(1,1)$ because $\mathcal{A}$ separates points. So $\mathcal{A}_{zw}$ is the linear span of $(1,0)$ or $(0,1)$. In either case, there exists $z_0 \in K$ such that $f(z_0) = 0$ for all $f \in \mathcal{A}$. There is only one such z_0 since $\mathcal{A}$ separates points, so if neither z nor w is z_0, then $\mathcal{A}_{zw} = \mathbb{R}^2$. Then Lemmas 12.1 and 12.2 imply that $\mathcal{A} = \{f \in C(K) : f(z_0) = 0\}$. If $\mathcal{A}$ contains constant functions, then this cannot be the case, so then $\mathcal{A}$ equals $C(K)$. □

The set of polynomials is an algebra that separates points and contains constants. Hence we get the following:

Corollary 12.5. *The restrictions of the real polynomial functions to K are dense in the space $C(K)$ of real-valued continuous functions.* □

Theorem 12.4 is false for the space $C(K)$ of complex-valued continuous functions. Nevertheless, there is a complex version of the Stone–Weierstraß theorem.

Theorem 12.6. *If $\mathcal{A}$ is a closed complex subalgebra of the space $C(K)$ of complex-valued continuous functions that separates points and is closed under complex conjugation, then either $\mathcal{A} = C(K)$, or $\mathcal{A} = \{f \in C(K) : f(z_0) = 0\}$ for some $z_0 \in K$.*

Proof. Since $\Re f = (f + \bar f)/2$ and $\Im f = (f - \bar f)/2i$, the set $\mathcal{A}_{\mathbb{R}}$ of real and imaginary parts of functions in $\mathcal{A}$ is a subalgebra of the space of real-valued continuous functions to which Theorem 12.4 applies. Since $\mathcal{A} = \{f + ig : f, g \in \mathcal{A}_{\mathbb{R}}\}$, the desired result follows. □

Corollary 12.7. *The set of all polynomials p of the form*

$$p(z,\overline{z}) = \sum_{|\alpha|\leq N} \lambda_\alpha z^{\alpha_1}\overline{z}^{\alpha_2},$$

where $\alpha = (\alpha_1,\alpha_2)$ is a multi-index in $\mathbb{N}^{2n}$, is dense in the space $C(K)$ of complex-valued continuous functions.

Proof. It is clear that the set of polynomials p of the above form separates points and is invariant under complex conjugation. Therefore we can apply Theorem 12.6. □

12.2 Commutators of the Bergman projection

Let Ω be a bounded pseudoconvex domain in $\mathbb{C}^n$, and let $A^2_{(0,1)}(\Omega)$ denote the space of all $(0,1)$-forms with holomorphic coefficients belonging to $L^2(\Omega)$. In Section 3.2, we showed that the canonical solution operator $S : A^2_{(0,1)}(\Omega) \longrightarrow L^2(\Omega)$ has the form

$$S(g)(z) = \int_\Omega K(z,w)\langle g(w), z-w\rangle \, d\lambda(w), \tag{12.1}$$

where K denotes the Bergman kernel of Ω, $g = \sum_{j=1}^n g_j \, d\overline{z}_j$ is a $(0,1)$-form belonging to $A^2_{(0,1)}(\Omega)$, and

$$\langle g(w), z-w\rangle = \sum_{j=1}^n g_j(w)(\overline{z}_j - \overline{w}_j)$$

for $z = (z_1,\dots,z_n)$ and $w = (w_1,\dots,w_n)$.

The restriction of the canonical solution operator to forms with holomorphic coefficients has many interesting aspects, which in most cases correspond to certain growth properties of the Bergman kernel. It is also of great interest to clarify to what extent the compactness of the restriction already implies the compactness of the original solution operator to $\overline{\partial}$. This is the case for convex domains; see [33]. There are many other examples of noncompactness where the obstruction already occurs for forms with holomorphic coefficients (see Chapter 14 and [68, 70]).

We define the operator

$$T : L^2_{(0,1)}(\Omega) \longrightarrow L^2(\Omega)$$

by

$$Tf(z) = \int_\Omega K(z,w)\langle f(w), z-w\rangle \, d\lambda(w), \tag{12.2}$$

where $f = \sum_{k=1}^n f_k \, d\overline{z}_k$ and $\langle f(w), z - w \rangle = \sum_{k=1}^n f_k(w)(\overline{z}_k - \overline{w}_k)$.

The operator T can be written as a sum of commutators

$$Tf = \sum_{k=1}^n [\overline{M_k}, P] f_k, \quad f = \sum_{k=1}^n f_k \, d\overline{z}_k, \tag{12.3}$$

where $\overline{M_k} v(z) = \overline{z}_k v(z)$, $v \in L^2(\Omega)$, $k = 1, \dots, n$, and P is the Bergman projection; see (1.5).

Let $\mathcal{P} : L^2_{(0,1)}(\Omega) \longrightarrow A^2_{(0,1)}(\Omega)$ be the orthogonal projection on the space of $(0,1)$-forms with holomorphic coefficients. We claim that

$$Tf = T\mathcal{P}f, \quad f \in L^2_{(0,1)}(\Omega).$$

It suffices to show that $Tg = 0$ for $g \perp A^2_{(0,1)}(\Omega)$:

$$\begin{aligned} Tg(z) &= -\sum_{k=1}^n P\overline{M_k} g_k(z) = -\sum_{k=1}^n \int_\Omega K(z, w) \overline{w}_k g_k(w) \, d\lambda(w) \\ &= -\sum_{k=1}^n \int_\Omega g_k(w) \, [K(w, z) w_k]^{-} \, d\lambda(w) = 0, \end{aligned}$$

because $w \mapsto K(w, z) w_k$ is holomorphic and $g_k \perp A^2(\Omega)$ for $k = 1, \dots, n$.

Now let S denote the canonical solution operator to $\overline{\partial}$ restricted to $A^2_{(0,1)}(\Omega)$. From (12.1) we have, for $f \in L^2_{(0,1)}(\Omega)$,

$$S(\mathcal{P}f) = T(\mathcal{P}f) = Tf. \tag{12.4}$$

Hence we have proved the following:

Theorem 12.8. *If $f \in L^2_{(0,1)}(\Omega)$, then $T(\mathcal{P}f) = Tf$. The operator S is compact as an operator from $A^2_{(0,1)}(\Omega)$ to $L^2(\Omega)$ if and only if the operator T is compact as an operator from $L^2_{(0,1)}(\Omega)$ to $L^2(\Omega)$.*

Remark 12.9. The adjoint operator $T^* : L^2(\Omega) \longrightarrow L^2_{(0,1)}(\Omega)$ is given by

$$T^*(g) = \sum_{k=1}^n [P, M_k] \, g \, d\overline{z}_k, \ g \in L^2(\Omega), \tag{12.5}$$

where $M_k v(z) = z_k v(z)$.

Here we have

$$T^*(I - P)(g) = T^*(g),$$

since

$$[P, M_k] \, Pg = PM_k Pg - M_k Pg = 0.$$

In a similar way, the following results can be proved.

Lemma 12.10.

(1) $PM_jP = M_jP$,

(2) $P\overline{M_j}P = P\overline{M_j}$.

Let

$$B^2_{(0,1)}(\Omega) = \{f \in L^2_{(0,1)}(\Omega) : f \in \ker \overline{\partial}\}.$$

Now suppose that Ω is a bounded pseudoconvex domain in $\mathbb{C}^n$. The $\overline{\partial}$-Neumann operator N can be viewed as an operator from $B^2_{(0,1)}(\Omega)$ to $B^2_{(0,1)}(\Omega)$. The operator

$$\overline{\partial}^* N : B^2_{(0,1)}(\Omega) \longrightarrow A^2(\Omega)^{\perp}$$

is the canonical solution operator to $\overline{\partial}$ (see Proposition 4.64).

Theorem 12.11. *If $f = \sum_{k=1}^n f_k d\overline{z}_k \in B^2_{(0,1)}(\Omega)$, then*

$$\mathcal{P}N\mathcal{P}f = \sum_{k=1}^n \left(\sum_{j=1}^n (PM_k\overline{M}_jPf_j - M_kP\overline{M}_jf_j) \right) d\overline{z}_k. \tag{12.6}$$

If $f = \sum_{k=1}^n f_k d\overline{z}_k \in A^2_{(0,1)}(\Omega)$, then

$$\mathcal{P}Nf = \sum_{k=1}^n [P, M_k] \left(\sum_{j=1}^n \overline{M}_j f_j \right) d\overline{z}_k. \tag{12.7}$$

Proof. First, we observe that for $f \in B^2_{(0,1)}(\Omega)$, we have

$$N\overline{\partial}\overline{\partial}^* Nf = N(I - \overline{\partial}^*\overline{\partial}N)f = Nf,$$

where we used the fact that

$$N : B^2_{(0,1)}(\Omega) \longrightarrow B^2_{(0,1)}(\Omega).$$

If $f \in A^2_{(0,1)}(\Omega)$, then by Theorem 12.8 it follows that

$$\overline{\partial}^* Nf = Tf.$$

Let $f \in A^2_{(0,1)}(\Omega)$, and let $g \in B^2_{(0,1)}(\Omega)$ with orthogonal decomposition $g = h + \tilde{h}$, where $h \in A^2_{(0,1)}(\Omega)$ and $\tilde{h} = (I - \mathcal{P})g$. Then

$$(g, N\overline{\partial}\overline{\partial}^* Nf) = (\overline{\partial}^* N(h + \tilde{h}), Tf) = (\overline{\partial}^* Nh, Tf) + (\overline{\partial}^* N\tilde{h}, Tf)$$

$$
\begin{aligned}
&= (Th, Tf) + (\overline{\partial}^* N\tilde{h}, Tf) = (Tg, Tf) + (\overline{\partial}^* N\tilde{h}, Tf) \\
&= (g, T^* Tf) + (\overline{\partial}^* N\tilde{h}, Tf).
\end{aligned}
$$

Since

$$
(\overline{\partial}^* N\tilde{h}, Tf) = (N\tilde{h}, \overline{\partial} Tf) = (N\tilde{h}, f) = (\tilde{h}, Nf),
$$

we obtain

$$
\begin{aligned}
(g, Nf) &= (g, N\overline{\partial}\overline{\partial}^* Nf) = (g, T^* Tf) + (\tilde{h}, Nf) \\
&= (g, T^* Tf) + ((I - \mathcal{P})g, Nf) = (g, T^* Tf) + (g, (I - \mathcal{P})Nf).
\end{aligned}
$$

Now since $g \in B^2_{(0,1)}(\Omega)$ was arbitrary, we get

$$
Nf = T^* Tf + Nf - \mathcal{P}Nf,
$$

and therefore

$$
\mathcal{P}Nf = T^* Tf.
$$

If we take into account that for $f \in B^2_{(0,1)}(\Omega)$, we have $Tf = T\mathcal{P}f$, we can now apply the last formula for $\mathcal{P}f$ and get

$$
\mathcal{P}N\mathcal{P}f = T^* Tf.
$$

It remains to compute $T^* T$. If $f \in B^2_{(0,1)}(\Omega)$, then

$$
\begin{aligned}
T^* Tf &= \sum_{k=1}^{n} [P, M_k]\left(\sum_{j=1}^{n} [\overline{M}_j, P] f_j \right) d\overline{z}_k \\
&= \sum_{k=1}^{n} \left(\sum_{j=1}^{n} (PM_k\overline{M}_j P - M_k P\overline{M}_j P - PM_k P\overline{M}_j + M_k P\overline{M}_j) f_j \right) d\overline{z}_k \\
&= \sum_{k=1}^{n} \left(\sum_{j=1}^{n} (PM_k\overline{M}_j Pf_j - M_k P\overline{M}_j f_j) \right) d\overline{z}_k,
\end{aligned}
$$

where we used Lemma 12.10.

If $f \in A^2_{(0,1)}(\Omega)$, then

$$
Pf_j = f_j,
$$

and we obtain the second formula in Theorem 12.11. □

Using the last results, we get a criterion for the compactness of commutators $[P, M_k]$.

Theorem 12.12. *Let Ω be a bounded pseudoconvex domain in $\mathbb{C}^n$. Then the following conditions are equivalent:*
(1) $N|_{A^2_{(0,1)}(\Omega)}$ *is compact;*
(2) $\overline{\partial}^* N|_{A^2_{(0,1)}(\Omega)}$ *is compact;*
(3) $[P, M_k]$ *is compact on* $L^2(\Omega)$ *for* $k = 1, \ldots, n$*;*
(4) $(I - P)\overline{M}_k P$ *is compact on* $L^2(\Omega)$ *for* $k = 1, \ldots, n$*;*
(5) $[M_\varphi, P]$ *is compact on* $L^2(\Omega)$ *for each continuous function* φ *on* $\overline{\Omega}$.

Proof. Let $S_1 = \overline{\partial}^* N_1 : B^2_{(0,1)}(\Omega) \longrightarrow A^2(\Omega)^\perp$ be the canonical solution operator to $\overline{\partial}$, and similarly $S_2 = \overline{\partial}^* N_2 : B^2_{(0,2)}(\Omega) \longrightarrow B^2_{(0,1)}(\Omega)^\perp$. Then

$$N_1 = S_1^* S_1 + S_2 S_2^*$$

(see Proposition 4.64). Since $S_2^*|_{A^2_{(0,1)}(\Omega)} = 0$, we have

$$N_1|_{A^2_{(0,1)}(\Omega)} = S_1^* S_1|_{A^2_{(0,1)}(\Omega)},$$

and (1) is equivalent to (2).

Now suppose that (2) holds. Then since the restriction of $\overline{\partial}^* N$ to $A^2_{(0,1)}(\Omega)$ is of the form

$$\overline{\partial}^* N f = \sum_{k=1}^{n} [\overline{M_k}, P] f_k,$$

where $f = \sum_{k=1}^n f_k \, dz_k \in A^2_{(0,1)}(\Omega)$, by Theorem 12.8 it follows that the operators $[\overline{M_k}, P]$ are compact on $L^2(\Omega)$. Since $[\overline{M_k}, P]^* = [P, M_k]$, we obtain property (3).

It is also clear by Theorem 12.8 that (3) implies (2).

Now suppose that (3) holds. It follows that $[\overline{M_k}, P]P$ is also compact, and since

$$[\overline{M_k}, P]P = \overline{M_k}P - P\overline{M_k}P = (I - P)\overline{M_k}P,$$

the Hankel operators $(I - P)\overline{M_k}P$ are compact. So we have shown that (3) implies (4).

Suppose that (4) holds. The Hankel operators $H_{z_j \overline{z_k}}$ with symbol $z_j \overline{z_k}$ can be written in the form

$$H_{z_j \overline{z_k}} = (I - P)M_j(P + (I - P))\overline{M_k}P = (I - P)M_j(I - P)\overline{M_k}P,$$

and hence it follows that $H_{z_j \overline{z_k}}$ is compact. Similarly, we can show that for any polynomial

$$p(z, \overline{z}) = \sum_{|\alpha| \le N} \lambda_\alpha z^{\alpha_1} \overline{z}^{\alpha_2},$$

where $\alpha = (\alpha_1, \alpha_2)$ is a multi-index in $\mathbb{N}^{2n}$, the corresponding Hankel operator $H_p = (I-P)M_pP$ is compact. Now let $\varphi \in C(\overline{\Omega})$. Then by Corollary 12.7 there exists a polynomial p of the above form such that

$$\|\varphi - p\|_\infty < \epsilon.$$

Hence

$$\|H_\varphi - H_p\| = \|(I-P)M_{\varphi-p}P\| \leq \|\varphi - p\|_\infty,$$

where $\|\cdot\|$ denotes the operator norm, and $\|\cdot\|_\infty$ denote the sup-norm on $\overline{\Omega}$.

Since the compact operators form a closed two-sided ideal in the operator norm (Corollary 2.4) and since for $g = g_1 + g_2$ with $g_1 \in A^2(\Omega)$ and $g_2 \in A^2(\Omega)^\perp$, we have

$$[M_\varphi, P]g = -H_{\overline{\varphi}}^* g_2 + H_\varphi g_1,$$

it follows that $[M_\varphi, P]$ is compact. □

Remark 12.13.

(a) If Ω is a bounded convex domain, then the compactness of

$$\overline{\partial}^* N|_{A^2_{(0,1)}(\Omega)}$$

implies already the compactness of $\overline{\partial}^* N$ on all of $L^2_{(0,1)}(\Omega)$ (see [33]), and hence in this case, property (1) of Theorem 12.12 can be replaced by the compactness of N on $L^2_{(0,1)}(\Omega)$, and property (2) of Theorem 12.12 can be replaced by the compactness of $\overline{\partial}^* N$ on $L^2_{(0,1)}(\Omega)$.

(b) For $n \geq 2$ and $1 \leq q \leq n$, let

$$B^2_{(0,q)}(\Omega) = \{f \in L^2_{(0,q)}(\Omega) : f \in \ker \overline{\partial}\}.$$

From Proposition 4.65 we know that the orthogonal projection

$$P_q : L^2_{(0,q)}(\Omega) \longrightarrow B^2_{(0,q)}(\Omega)$$

can be written in the form

$$P_q = I - \overline{\partial}^* N_{q+1} \overline{\partial}.$$

Recently, Celik and Sahutoglu [15] showed that for $2 \leq q \leq n$, the following statements are equivalent:

(1) N_q is compact on $L^2_{(0,q)}(\Omega)$,

(2) $\overline{\partial}^* N_q$ is compact on $B^2_{(0,q)}(\Omega)$,

(3) $[M_\varphi, P_q]$ is compact on $B^2_{(0,q)}(\Omega)$ for each $\varphi \in C(\overline{\Omega})$.

This is mainly due to the fact that $B^2_{(0,q)}(\Omega)$ is a much larger space than the space of $(0,q)$-forms with holomorphic coefficients belonging to $L^2(\Omega)$.

Whether the same is true for $q = 1$ is still not known; compare with Theorem 12.12, where in the first statement, we have the compactness only on $A^2_{(0,1)}(\Omega)$.

12.3 Notes

The proof of the Stone–Weierstraß theorem stems from [29]. Salinas [86] showed that the compactness of the canonical solution operator $\overline{\partial}^* N_{q+1}$ restricted to forms with holomorphic coefficients is a consequence of the compactness of the commutators $[\overline{M}_j, P_q]$, $1 \leq j \leq n$. The compactness of N_q for $1 \leq q \leq n$ can also be characterized in terms of C^* algebras (see [86, 87] and [15]). Catlin and D'Angelo [14] used the compactness of the commutators $[M_\varphi, P]$ in conjunction with a complex-variable analogue of Hilbert's 17th problem. They showed that the compactness of N_1 implies that the commutators $[M, P]$ are compact for all tangential pseudodifferential operators M of order 0. Theorems 12.11 and 12.12 are from [41].

13 Compact resolvents

In this chapter, we return to the differential operators introduced in Chapter 10 and discuss the question whether these operators are with compact resolvents. For this purpose, we use the general characterization of the compactness of the $\overline{\partial}$-Neumann operator.

13.1 Schrödinger operators

Here we characterize the compactness of the $\overline{\partial}$-Neumann operator N_φ on $L^2(\mathbb{C}, e^{-\varphi})$, which was originally done in [46] using methods from Schrödinger operators and later in [74] using estimates of the Bergman kernel in $A^2(\mathbb{C}, e^{-\varphi})$. We follow the method from [46], showing that the necessary condition for compactness (11.22) is also sufficient for $n = 1$ if we suppose that the Laplacian of the weight function belongs to the reverse Hölder class $B_2(\mathbb{R}^2)$ consisting of positive and almost nonzero everywhere L^2 functions V for which there exists a constant $C > 0$ such that

$$\left(\frac{1}{|B|}\int_B V^2\,d\lambda\right)^{1/2} \leq C\left(\frac{1}{|B|}\int_B V\,d\lambda\right) \tag{13.1}$$

for any ball B in $\mathbb{R}^2$. Note that any positive (nonzero) polynomial is in $B_2(\mathbb{R}^2)$. Before we proceed characterizing the compactness of N_φ on $L^2(\mathbb{C}, e^{-\varphi})$, we collect important properties of functions belonging to the reverse Hölder class.

A function $w(x) \geq 0$ is called an A_1 weight if

$$[w]_{A_1} = \sup_{Q \text{ cubes in } \mathbb{R}^2}\left(\frac{1}{|Q|}\int_Q w(x)\,d\lambda(x)\right)\|w^{-1}\|_{L^\infty(Q)} < \infty;$$

for $1 < p < \infty$, a weight w is said to be of class A_p if

$$[w]_{A_p} = \sup_{Q \text{ cubes in } \mathbb{R}^2}\left(\frac{1}{|Q|}\int_Q w(x)\,d\lambda(x)\right)\left(\frac{1}{|Q|}\int_Q w(x)^{-\frac{1}{p-1}}\,d\lambda(x)\right)^{p-1} < \infty;$$

a weight w is called an A_∞ weight if

$$[w]_{A_\infty} = \sup_{Q \text{ cubes in } \mathbb{R}^2}\left(\frac{1}{|Q|}\int_Q w(x)\,d\lambda(x)\right)\exp\left(\frac{1}{|Q|}\int_Q \log w(x)^{-1}\,d\lambda(x)\right) < \infty;$$

the quantity $[w]_{A_p}$ is called the A_p Muckenhoupt characteristic constant of w. It follows that for all $1 \leq p < \infty$, we have

https://doi.org/10.1515/9783111182926-013

$$[w]_{A_\infty} \le [w]_{A_p} \quad \text{and} \quad \bigcup_{1\le p<\infty} A_p \subseteq A_\infty.$$

Using Theorem 7.3.3 of [37], we in fact see that

$$\bigcup_{1\le p<\infty} A_p = A_\infty$$

and that $w \in A_\infty$ if and only if the reverse Hölder condition holds for w, i. e., there exist $0 < C_1, \epsilon < \infty$ such that for all cubes Q, we have

$$\left(\frac{1}{|Q|}\int_Q w(x)^{1+\epsilon}\,d\lambda(x)\right)^{\frac{1}{1+\epsilon}} \le \frac{C_1}{|Q|}\int_Q w(x)\,d\lambda(x).$$

In [37], we find the following properties of functions in A_∞, which will be important for the compactness of N_φ.

Lemma 13.1. *Let $w \in A_\infty$. Then*
(1) $[Cw]_{A_\infty} = [w]_{A_\infty}$ *for all* $C > 0$;
(2) $[w]_{A_\infty} \ge 1$;
(3) $[w]_{A_\infty} = \sup\{\frac{w(Q)}{\int_Q |f(x)|w(x)d\lambda(x)}\exp(\frac{1}{|Q|}\int_Q \log|f(x)|d\lambda(x))\}$, *where the supremum is taken over all cubes Q in $\mathbb{R}^2$ and all functions f such that* $\log|f| \in L^1(Q)$ *and* $\int_Q |f|wd\lambda > 0$, *and where* $w(Q) = \int_Q w\,d\lambda$;
(4) *For all $C > 1$ and all cubes Q, we have*

$$w(CQ) \le 2^{C^2}[w]_{A_\infty}^{C^2}\,w(Q),$$

and we say that the measure $wd\lambda$ is doubling.

Proof. (1) immediately follows from the definition. To prove (2), we have to recall the consequence of Jensen's inequality

$$\left(\int_X |h(t)|\,d\mu(t)\right)^{1/q} \ge \exp\left(\int_X \log|h(t)|\,d\mu(t)\right),$$

which holds for all measurable functions h on a probability space (X, μ) and all $0 < q < \infty$. To prove (3), observe that by taking $f = w^{-1}$ the expression on the right of (3) is at least as large as $[w]_{A_\infty}$. From Jensen's inequality we get

$$\exp\left(\frac{1}{|Q|}\int_Q \log(|f(x)|w(x))\,d\lambda(x)\right) \le \frac{1}{|Q|}\int_Q |f(x)|w(x)\,d\lambda(x),$$

which can be written as

$$\frac{w(Q)}{\int_Q |f|w\,d\lambda}\exp\Big(\frac{1}{Q}\int\limits_Q \log|f|\,d\lambda\Big) \le \frac{w(Q)}{|Q|}\exp\Big(-\frac{1}{|Q|}\int\limits_Q \log|w|\,d\lambda\Big)$$

when f does not vanish almost everywhere on Q, and we get (3).

To show (4), we apply (3) to the cube CQ instead of Q and to the function $f(x) = c$ on Q and $f(x) = 1$ on $\mathbb{R}^2 \setminus Q$, where c is chosen such that $c^{1/C^n} = 2[w]_{A_\infty}$. We obtain

$$\frac{w(CQ)}{w(CQ \setminus Q) + cw(Q)}\exp\Big(\frac{\log c}{C^n}\Big) \le [w]_{A_\infty},$$

which implies (4). □

Lemma 13.2.

(a) *Let $w \in A_\infty$. Then there exists $\epsilon > 0$, independent of Q, such that $E = \{x \in Q : w(x) > \epsilon\,\mathrm{av}_Q w\}$ satisfies $|E| \ge \frac{1}{2}|Q|$, where $\mathrm{av}_Q w = |Q| \int_Q w\,d\lambda$.*

(b) *If $w \in A_\alpha$ for $1 \le \alpha < \infty$, then there exists a constant $C > 0$ such that*

$$\frac{\mathrm{av}_E w}{\mathrm{av}_Q w} \ge C\Big(\frac{|E|}{|Q|}\Big)^{\alpha-1}$$

for each measurable subset E of Q.

Proof. (a) We show that there exists $\epsilon > 0$, independent of Q, such that

$$\big|\{x \in Q : w(x) \le \epsilon\,\mathrm{av}_Q w\}\big| \le \frac{1}{2}|Q|,$$

which implies (a). By Lemma 13.1(1) and the log-term in the definition of the characteristic $[w]_{A_\infty}$ we can assume that $\int_Q \log w\,d\lambda = 0$. Hence we have $\mathrm{av}_Q w \le [w]_{A_\infty}$, and we get

$$\begin{aligned}
\big|\{x \in Q : w(x) \le \epsilon\,\mathrm{av}_Q w\}\big| &\le \big|\{x \in Q : w(x) \le \epsilon\,[w]_{A_\infty}\}\big| \\
&= \big|\{x \in Q : \log(1 + w(x)^{-1}) \ge \log(1 + (\epsilon[w]_{A_\infty})^{-1})\}\big| \\
&\le \frac{1}{\log(1 + (\epsilon[w]_{A_\infty})^{-1})}\int\limits_Q \log\frac{1 + w(x)}{w(x)}\,d\lambda(x) \\
&= \frac{1}{\log(1 + (\epsilon[w]_{A_\infty})^{-1})}\int\limits_Q \log(1 + w(x))\,d\lambda(x) \\
&\le \frac{1}{\log(1 + (\epsilon[w]_{A_\infty})^{-1})}\int\limits_Q w(x)\,d\lambda(x) \\
&\le \frac{[w]_{A_\infty}|Q|}{\log(1 + (\epsilon[w]_{A_\infty})^{-1})} \\
&= \frac{1}{2}|Q|,
\end{aligned}$$

where $\epsilon = [w]_{A_\infty}^{-1}(e^{2[w]_{A_\infty}} - 1)^{-1}$.

To prove (b), we apply Hölder's inequality with exponents p and p' such that $\frac{1}{p} + \frac{1}{p'} = 1$, and for a suitable measurable function f, we get that

$$\begin{aligned}
\left(\frac{1}{|Q|}\int_Q |f(x)|\,d\lambda(x)\right)^p &= \left(\frac{1}{|Q|}\int_Q |f(x)|w(x)^{1/p}w(x)^{-1/p}\,d\lambda(x)\right)^p \\
&\le \frac{1}{|Q|^p}\left(\int_Q |f(x)|^p w(x)\,d\lambda(x)\right)\left(\int_Q w(x)^{-p'/p}\,d\lambda(x)\right)^{p/p'} \\
&= \left(\frac{1}{w(Q)}\int_Q |f(x)|^p w(x)\,d\lambda(x)\right)\left(\frac{1}{|Q|}\int_Q w(x)\,d\lambda(x)\right) \\
&\quad\times \left(\frac{1}{|Q|}\int_Q w(x)^{-\frac{1}{p-1}}\,d\lambda(x)\right)^{p-1} \\
&\le [w]_{A_p}\frac{1}{w(Q)}\int_Q |f(x)|^p w(x)\,d\lambda(x).
\end{aligned}$$

Now take for f the characteristic function of E. Then

$$\left(\frac{|E|}{|Q|}\right)^p \le [w]_{A_p}\frac{w(E)}{w(Q)},$$

which proves (b). □

Theorem 13.3. *Let φ be a subharmonic $\mathcal{C}^2$-function such that $\triangle\varphi(z) > \epsilon > 0$ and suppose that $\triangle\varphi \in B_2(\mathbb{R}^2)$. The $\overline{\partial}$-Neumann operator N_φ is compact on $L^2(\mathbb{C}, e^{-\varphi})$ if and only if*

$$\lim_{|z|\to\infty}\int_{Q_z} \triangle\varphi\,d\lambda = +\infty, \tag{13.2}$$

where $Q_z = \{w \in \mathbb{C} : |w - z| < 1\}$.

Proof. Suppose that N_φ is compact. In Chapter 10, we showed that $\Box_\varphi = e^{\varphi/2}\overline{D}\,\overline{D}^*\,e^{-\varphi/2}$ and that $\overline{D}\,\overline{D}^* = -\frac{1}{4}\triangle_A + \frac{1}{8}\triangle\varphi$. Now Theorem 11.20 implies that

$$\lim_{|z|\to\infty}\int_{Q_z} (\triangle\varphi)^2\,d\lambda = +\infty. \tag{13.3}$$

Since $\triangle\varphi \in B_2(\mathbb{R}^2)$, we get (13.2).

For the other direction, we first use a version of the diamagnetic property for Schrödinger operators (see, for example, [67]) saying that $H(A, \triangle\varphi) = -\triangle_A + \triangle\varphi$ has a compact resolvent if $H(0, \triangle\varphi) = -\triangle + \triangle\varphi$ has a compact resolvent. To see this, let

$-\triangle + \triangle\varphi$ be with compact resolvent. Then by Theorem 11.9 there exists a continuous function $\Lambda : \mathbb{R}^n \longrightarrow \mathbb{R}$ such that $\Lambda(x) \to \infty$ as $|x| \to \infty$ and

$$(H(0, \triangle\varphi)\phi, \phi) \geq \int_{\mathbb{R}^n} \Lambda(x)|\phi(x)|^2 \, d\lambda(x) \tag{13.4}$$

for all $\phi \in C_0^\infty(\mathbb{R}^n)$. Kato's inequality implies that

$$(H(A, \triangle\varphi)\phi, \phi) \geq (H(0, \triangle\varphi)|\phi|, |\phi|);$$

see the proof of Proposition 10.11. So we get

$$(H(A, \triangle\varphi)\phi, \phi) \geq (H(0, \triangle\varphi)|\phi|, |\phi|) \geq (\Lambda|\phi|, |\phi|) = (\Lambda\phi, \phi),$$

and Theorem 11.9 gives that $H(A, \triangle\varphi)$ has a compact resolvent.

Now, again by Theorem 11.9(c), it suffices to show that

$$e_{0,\triangle\varphi}(Q_z) \to \infty \tag{13.5}$$

as $|z| \to \infty$. For this purpose, we have to estimate

$$\frac{\int_{Q_z} (H(0, \triangle\varphi)\phi)\overline{\phi} \, d\lambda}{\int_{Q_z} |\phi|^2 \, d\lambda} = \frac{\int_{Q_z} (|\nabla\phi|^2 + \triangle\varphi|\phi|^2) \, d\lambda}{\int_{Q_z} |\phi|^2 \, d\lambda}.$$

Without loss of generality, we can consider cubes instead of balls when using the following improved version of the Fefferman–Phong lemma as given in [2].

Lemma 13.4. *Let $V \in A_\infty$. Then there exist constants $C_V > 0$ and $\beta_V \in (0,1)$ such that for all cubes Q with side length R and for all $u \in C_0^\infty(Q)$,*

$$C_V \frac{m_\beta(R^2 \mathrm{av}_Q V)}{R^2} \int |u|^2 \, d\lambda \leq \int (|\nabla u|^2 + V|u|^2) \, d\lambda, \tag{13.6}$$

where $m_\beta(t) = t$ for $t \leq 1$ and $m_\beta(t) = t^{\beta_V}$ for $t \geq 1$.

We apply this lemma for $R = 1$ and $V = \triangle\varphi$. Then by (13.6) we obtain

$$e_{0,\triangle\varphi}(Q_z) \geq C_V \mathrm{av} V_{Q_z}^{\beta_V}$$

if $\mathrm{av} V_{Q_z} = \int_{Q_z} \triangle\varphi \, d\lambda \geq 1$. Hence (13.2) implies (13.5), and we are done. □

It remains to prove Lemma 13.4. From (11.12) we have

$$|u(x) - u(y)|^2 \leq |x - y|^2 \int_0^1 |\nabla u(x + t(y - x))|^2 \, dt,$$

and hence

$$\int_Q \int_Q |u(x) - u(y)|^2 \, d\lambda(x) d\lambda(y) \le R^4 \int_Q |\nabla u(x)|^2 \, d\lambda(x).$$

In addition, we also have

$$R^2 \int_Q V(x)|u(x)|^2 \, d\lambda(x) = \int_Q \int_Q V(x)|u(x)|^2 \, d\lambda(x) d\lambda(y),$$

so we get

$$\begin{aligned}\int_Q (|\nabla u|^2 + V|u|^2) \, d\lambda &\ge R^{-4} \int_Q \int_Q |u(x) - u(y)|^2 \, d\lambda(x) d\lambda(y) \\ &\quad + R^{-2} \int_Q \int_Q V(y)|u(y)|^2 \, d\lambda(x) d\lambda(y) \\ &\ge R^{-2} \int_Q \int_Q \min(V(y), R^{-2})(|u(x) - u(y)|^2 + |u(y)|^2) \, d\lambda(x) d\lambda(y) \\ &\ge \frac{1}{2} \mathrm{av}_Q[\min(V, R^{-2})] \int_Q |u|^2 \, d\lambda,\end{aligned}$$

where we used the inequality

$$|u(x) - u(y)|^2 + |u(y)|^2 \ge \frac{1}{2} |u(x)|^2.$$

By Lemma 13.2(a) there exists $\epsilon > 0$, independent of Q, such that $E = \{x \in Q : V(x) > \epsilon \mathrm{av}_Q V\}$ satisfies $|E| \ge \frac{1}{2}|Q|$. Hence

$$\mathrm{av}_Q[\min(V, R^{-2})] \ge \frac{1}{2} \min(\epsilon \mathrm{av}_Q V, R^{-2}).$$

So we get the desired inequality if $R^2 \mathrm{av}_Q V \le 1$.

It remains to analyze the case $R^2 \mathrm{av}_Q V \ge 1$. For this purpose, we subdivide Q in a dyadic manner and stop for the first time when $R(Q')^2 \mathrm{av}_{Q'} V < 1$. In this way, we obtain a collection $\{Q_i\}$ of strict dyadic subcubes of Q that are maximal for the property $R_i^2 \mathrm{av}_{Q_i} V < 1$. Since, by Lemma 13.1 (4), $w\, d\lambda$ is a doubling measure and the ancestor $\hat{Q}_i$ of Q_i satisfies $(2R_i)^2 \mathrm{av}_{\hat{Q}_i} V \ge 1$, there exists a constant $A > 0$ such that

$$R_i^2 \mathrm{av}_{Q_i} V \ge A. \tag{13.7}$$

This implies that the Q_i form a disjoint covering of Q up to a set of null measure. Now we get

$$\begin{aligned}\int_Q (|\nabla u|^2 + V|u|^2)\,d\lambda &= \sum_i \int_{Q_i} (|\nabla u|^2 + V|u|^2)\,d\lambda\\ &\ge C' \sum_i \min(\mathrm{av}_{Q_i} V, R_i^{-2}) \int_{Q_i} |u|^2\,d\lambda\\ &\ge AC' \sum_i R_i^{-2} \int_{Q_i} |u|^2\,d\lambda\\ &\ge AC' \min_i \left(\frac{R}{R_i}\right)^2 R^{-2} \int_{Q_i} |u|^2\,d\lambda,\end{aligned}$$

where we used (13.7). By Lemma 13.2(b) there exists $\alpha \ge 1$ such that for any cube Q and any measurable subset E of Q, we have

$$\left(\frac{\mathrm{av}_E V}{\mathrm{av}_Q V}\right) \ge C\left(\frac{|E|}{|Q|}\right)^{\alpha-1}.$$

We apply this to $E = Q_i$ and get

$$\begin{aligned}\left(\frac{R}{R_i}\right)^2 &= \frac{R^2 \mathrm{av}_Q V}{R_i^2 \mathrm{av}_{Q_i} V}\,\frac{\mathrm{av}_{Q_i} V}{\mathrm{av}_Q V}\\ &\ge R^2\,\mathrm{av}_Q V\,\frac{\mathrm{av}_{Q_i} V}{\mathrm{av}_Q V}\\ &\ge CR^2\,\mathrm{av}_Q V\left(\frac{R_i}{R}\right)^{2(\alpha-1)}.\end{aligned}$$

Finally, this implies

$$\min_i \left(\frac{R}{R_i}\right)^2 \ge C(R^2 \mathrm{av}_Q V)^{\beta}$$

with $\beta = 1/\alpha$, which proves the lemma.

Corollary 13.5. *Let φ be a subharmonic C^2-function such that*

$$\Delta\varphi(z) > \epsilon > 0 \text{ and } \Delta\varphi \in B_2(\mathbb{R}^2).$$

The canonical solution operator S_φ to $\bar\partial$ is compact on $L^2(\mathbb{C}, e^{-\varphi})$ if and only if $\lim_{|z|\to\infty} \int_{Q_z} \Delta\varphi\,d\lambda = +\infty$.

Proof. Since

$$N_\varphi = S_\varphi^* S_\varphi,$$

the result follows from Theorems 13.3 and 2.5. □

Remark 13.6. Using the notations of Chapter 8 and the assumptions on φ, we can express the last theorem in the following way: Let φ be a subharmonic C^2-function such that

$$\triangle\varphi(z) > \epsilon > 0 \quad \text{and} \quad \triangle\varphi \in B_2(\mathbb{R}^2).$$

The Schrödinger operator with magnetic field

$$\mathcal{S} = \frac{1}{4}(-\triangle_A + B),$$

where

$$\Delta_A = \left(\frac{\partial}{\partial x} + \frac{i}{2}\frac{\partial\varphi}{\partial y}\right)^2 + \left(\frac{\partial}{\partial y} - \frac{i}{2}\frac{\partial\varphi}{\partial x}\right)^2 \quad \text{and} \quad B = \frac{1}{2}\triangle\varphi,$$

has a compact resolvent if and only if $\lim_{|z|\to\infty}\int_{Q_z}\triangle\varphi\, d\lambda = +\infty$.

13.2 Dirac and Pauli operators

We return to the Dirac and Pauli operators related to the weight function φ:

$$\mathcal{D} = \left(-i\frac{\partial}{\partial x} - A_1\right)\sigma_1 + \left(-i\frac{\partial}{\partial y} - A_2\right)\sigma_2,$$

where $A_1 = -\frac{1}{2}\frac{\partial\varphi}{\partial y}$, $A_2 = \frac{1}{2}\frac{\partial\varphi}{\partial x}$ and

$$\sigma_1 = \begin{pmatrix} 0 & 1 \\ 1 & 0 \end{pmatrix}, \quad \sigma_2 = \begin{pmatrix} 0 & -i \\ i & 0 \end{pmatrix}.$$

The square of $\mathcal{D}$ is diagonal with the Pauli operators $P_\pm$ on the diagonal:

$$\mathcal{D}^2 = \begin{pmatrix} P_- & 0 \\ 0 & P_+ \end{pmatrix},$$

where

$$P_\pm = \left(-i\frac{\partial}{\partial x} - A_1\right)^2 + \left(-i\frac{\partial}{\partial y} - A_2\right)^2 \pm B = -\Delta_A \pm B$$

with $B = \frac{1}{2}\triangle\varphi$.

Theorem 13.7. *Suppose that* $|z|^2\triangle\varphi(z) \to +\infty$ *as* $|z| \to \infty$. *Then the corresponding Dirac operator* $\mathcal{D}$ *has a noncompact resolvent.*

Proof. By Proposition 9.40, $\mathcal{D}^2$ has a compact resolvent if and only if $\mathcal{D}$ has a compact resolvent. Suppose that $\mathcal{D}$ has a compact resolvent. Since

$$\mathcal{D}^2 = \begin{pmatrix} P_- & 0 \\ 0 & P_+ \end{pmatrix},$$

this would imply that both $P_\pm$ have compact resolvents.

We know from (10.30) that

$$P_- = 4\overline{D}^* \,\overline{D} = 4e^{-\varphi/2}\,\overline{\partial}_\varphi^*\,\overline{\partial}\, e^{\varphi/2}$$

and that P_- is a nonnegative self-adjoint operator. It follows from Theorem 8.11 that the space of entire functions $A^2(\mathbb{C}, e^{-\varphi})$ is of infinite dimension. This means that 0 belongs to the essential spectrum of P_-. Hence, by Proposition 9.39, P_- fails to have a compact resolvent, and we arrive at a contradiction. □

A similar conclusion can be drawn in several variables for the Witten Laplacian

$$\Delta_\varphi^{(0,0)} = \overline{D}_1^*\overline{D}_1 = e^{-\varphi/2}\,\overline{\partial}_\varphi^*\,\overline{\partial}\, e^{\varphi/2};$$

if $\lim_{|z|\to\infty} |z|^2\mu_\varphi(z) = +\infty$, then $\Delta_\varphi^{(0,0)}$ fails to have a compact resolvent (μ_φ is the lowest eigenvalue of the Levi matrix M_φ).

Remark 13.8. The notion of a doubling measure appeared already in Lemma 13.1(4). Recall that a nonnegative Borel measure μ on $\mathbb{C}$ is doubling if there exists a constant $C > 0$ such that for all $z \in \mathbb{C}$ and $r > 0$,

$$\mu(D(z,r)) \le C\mu(D(z,r/2)). \tag{13.8}$$

It can be shown that

$$\mu(D(z,2r)) \ge (1 + C^{-3})\mu(D(z,r)) \tag{13.9}$$

for all $z \in \mathbb{C}$ and $r > 0$; in particular, $\mu(\mathbb{C}) = \infty$, unless $\mu(\mathbb{C}) = 0$ (see [90]).

Examples: If $p(z,\overline{z})$ is a polynomial on $\mathbb{C}$ of degree d, then

$$d\mu(z) = |p(z,\overline{z})|^a\, d\lambda(z), \quad a > -\frac{1}{d},$$

is a doubling measure on $\mathbb{C}$; see [90]. If V belongs to the reverse Hölder class $B_2(\mathbb{R}^2)$ or to A_∞, then $V(x)d\lambda(x)$ is a doubling measure.

Let $\varphi : \mathbb{C} \longrightarrow \mathbb{R}_+$ be a subharmonic $\mathcal{C}^2$-function. Suppose that $d\mu = \Delta\varphi\, d\lambda$ is a nontrivial doubling measure.

We show that the weighted space of entire functions

$$A^2(\mathbb{C}, e^{-\varphi}) = \left\{ f \text{ entire} : \|f\|_\varphi^2 = \int_{\mathbb{C}} |f|^2 e^{-\varphi}\, d\lambda < \infty \right\}$$

is of infinite dimension.

We know that

$$\mu(\mathbb{C}) = \int_{\mathbb{C}} \triangle\varphi\, d\lambda = \infty. \tag{13.10}$$

Let $K_\varphi(z, w)$ denote the Bergman kernel of $A^2(\mathbb{C}, e^{-\varphi})$. So if $f \in A^2(\mathbb{C}, e^{-\varphi})$, then

$$f(z) = \int_{\mathbb{C}} K_\varphi(z, w) f(w) e^{-\varphi(w)}\, d\lambda(w). \tag{13.11}$$

We claim that

$$\int_{\mathbb{C}} K_\varphi(z, z) e^{-\varphi(z)}\, d\lambda(z) = \sum_{k=1}^{\infty} \|f_k\|_\varphi^2 \tag{13.12}$$

for any complete orthonormal basis $(f_k)_k$ of $A^2(\mathbb{C}, e^{-\varphi})$, finite or infinite.

The partial sums

$$F_N(z) = \sum_{k=1}^{N} |f_k(z)|^2$$

form a pointwise nondecreasing sequence: $F_N(z) \leq F_{N+1}(z)$ for all $z \in \mathbb{C}$ and $N \in \mathbb{N}$. In addition, we have

$$\lim_{N\to\infty} F_N(z) = \sum_{k=1}^{\infty} f_k(z)\overline{f_k(z)} = K_\varphi(z, z);$$

see (1.9). The monotone convergence theorem implies that

$$\sum_{k=1}^{\infty} \|f_k(z)\|_\varphi^2 = \int_{\mathbb{C}} \sum_{k=1}^{\infty} |f_k(z)|^2 e^{-\varphi(z)}\, d\lambda(z) = \int_{\mathbb{C}} K_\varphi(z, z) e^{-\varphi(z)}\, d\lambda(z). \tag{13.13}$$

Now we use the following pointwise inequality for the Bergman kernel, which is due to Marzo and Ortega-Cerdà ([74]): there exists a constant $C > 0$ such that

$$C^{-1} \triangle \varphi(z) \leq K_\varphi(z, z) e^{-\varphi(z)} \leq C \triangle \varphi(z) \tag{13.14}$$

for each $z \in \mathbb{C}$. From this we conclude that $A^2(\mathbb{C}, e^{-\varphi})$ is of infinite dimension.

Relations (13.10) and (13.13) imply that the corresponding Dirac operator $\mathcal{D}$ fails to be with compact resolvent if φ is a subharmonic function such that $d\mu = \triangle\varphi\, d\lambda$ is a nontrivial doubling measure.

13.3 Notes

Iwatsuka [59] and Avron, Herbst, and Simon [3] were the first to obtain results on the problem of compact resolvence of Schrödinger operators with magnetic field.

Theorem 13.3 was inspired by the work of Iwatsuka. Christ [17] initiated the study of weighted spaces of entire functions, where the weight φ has the property that $d\mu = \triangle\varphi\, d\lambda$ is a nontrivial doubling measure.

There is another approach to characterize the compactness of N_φ by means of estimates of the Bergman kernel of $A^2(\mathbb{C}, e^{-\varphi})$ [74]. The results there read as follows: Suppose that $\triangle\varphi$ defines a doubling measure. Then N_φ is compact if and only if $\int_{\mathbb{B}(z,1)} \triangle\varphi\, d\lambda \to \infty$ as $|z| \to \infty$, where $\mathbb{B}(z,1) = \{w : |w - z| < 1\}$. Witten Laplacians, as they appear in the context of semiclassical analysis, are studied in [50, 53] and [61]. A thorough treatment of Dirac and Pauli operators can be found in [21, 54] and [92].

14 Spectrum of $\square$ on the Fock space

In this chapter, we investigate the spectral properties of the $\square$-operator for the Fock space using solutions of the corresponding partial differential equations without boundary conditions. It is important that in this case the $\square$-operator acts diagonally, which is related to the fact that the weight function is decoupled, a property that turns out to be an obstruction for compactness (see Chapter 15). We show that the spectrum of the $\square$-operator for the Fock space consists of positive integers of infinite multiplicity and indicates consequences of this result related to the essential spectrum.

14.1 The general setting

We consider the weighted $\overline{\partial}$-complex

$$L^2_{(0,q-1)}(\mathbb{C}^n, e^{-\varphi}) \underset{\overline{\partial}^*_\varphi}{\overset{\overline{\partial}}{\rightleftarrows}} L^2_{(0,q)}(\mathbb{C}^n, e^{-\varphi}) \underset{\overline{\partial}^*_\varphi}{\overset{\overline{\partial}}{\rightleftarrows}} L^2_{(0,q+1)}(\mathbb{C}^n, e^{-\varphi}),$$

where for $(0,q)$-forms $u = \sum'_{|J|=q} u_J \, d\overline{z}_J$ with coefficients in $\mathcal{C}_0^\infty(\mathbb{C}^n)$, we have

$$\overline{\partial} u = \sum_{|J|=q}{}' \sum_{j=1}^{n} \frac{\partial u_J}{\partial \overline{z}_j} \, d\overline{z}_j \wedge d\overline{z}_J$$

and

$$\overline{\partial}^*_\varphi u = -\sum_{|K|=q-1}{}' \sum_{k=1}^{n} \delta_k u_{kK} \, d\overline{z}_K,$$

where $\delta_k = \frac{\partial}{\partial z_k} - \frac{\partial \varphi}{\partial z_k}$.

The complex Laplacian on $(0,q)$-forms is defined as

$$\square_{\varphi,q} := \overline{\partial}\,\overline{\partial}^*_\varphi + \overline{\partial}^*_\varphi \overline{\partial}.$$

The operator $\square_{\varphi,q}$ is self-adjoint and positive, which means that

$$(\square_{\varphi,q} f, f)_\varphi \geq 0 \quad \text{for } f \in \operatorname{dom}(\square_\varphi).$$

The associated Dirichlet form is defined as

$$Q_\varphi(f,g) = (\overline{\partial} f, \overline{\partial} g)_\varphi + (\overline{\partial}^*_\varphi f, \overline{\partial}^*_\varphi g)_\varphi \tag{14.1}$$

for $f, g \in \operatorname{dom}(\overline{\partial}) \cap \operatorname{dom}(\overline{\partial}^*_\varphi)$. The weighted $\overline{\partial}$-Neumann operator $N_{\varphi,q}$ is – if it exists – the bounded inverse of $\square_{\varphi,q}$.

https://doi.org/10.1515/9783111182926-014

We indicate that a square-integrable (0, 1)-form $f = \sum_{j=1}^n f_j \, d\overline{z}_j$ belongs to $\operatorname{dom}(\overline{\partial}_\varphi^*)$ if and only if

$$e^{\varphi} \sum_{j=1}^{n} \frac{\partial}{\partial z_j}(f_j e^{-\varphi}) \in L^2(\mathbb{C}^n, e^{-\varphi}),$$

where the derivative is to be taken in the sense of distributions, and that the forms with coefficients in $C_0^\infty(\mathbb{C}^n)$ are dense in $\operatorname{dom}(\overline{\partial}) \cap \operatorname{dom}(\overline{\partial}_\varphi^*)$ in the graph norm $f \mapsto (\|\overline{\partial} f\|_\varphi^2 + \|\overline{\partial}_\varphi^* f\|_\varphi^2)^{\frac{1}{2}}$ (see Chapter 6 and [36]).

We consider the Levi matrix

$$M_\varphi = \left(\frac{\partial^2 \varphi}{\partial z_j \partial \overline{z}_k} \right)_{jk}$$

of φ and suppose that the sum s_q of any q (equivalently, the smallest q) eigenvalues of M_φ satisfies

$$\liminf_{|z|\to\infty} s_q(z) > 0. \tag{14.2}$$

In Proposition 6.7, we showed that (14.2) implies that there exists a continuous linear operator

$$N_{\varphi,q} : L^2_{(0,q)}(\mathbb{C}^n, e^{-\varphi}) \longrightarrow L^2_{(0,q)}(\mathbb{C}^n, e^{-\varphi})$$

such that $\Box_{\varphi,q} \circ N_{\varphi,q} u = u$ for all $u \in L^2_{(0,q)}(\mathbb{C}^n, e^{-\varphi})$.

If the sum s_q of any q (equivalently, the smallest q) eigenvalues of M_φ satisfies

$$\lim_{|z|\to\infty} s_q(z) = \infty, \tag{14.3}$$

then the $\overline{\partial}$-Neumann operator $N_{\varphi,q} : L^2_{(0,q)}(\mathbb{C}^n, e^{-\varphi}) \longrightarrow L^2_{(0,q)}(\mathbb{C}^n, e^{-\varphi})$ is compact (see Proposition 11.16).

To find the canonical solution to $\overline{\partial} f = u$, where $u \in L^2_{(0,1)}(\mathbb{C}^n, e^{-\varphi})$ is a given (0, 1)-form such that $\overline{\partial} u = 0$, we can take $f = \overline{\partial}_\varphi^* N_{\varphi,1} u$, which will also satisfy $f \perp \ker \overline{\partial}$.

If the weight function is $\varphi(z) = |z|^2$, then the corresponding Levi matrix M_φ is the identity. The space $A^2(\mathbb{C}^n, e^{-|z|^2})$ of entire functions belonging to $L^2(\mathbb{C}^n, e^{-|z|^2})$ is the Fock space, which plays an important role in quantum mechanics. In this case,

$$\Box_{\varphi,0} u = \overline{\partial}_\varphi^* \overline{\partial} u = -\frac{1}{4} \triangle u + \sum_{j=1}^{n} \overline{z}_j u_{\overline{z}_j} \tag{14.4}$$

for $u \in \operatorname{dom}(\Box_{\varphi,0}) \subseteq L^2(\mathbb{C}^n, e^{-|z|^2})$, and

$$\square_{\varphi,n}\, u = \overline{\partial}\,\overline{\partial}_{\varphi}^{*} u = -\frac{1}{4}\,\triangle\, u + \sum_{j=1}^{n} \overline{z}_j\, u_{\overline{z}_j} + n\, u \tag{14.5}$$

for $u \in \operatorname{dom}(\square_{\varphi,n}) \subseteq L^2_{(0,n)}(\mathbb{C}^n, e^{-|z|^2})$.

For $n > 1$ and $1 \le q \le n-1$, the $\overline{\partial}$-Neumann Laplacian $\square_{\varphi,q}$ acts diagonally (see Chapter 10): for

$$u = \sum_{|J|=q}{}' u_J\, d\overline{z}_J \in \operatorname{dom}(\square_{\varphi,q}) \subseteq L^2_{(0,q)}(\mathbb{C}^n, e^{-|z|^2}),$$

we have

$$\square_{\varphi,q}\, u = (\overline{\partial}\,\overline{\partial}_{\varphi}^{*} + \overline{\partial}_{\varphi}^{*}\,\overline{\partial})\, u = \sum_{|J|=q}{}' \left(-\frac{1}{4}\,\triangle\, u_J + \sum_{j=1}^{n} \overline{z}_j\, u_{J\overline{z}_j} + q\, u_J \right) d\overline{z}_J. \tag{14.6}$$

14.2 Determination of the spectrum

To determine the spectrum of $\square_{\varphi,q}$ for $\varphi(z) = |z|^2$, we use Lemma 9.33.

For simplicity, to explain the general method, we start with the complex one-dimensional case. Looking for the eigenvalues μ of $\square_{\varphi,0}$, we have to consider

$$\square_{\varphi,0}\, u = -u_{z\overline{z}} + \overline{z} u_{\overline{z}} = \mu u. \tag{14.7}$$

It is clear that the space $A^2(\mathbb{C}^n, e^{-|z|^2})$ is contained in the eigenspace of the eigenvalue $\mu = 0$.

For any positive integer k, the antiholomorphic monomial $\overline{z}^k$ is an eigenfunction for the eigenvalue $\mu = k$.

In the following, we denote $\mathbb{N}_0 = \mathbb{N} \cup \{0\}$.

Lemma 14.1. *Let $n = 1$. For $k \in \mathbb{N}_0$ and $m \in \mathbb{N}$, the functions*

$$u_{k,m}(z, \overline{z}) = \overline{z}^{k+m}\, z^m + \sum_{j=1}^{m} \frac{(-1)^j (k+m)!\, m!}{j!\,(k+m-j)!\,(m-j)!}\, \overline{z}^{k+m-j}\, z^{m-j} \tag{14.8}$$

are eigenfunctions for the eigenvalue $k+m$ of the operator $\square_{\varphi,0}\, u = -u_{z\overline{z}} + \overline{z} u_{\overline{z}}$.

For $k \in \mathbb{N}$ and $m \in \mathbb{N}_0$, the functions

$$v_{k,m}(z, \overline{z}) = \overline{z}^{k}\, z^{k+m} + \sum_{j=1}^{k} \frac{(-1)^j (k+m)!\, k!}{j!\,(k+m-j)!\,(k-j)!}\, \overline{z}^{k-j}\, z^{k+m-j} \tag{14.9}$$

are eigenfunctions for the eigenvalue k of the operator $\square_{\varphi,0}\, u = -u_{z\overline{z}} + \overline{z} u_{\overline{z}}$.

Proof. To prove (14.8), we set

$$u_{k,m}(z,\overline{z}) = \overline{z}^{k+m}\, z^m + a_1\overline{z}^{k+m-1}\, z^{m-1} + a_2\overline{z}^{k+m-2}\, z^{m-2} + \cdots + a_{m-1}\overline{z}^{k+1}\, z + a_m\overline{z}^k$$

and compute

$$\begin{aligned}&\frac{\partial^2}{\partial z\partial\overline{z}}\, u_{k,m}(z,\overline{z})\\ &\quad = (k+m)m\overline{z}^{k+m-1}\, z^{m-1} + a_1(k+m-1)(m-1)\overline{z}^{k+m-2}\, z^{m-2} + \cdots + a_{m-1}(k+1)\overline{z}^k\end{aligned}$$

and

$$\begin{aligned}&\overline{z}\,\frac{\partial}{\partial\overline{z}}\, u_{k,m}(z,\overline{z})\\ &\quad = (k+m)\overline{z}^{k+m}\, z^m + a_1(k+m-1)\overline{z}^{k+m-1}\, z^{m-1} + \cdots + a_{m-1}(k+1)\overline{z}^{k+1}\, z + a_m k\overline{z}^k,\end{aligned}$$

which implies that the function $u_{k,m}$ is an eigenfunction for the eigenvalue $\mu = k + m$, and from (14.7) we obtain, comparing the highest exponents of $\overline{z}$ and z,

$$(k+m)m - a_1(k+m-1) = -(k+m)a_1,$$

and hence $a_1 = -(k+m)m$. Comparing the next lower exponents, we get

$$a_1(k+m-1)(m-1) - a_2(k+m-2) = -a_2(k+m)$$

and $a_2 = \frac{1}{2}(k+m)(k+m-1)m(m-1)$. In general, we find that for $j = 1, 2, \ldots, m$,

$$a_j = \frac{(-1)^j (k+m)!\, m!}{j!\,(k+m-j)!\,(m-j)!},$$

which proves (14.8).

To show (14.9), we set

$$v_{k,m}(z,\overline{z}) = \overline{z}^k\, z^{k+m} + b_1\overline{z}^{k-1}\, z^{k+m-1} + b_2\overline{z}^{k-2}\, z^{k+m-2} + \cdots + b_{k-1}\overline{z}\, z^{m+1} + b_k z^m$$

and compute

$$\begin{aligned}&\frac{\partial^2}{\partial z\partial\overline{z}}\, v_{k,m}(z,\overline{z})\\ &\quad = k(k+m)\overline{z}^{k-1}\, z^{k+m-1} + b_1(k-1)(k+m-1)\overline{z}^{k-2}\, z^{k+m-2} + \cdots + b_{k-1}\overline{z}\, z^{m+1}\end{aligned}$$

and

$$\overline{z}\,\frac{\partial}{\partial\overline{z}}\, v_{k,m}(z,\overline{z}) = k\overline{z}^k\, z^{k+m} + b_1(k-1)\overline{z}^{k-1}\, z^{k+m-1} + \cdots + b_{k-1}\overline{z}\, z^{m+1},$$

which implies that the function $v_{k,m}$ is an eigenfunction for the eigenvalue $\mu = k$, for each $m \in \mathbb{N}$, and from (14.7) we obtain, comparing the highest exponents of $\overline{z}$ and z,

$$k(k+m) - b_1(k-1) = -kb_1,$$

and hence $b_1 = -(k+m)k$. Comparing the next lower exponents, we get

$$b_1(k-1)(k+m-1) - b_2(k-2) = -b_2 k$$

and $b_2 = \frac{1}{2}(k+m)(k+m-1)k(k-1)$. In general, we find that for $j = 1, 2, \dots, k$,

$$b_j = \frac{(-1)^j (k+m)!\, k!}{j!\,(k+m-j)!\,(k-j)!},$$

which proves (14.9). □

Now we are able to prove the following:

Theorem 14.2. *Let $n = 1$ and $\varphi(z) = |z|^2$. The spectrum of $\Box_{\varphi,0}$ consists of all nonnegative integers $\{0, 1, 2, \dots\}$ each of which is of infinite multiplicity, so 0 is the bottom of the essential spectrum. The spectrum of $\Box_{\varphi,1}$ consists of all positive integers $\{1, 2, 3, \dots\}$ each of which is of infinite multiplicity.*

Proof. We already know that the whole Bergman space $A^2(\mathbb{C}, e^{-|z|^2})$ is contained in the eigenspace of the eigenvalue 0 of the operator $\Box_{\varphi,0}$ and for any positive integer k, the antiholomorphic monomial $\overline{z}^k$ is an eigenfunction for the eigenvalue $\mu = k$. In addition, all functions of the form $\overline{z}^\nu z^\kappa$ with $\nu, \kappa \in \mathbb{N}_0$ can be expressed as linear combinations of functions of the form (14.8) and (14.9). For a fixed $k \in \mathbb{N}$, the functions of the form (14.9) have infinite multiplicity as the parameter $m \in \mathbb{N}_0$ is free. So these eigenvalues are of infinite multiplicity. All the eigenfunctions considered so far yield a complete orthogonal basis of $L^2(\mathbb{C}, e^{-|z|^2})$, since the Hermite polynomials

$$\{H_0(x)H_k(y), H_1(x)H_{k-1}(y), \dots, H_k(x)H_0(y)\}$$

for $k = 0, 1, 2, \dots$ form a complete orthogonal system in $L^2(\mathbb{R}^2, e^{-x^2-y^2})$ (see, for instance, [29]), and since $x = 1/2(z + \overline{z})$, $y = i/2(\overline{z} - z)$, we can apply Lemma 9.33 and obtain that the spectrum of $\Box_{\varphi,0}$ is $\mathbb{N}_0$.

The statement for the spectrum of $\Box_{\varphi,1}$ follows from (14.5). □

For several variables, we can adopt the method from above to obtain the following result.

Theorem 14.3. *Let $n > 1$, $\varphi(z) = |z_1|^2 + \cdots + |z_n|^2$, and $0 \le q \le n$. The spectrum of $\Box_{\varphi,q}$ consists of all integers $\{q, q+1, q+2, \dots\}$ each of which is of infinite multiplicity.*

Proof. Recall that the $\overline{\partial}$-Neumann Laplacian $\square_{\varphi,q}$ acts diagonally and that

$$\square_{\varphi,q}\, u = \sum_{|J|=q}{}' \left(-\frac{1}{4} \,\triangle\, u_J + \sum_{j=1}^{n} \overline{z}_j\, u_{J\overline{z}_j} + q\, u_J \right) d\overline{z}_J .$$

The factor q in the last formula is responsible for the fact that the eigenvalues start with q, which can be seen, in each component separately, by

$$-\frac{1}{4} \,\triangle\, u_J + \sum_{j=1}^{n} \overline{z}_j\, u_{J\overline{z}_j} = (\mu - q)\, u_J .$$

Now let $k_1, k_2, \ldots, k_n \in \mathbb{N}_0$ and $m_1, m_2, \ldots, m_n \in \mathbb{N}$. Then the function

$$u_{k_1,m_1}(z_1,\overline{z}_1)\, u_{k_2,m_2}(z_2,\overline{z}_2) \ldots\, u_{k_n,m_n}(z_n,\overline{z}_n)$$

is the component of an eigenfunction for the eigenvalue $\sum_{j=1}^{n}(k_j + m_j)$ of the operator $\square_{\varphi,q}$, which follows from (14.6) and (14.8).

Similarly, it follows from (14.6) and (14.9) that for $k_1, k_2, \ldots, k_n \in \mathbb{N}$ and $m_1, m_2, \ldots, m_n \in \mathbb{N}_0$, the function

$$v_{k_1,m_1}(z_1,\overline{z}_1)\, v_{k_2,m_2}(z_2,\overline{z}_2) \ldots\, v_{k_n,m_n}(z_n,\overline{z}_n)$$

is an eigenfunction for the eigenvalue $\sum_{j=1}^{n} k_j$.

All other possible n-fold products with factors u_{k_j,m_j} or v_{k_j,m_j} (also mixed) appear as eigenfunctions of $\square_{\varphi,q}$.

From this we obtain that all expressions of the form $z_1^{\alpha_1}\,\overline{z}_1^{\beta_1} \ldots z_n^{\alpha_n}\,\overline{z}_n^{\beta_n}$ for arbitrary $\alpha_j, \beta_j \in \mathbb{N}_0$, $j = 1, \ldots, n$, can be written as linear combinations of eigenfunctions of $\square_{\varphi,q}$, which proves that all these eigenfunctions constitute a complete basis in $L^2_{(0,q)}(\mathbb{C}^n, e^{-|z|^2})$; see the proof of Theorem 14.2. So we can again apply Lemma 9.33. □

Remark 14.4.
(i) Since in all cases the essential spectrum is nonempty, the corresponding $\square_{\varphi,q}$ operator fails to be with compact resolvent (see Proposition 9.39).
(ii) For the weight function

$$\varphi(z) = \left(|z_1|^2 + |z_2|^2 + \cdots + |z_n|^2\right)^{\alpha} + |z|^2 \quad \text{for } \alpha > 1,$$

the situation is completely different: The operators $\square_{\varphi,q}$, $1 \le q \le n$, are with compact resolvent (see Proposition 11.16), so the essential spectrum must be empty.

We can use the results from above to settle the corresponding questions for the so-called Witten Laplacian, which is defined on $L^2(\mathbb{C}^n)$.

For this purpose, we set $Z_k = \frac{\partial}{\partial \overline{z}_k} + \frac{1}{2}\frac{\partial \varphi}{\partial \overline{z}_k}$ and $Z_k^* = -\frac{\partial}{\partial z_k} + \frac{1}{2}\frac{\partial \varphi}{\partial z_k}$ and consider $(0,q)$-forms $h = \sum'_{|J|=q} h_J \, d\overline{z}_J$, where $\sum'$ means that we sum up only increasing multi-indices $J = (j_1, \dots, j_q)$, and where $d\overline{z}_J = d\overline{z}_{j_1} \wedge \cdots \wedge d\overline{z}_{j_q}$. We define

$$\overline{D}_{q+1} h = \sum_{k=1}^{n} \sum_{|J|=q}{}' Z_k(h_J)\, d\overline{z}_k \wedge d\overline{z}_J$$

and

$$\overline{D}_q^* h = \sum_{k=1}^{n} \sum_{|J|=q}{}' Z_k^*(h_J)\, d\overline{z}_k \rfloor d\overline{z}_J,$$

where $d\overline{z}_k \rfloor d\overline{z}_J$ denotes the contraction or interior multiplication by $d\overline{z}_k$, i. e., we have

$$\langle \alpha, d\overline{z}_k \rfloor d\overline{z}_J \rangle = \langle d\overline{z}_k \wedge \alpha, d\overline{z}_J \rangle$$

for each $(0, q-1)$-form α.

The complex Witten Laplacian on $(0,q)$-forms is then given by

$$\Delta_\varphi^{(0,q)} = \overline{D}_q \overline{D}_q^* + \overline{D}_{q+1}^* \overline{D}_{q+1}$$

for $q = 1, \dots, n-1$.

The general $\overline{D}$-complex has the form

$$L^2_{(0,q-1)}(\mathbb{C}^n) \underset{\overline{D}_q^*}{\overset{\overline{D}_q}{\rightleftarrows}} L^2_{(0,q)}(\mathbb{C}^n) \underset{\overline{D}_{q+1}^*}{\overset{\overline{D}_{q+1}}{\rightleftarrows}} L^2_{(0,q+1)}(\mathbb{C}^n).$$

It follows that

$$\overline{D}_{q+1} \Delta_\varphi^{(0,q)} = \Delta_\varphi^{(0,q+1)} \overline{D}_{q+1} \quad \text{and} \quad \overline{D}_{q+1}^* \Delta_\varphi^{(0,q+1)} = \Delta_\varphi^{(0,q)} \overline{D}_{q+1}^*.$$

We remark that

$$\overline{D}_q^* h = \sum_{k=1}^{n} \sum_{|J|=q}{}' Z_k^*(h_J)\, d\overline{z}_k \rfloor d\overline{z}_J = \sum_{|K|=q-1}{}' \sum_{k=1}^{n} Z_k^*(h_{kK})\, d\overline{z}_K.$$

In particular, for a function $v \in L^2(\mathbb{C}^n)$, we get

$$\Delta_\varphi^{(0,0)} v = \overline{D}_1^* \overline{D}_1 v = \sum_{j=1}^{n} Z_j^* Z_j(v), \tag{14.10}$$

and for a $(0,1)$-form $g = \sum_{\ell=1}^{n} g_\ell \, d\overline{z}_\ell \in L^2_{(0,1)}(\mathbb{C}^n)$, we obtain

$$\Delta_\varphi^{(0,1)} g = (\overline{D}_1 \overline{D}_1^* + \overline{D}_2^* \overline{D}_2) g = (\Delta_\varphi^{(0,0)} \otimes I) g + M_\varphi g, \tag{14.11}$$

where we set

$$M_\varphi g = \sum_{j=1}^{n}\left(\sum_{k=1}^{n} \frac{\partial^2 \varphi}{\partial z_k \partial \overline{z}_j} g_k\right) d\overline{z}_j$$

and

$$(\Delta_\varphi^{(0,0)} \otimes I)\, g = \sum_{k=1}^{n} \Delta_\varphi^{(0,0)} g_k \, d\overline{z}_k.$$

In general, we have that

$$\Delta_\varphi^{(0,q)} = e^{-\varphi/2} \, \Box_{\varphi,q} \, e^{\varphi/2} \tag{14.12}$$

for $q = 0, 1, \ldots, n$.

In our case $\varphi(z) = |z_1|^2 + \cdots + |z_n|^2$, we get

$$\Delta_\varphi^{(0,q)} h = \sum_{|J|=q}{}' \left(-\frac{1}{4} \triangle h_J + \frac{1}{2} \sum_{j=1}^{n} (\overline{z}_j h_{J\overline{z}_j} - z_j h_{Jz_j}) + \frac{1}{4} |z|^2 h_J + \left(q - \frac{n}{2}\right) h_J \right) d\overline{z}_J \tag{14.13}$$

for

$$h = \sum_{|J|=q}{}' h_J \, d\overline{z}_J \in \mathrm{dom}(\Delta_\varphi^{(0,q)}) \subseteq L^2_{(0,q)}(\mathbb{C}^n).$$

Using (14.12) and Lemma 9.31, we get that $\Delta_\varphi^{(0,q)}$ and $\Box_{\varphi,q}$ have the same spectrum. Hence by Theorem 14.3 we obtain the following:

Theorem 14.5. *Let $\varphi(z) = |z_1|^2 + \cdots + |z_n|^2$ and $0 \le q \le n$. The spectrum of the Witten Laplacian $\Delta_\varphi^{(0,q)}$ consists of all integers $\{q, q+1, q+2, \ldots\}$ each of which is of infinite multiplicity.*

14.3 Notes

Only in a few cases it is possible to determine the exact form of the spectrum of a □-operator. Folland [28] described it for the unit ball in $\mathbb{C}^n$ using spherical harmonics; in this case the spectrum is discrete consisting of eigenvalues of finite multiplicity. With the help of Bessel functions, Fu [32] showed that the spectrum of the □-operator for a polycylinder contains eigenvalues of finite and infinite multiplicity.

The spectrum of $\Delta_\varphi^{(0,0)}$, even in a more general form, was calculated by Ma and Marinescu [72, 73]. They directly use Hermite polynomials, the fact that the Hermite polynomials form a complete orthogonal system of the weighted space $L^2(\mathbb{R}, e^{-x^2})$, and general properties of the Kodaira Laplacian.

15 Obstructions to compactness

In this chapter, we give some examples of domains or weights for which the corresponding $\overline{\partial}$-Neumann operator or the canonical solution operator to $\overline{\partial}$ fails to be compact. For this purpose, we will mainly use the characterization of compactness from Chapter 11. The standard examples for noncompactness are polydiscs and so-called decoupled weights.

15.1 The bidisc

First, we consider the canonical solution operator to $\overline{\partial}$ for the bidisc $\mathbb{D} \times \mathbb{D}$ (see [68]). We know from Chapter 1 that the monomials

$$\varphi_n(z) = \sqrt{\frac{n+1}{\pi}}\, z^n, \quad n = 0, 1, 2, \dots,$$

constitute a complete orthonormal system in $A^2(\mathbb{D})$. Consider the following $(0,1)$-forms α_n in $L^2_{(0,1)}(\mathbb{D} \times \mathbb{D})$ with holomorphic coefficients:

$$\alpha_n(z_1, z_2) = \varphi_n(z_1)\, d\overline{z}_2.$$

They are $\overline{\partial}$-closed, and their norms in $L^2_{(0,1)}(\mathbb{D} \times \mathbb{D})$ are

$$\|\alpha_n\| = \sqrt{\pi}, \quad n = 0, 1, 2, \dots.$$

The canonical solution to $\overline{\partial} u = \alpha_n$ is given by

$$u_n(z_1, z_2) = \varphi_n(z_1)\, \overline{z}_2;$$

this means that $u_n \in A^2(\mathbb{D} \times \mathbb{D})^\perp$, which follows by the fact that for each $h \in A^2(\mathbb{D} \times \mathbb{D})$, we have

$$\int_{\mathbb{D}\times\mathbb{D}} u_n(z_1, z_2)\, \overline{h(z_1, z_2)}\, d\lambda(z_1, z_2) = \int_{\mathbb{D}} \varphi_n(z_1) \left(\int_{\mathbb{D}} z_2 h(z_1, z_2)\, d\lambda(z_2) \right)^{-} d\lambda(z_1),$$

where the inner integral vanishes by Cauchy's theorem applied to the holomorphic function $z_2 \mapsto z_2 h(z_1, z_2)$. Finally,

$$\|u_n\| = \sqrt{\frac{\pi}{2}} \quad \text{and} \quad u_n \perp u_m \quad \text{if } n \neq m,$$

which follows from

$$(u_n, u_m) = \int_{\mathbb{D}} |z_2|^2\, d\lambda(z_2) \int_{\mathbb{D}} \varphi_n(z_1) \overline{\varphi_m(z_1)}\, d\lambda(z_1)$$

https://doi.org/10.1515/9783111182926-015

and $(\varphi_n, \varphi_m) = \delta_{n,m}$.

Thus $\{u_n\}$ has no convergent subsequence in $L^2(\mathbb{D} \times \mathbb{D})$. This shows that the canonical solution operator $\overline{\partial}^* N$ to $\overline{\partial}$ fails to be compact.

Further obstructions to compactness on weakly pseudoconvex domains can be found in [33, 34] and [91].

15.2 Weighted spaces

We continue to calculate the integrals in (11.30) for the weight $\varphi(z) = |z|^\alpha$ in $\mathbb{C}$. We set $\beta = \alpha/2$ and $u_k(z) = z^k$ for $k \in \mathbb{N}$. The left-hand side of (11.30) is

$$\int_{\mathbb{C}\setminus B_R} |u_k(z)|^2 e^{-|z|^\alpha} \, d\lambda(z) = 2\pi \int_R^\infty r^{2k+1} e^{-r^\alpha} \, dr;$$

we indicate that

$$\int_0^\infty r^{2k+1} e^{-r^\alpha} \, dr = \frac{1}{2\beta} \Gamma\left(\frac{k}{\beta} + \frac{1}{\beta}\right).$$

The right-hand side of (11.30) reads

$$\begin{aligned}
\int_{\mathbb{C}} |\overline{\partial}_\varphi^* u_k(z)|^2 e^{-|z|^\alpha} \, d\lambda(z) &= \int_{\mathbb{C}} |-kz^{k-1} + \beta z^{\beta+k-1} \overline{z}^\beta|^2 \, e^{-|z|^\alpha} \, d\lambda(z) \\
&= 2\pi \int_0^\infty (k^2 r^{2k-1} - 2k\beta r^{2\beta+2k-1} + \beta^2 r^{4\beta+2k-1}) \, e^{-r^{2\beta}} \, dr \\
&= 2\pi \left[\frac{k^2}{2\beta} \Gamma\left(\frac{k}{\beta}\right) - k\,\Gamma\left(\frac{k}{\beta} + 1\right) + \frac{\beta}{2} \Gamma\left(\frac{k}{\beta} + 2\right)\right] \\
&= \pi\beta\,\Gamma\left(\frac{k}{\beta} + 1\right).
\end{aligned}$$

If $\alpha = 2$, then it follows that condition (11.30) is not satisfied. For this purpose, we consider the integral

$$\int_R^\infty r^{2k+1} e^{-r^2} \, dr,$$

and substituting $r^2 = s$, we obtain

$$\int_R^\infty r^{2k+1} e^{-r^2} \, dr = \int_{R^2}^\infty s^k e^{-s} \, ds.$$

Now k-times integrating by parts, we get

$$\int_{R^2}^{\infty} s^k e^{-s}\, ds = e^{-R^2} R^{2k} + k \int_{R^2}^{\infty} s^{k-1} e^{-s}\, ds = e^{-R^2}\, k! \sum_{j=0}^{k} \frac{R^{2j}}{j!}.$$

Observe that for $\beta = 1$, we have

$$\Gamma\left(\frac{k}{\beta} + 1\right) = \Gamma\left(\frac{k}{\beta} + \frac{1}{\beta}\right) = k!.$$

There is $\epsilon_0 > 0$ such that for each $R > 0$, there exists $k \in \mathbb{N}$ such that

$$e^{-R^2} \sum_{j=0}^{k} \frac{R^{2j}}{j!} > \epsilon_0,$$

so that condition (11.30) is not satisfied for $\alpha = 2$. This means that the $\overline{\partial}$-Neumann operator N_φ on $L^2(\mathbb{C}, e^{-|z|^2})$ fails to be compact, and as $N_\varphi = S^*S$, where S is the canonical solution operator to $\overline{\partial}$, the canonical solution operator S also fails to be compact (compare Theorem 2.22).

Another proof of this fact uses spectral theory: from (10.3) we know that

$$\Box_\varphi u = \overline{\partial}\,\overline{\partial}_\varphi^* u = -\frac{\partial^2 u}{\partial z \partial \overline{z}} + \overline{z}\frac{\partial u}{\partial \overline{z}} + u,$$

and hence it follows immediately that the whole space $A^2(\mathbb{C}, e^{-|z|^2})$ is a subspace of the eigenspace to the eigenvalue 1 of the operator $\Box_\varphi$, which means that the essential spectrum of $\Box_\varphi$ is nonempty and N_φ fails to be compact by Proposition 9.39.

In the next examples, we consider the decoupled $\mathcal{C}^2$-weights

$$\varphi(z_1, z_2, \ldots, z_n) = \varphi_1(z_1) + \varphi_2(z_2) + \cdots + \varphi_n(z_n)$$

and follow an idea of Schneider [88] (see also the results in Chapter 3).

Proposition 15.1. *Suppose that $n \geq 2$ and that there exists ℓ such that $A^2(\mathbb{C}, e^{-\varphi_\ell})$ is infinite dimensional. Suppose also that $1 \in L^2(\mathbb{C}, e^{-\varphi_j})$ for all j. Suppose finally that $\overline{z}_k \in L^2(\mathbb{C}, e^{-\varphi_k})$ for some $k \neq \ell$. Then the canonical solution operator to $\overline{\partial}$ fails to be compact even on the space $A^2_{(0,1)}(\mathbb{C}^n, e^{-\varphi})$.*

Proof. Let P_k denote the Bergman projection

$$P_k : L^2(\mathbb{C}, e^{-\varphi_k}) \longrightarrow A^2(\mathbb{C}, e^{-\varphi_k}).$$

It is clear that the function $(\overline{z}_k - P_k\overline{z}_k)$ is not zero. Let $(f_\nu)_\nu$ be an infinite orthonormal system in $A^2(\mathbb{C}, e^{-\varphi_\ell})$ and define

$$h_\nu(z) := f_\nu(z_\ell)(\overline{z}_k - P_k\overline{z}_k).$$

Then $(h_\nu)_\nu$ is an orthogonal family in $A^2(\mathbb{C}^n, e^{-\varphi})^\perp$. To see this, let $g \in A^2(\mathbb{C}^n, e^{-\varphi})$ and note that $(v, P_k\overline{z}_k)_{\varphi_k} = (v, \overline{z}_k)_{\varphi_k}$ for $v \in A^2(\mathbb{C}, e^{-\varphi_k})$, which implies that $(g, h_\nu)_\varphi = 0$. In addition, we have $\overline{\partial} h_\nu = f_\nu(z_\ell) d\overline{z}_k$.

Hence $(\overline{\partial} h_\nu)_\nu$ constitutes a bounded sequence in $A^2_{(0,1)}(\mathbb{C}^n, e^{-\varphi})$, and for the canonical solution operator S, we have $S(f_\nu(z_\ell)d\overline{z}_k) = h_\nu$, and since $(h_\nu)_\nu$ is an orthogonal family, it has no convergent subsequence, which implies the result. □

Remark 15.2. If the conditions of Proposition 15.1 are satisfied, then the corresponding $\overline{\partial}$-Neumann operator $N_{\varphi,1}$ also fails to be compact, which follows from Proposition 11.27.

In the following example, we consider the $\overline{\partial}$-Neumann operator $N_{\varphi,1}$ for a decoupled weight φ.

Example. Let $\varphi(z_1, z_2) = |z_1|^2 + |z_2|^2$ and consider the corresponding $\overline{\partial}$-Neumann operator $N_{\varphi,1}$. We will investigate the following sequence of $(0,1)$-forms:

$$u_k(z_1, z_2) = \psi_k(z_1)\, d\overline{z}_2,$$

where $\psi_k(z_1) = \frac{z_1^k}{\sqrt{\pi k!}}$ for $k \in \mathbb{N}$. It follows that $\overline{\partial} u_k = 0$ for each $k \in \mathbb{N}$ and

$$\overline{\partial}^*_\varphi u_k(z_1, z_2) = \overline{z}_2 \psi_k(z_1).$$

This implies

$$\Box_{\varphi,1}\, u_k = u_k \quad \text{and} \quad N_{\varphi,1}\, u_k = u_k$$

for each $k \in \mathbb{N}$. The set $\{u_k \,:\, k \in \mathbb{N}\}$ is a bounded set of mutually orthogonal $(0,1)$-forms in $L^2_{(0,1)}(\mathbb{C}^n, e^{-\varphi})$. As $N_{\varphi,1}\, u_k = u_k$, it follows that $N_{\varphi,1}$ fails to be compact.

The following computation shows that condition (11.30) is not satisfied for the $(0,1)$-forms u_k, where we consider $\int_{\mathbb{C}^2\setminus Q_R}$ instead of $\int_{\mathbb{C}^2\setminus \mathbb{B}_R}$, where

$$Q_R = \{(z_1, z_2) \,:\, |z_1| < R,\, |z_2| < R\}.$$

We have

$$\int_{\mathbb{C}^2\setminus Q_R} |u_k(z_1, z_2)|^2 e^{-|z_1|^2 - |z_2|^2}\, d\lambda(z_1, z_2) = \frac{4\pi}{k!} \int_0^R \left(\int_R^\infty r_1^{2k+1} e^{-r_1^2}\, dr_1 \right) r_2\, e^{-r_2^2}\, dr_2$$
$$+ \frac{4\pi}{k!} \int_R^\infty \left(\int_0^\infty r_1^{2k+1} e^{-r_1^2}\, dr_1 \right) r_2\, e^{-r_2^2}\, dr_2.$$

After the substitution $r_1^2 = s$, the first integral is equal to

$$\frac{2\pi}{k!}\int_0^R \left(\int_{R^2}^{\infty} s^k e^{-s}\, ds\right) r_2\, e^{-r_2^2}\, dr_2.$$

As in the above example, we get

$$\int_{R^2}^{\infty} s^k e^{-s}\, ds = e^{-R^2}\, k! \sum_{j=0}^{k} \frac{R^{2j}}{j!},$$

and finally substituting $r_2^2 = t$,

$$\begin{aligned}\frac{2\pi}{k!}\int_0^R \left(\int_{R^2}^{\infty} s^k e^{-s}\, ds\right) r_2\, e^{-r_2^2}\, dr_2 &= \pi e^{-R^2} \sum_{j=0}^{k} \frac{R^{2j}}{j!} \int_0^{R^2} e^{-t}\, dt \\ &= \pi e^{-R^2} \sum_{j=0}^{k} \frac{R^{2j}}{j!} (1 - e^{-R^2}).\end{aligned}$$

On the right-hand side of (11.30), we only have the term

$$\begin{aligned}&\frac{1}{\pi k!} \int_{\mathbb{C}^2} |z_1|^{2k}\, |\overline{z}_2|^2 e^{-|z_1|^2 - |z_2|^2}\, d\lambda(z_1, z_2) \\ &\quad = \frac{4\pi}{k!} \int_0^{\infty} r_1^{2k+1} e^{-r_1^2}\, dr_1 \int_0^{\infty} r_2^3 e^{-r_2^2}\, dr_2 \\ &\quad = \pi.\end{aligned}$$

This implies

$$\int_{\mathbb{C}^2 \setminus Q_R} |u_k(z_1, z_2)|^2 e^{-|z_1|^2 - |z_2|^2}\, d\lambda(z_1, z_2) \geq \pi e^{-R^2} \sum_{j=0}^{k} \frac{R^{2j}}{j!} (1 - e^{-R^2}).$$

There is $\epsilon_0 > 0$ such that for each $R > 0$, there exists $k \in \mathbb{N}$ such that

$$e^{-R^2} \sum_{j=0}^{k} \frac{R^{2j}}{j!} (1 - e^{-R^2}) > \epsilon_0,$$

so that condition (11.30) is not satisfied.

Finally, we discuss the compactness of $N_{\varphi,1}$ and $N_{\varphi,2}$ in $\mathbb{C}^2$ in a more general setting: let $\varphi(z_1, z_2) = \varphi_1(z_1) + \varphi_2(z_2)$. The eigenvalues of the Levi matrix are

$$\frac{\partial^2 \varphi_1}{\partial z_1 \partial \overline{z}_1} \quad \text{and} \quad \frac{\partial^2 \varphi_2}{\partial z_2 \partial \overline{z}_2}.$$

If the $(0,1)$-form $u = u_1 d\overline{z}_1 + u_2 d\overline{z}_2$ belongs to $\mathrm{dom}(\Box_{\varphi,1})$, then

$$\Box_{\varphi,1} u = \left(-\frac{\partial^2 u_1}{\partial z_1 \partial \overline{z}_1} - \frac{\partial^2 u_1}{\partial z_2 \partial \overline{z}_2} + \frac{\partial \varphi_1}{\partial z_1} \frac{\partial u_1}{\partial \overline{z}_1} + \frac{\partial \varphi_2}{\partial z_2} \frac{\partial u_1}{\partial \overline{z}_2} + \frac{\partial^2 \varphi_1}{\partial z_1 \partial \overline{z}_1} u_1 \right) d\overline{z}_1$$
$$+ \left(-\frac{\partial^2 u_2}{\partial z_1 \partial \overline{z}_1} - \frac{\partial^2 u_2}{\partial z_2 \partial \overline{z}_2} + \frac{\partial \varphi_1}{\partial z_1} \frac{\partial u_2}{\partial \overline{z}_1} + \frac{\partial \varphi_2}{\partial z_2} \frac{\partial u_2}{\partial \overline{z}_2} + \frac{\partial^2 \varphi_2}{\partial z_2 \partial \overline{z}_2} u_2 \right) d\overline{z}_2,$$

and for $V = v\, d\overline{z}_1 \wedge d\overline{z}_2 \in \mathrm{dom}(\Box_{\varphi,2})$, we have that $\Box_{\varphi,2} V$ equals

$$\left(-\frac{\partial^2 v}{\partial z_1 \partial \overline{z}_1} - \frac{\partial^2 v}{\partial z_2 \partial \overline{z}_2} + \frac{\partial \varphi_1}{\partial z_1} \frac{\partial v}{\partial \overline{z}_1} + \frac{\partial \varphi_2}{\partial z_2} \frac{\partial v}{\partial \overline{z}_2} + \frac{\partial^2 \varphi_1}{\partial z_1 \partial \overline{z}_1} v + \frac{\partial^2 \varphi_2}{\partial z_2 \partial \overline{z}_2} v \right) d\overline{z}_1 \wedge d\overline{z}_2.$$

Now suppose that $A^2(\mathbb{C}, e^{-\varphi_1})$ is infinite dimensional, that $1 \in L^2(\mathbb{C}, e^{-\varphi_j})$ for $j = 1, 2$, that $\overline{z}_2 \in L^2(\mathbb{C}, e^{-\varphi_2})$, and finally that

$$\frac{\partial^2 \varphi_1(z_1)}{\partial z_1 \partial \overline{z}_1} + \frac{\partial^2 \varphi_2(z_2)}{\partial z_2 \partial \overline{z}_2} \to \infty \quad \text{as } |z_1|^2 + |z_2|^2 \to \infty.$$

Then $N_{\varphi,2}$ is compact, but $N_{\varphi,1}$ fails to be compact.

Our assumptions imply that $N_{\varphi,2}$ is compact by Proposition 11.16. In addition, we have that

$$N_{\varphi,2} = S_2^* S_2,$$

where S_2 is the canonical solution operator for $\overline{\partial}$ for $(0,2)$-forms. Hence S_2 is also compact. Now suppose that $N_{\varphi,1}$ is compact. Since

$$N_{\varphi,1} = S_1^* S_1 + S_2 S_2^*,$$

this would imply that S_1 is compact, contradicting Proposition 15.1. We get the same conclusion if we apply Proposition 11.27.

The above assumptions are all satisfied, for instance, for the weight functions

$$\varphi(z_1, z_2) = |z_1|^{2k} + |z_2|^{2k}, \quad k = 2, 3, \ldots.$$

15.3 Notes

A general characterization of compactness of the $\overline{\partial}$-Neumann operator for smoothly bounded domains in terms of geometric properties of the boundary of the domain is not known. Christ and Fu [18] showed: If $\Omega \subset \mathbb{C}^2$ is a smoothly bounded pseudoconvex Hartogs domain (i. e., with $(z, w) \in \Omega$, we have $(z, e^{i\theta} w) \in \Omega$ for every $\theta \in \mathbb{R}$), then the $\overline{\partial}$-Neumann operator N is compact if and only if property (P) holds. Also, analytic discs in the boundary are an obstruction to compactness. However, there are examples of Hartogs domains that fail property (P) although their boundaries contain no analytic discs.

In the case of a weighted space $L^2(\mathbb{C}, e^{-\varphi})$, the compactness of the $\overline{\partial}$-Neumann operator can be characterized in terms of the weight function (see Theorem 13.3); a different approach to this problem, using estimates of the Bergman kernel, is contained in [74]. An analogous characterization for the compactness of the $\overline{\partial}$-Neumann operator on $L^2(\mathbb{C}^n, e^{-\varphi})$, $n \geq 2$, has also not been found up to now; see [7] for various partial results.

16 The ∂-complex on the Segal–Bargmann space

We use the powerful classical methods of the $\overline{\partial}$-complex based on the theory of unbounded densely defined operators on Hilbert spaces (see Chapter 4 or [91]) to study the ∂-complex on an appropriate Hilbert space. The main difference to the classical theory is that the underlying Hilbert space is now not an L^2-space but a closed subspace of an L^2-space, the Segal–Bargmann space of entire functions $A^2(\mathbb{C}^n, e^{-|z|^2})$. It is well known that the differentiation with respect to z_j defines an unbounded operator on $A^2(\mathbb{C}^n, e^{-|z|^2})$. We consider the operator

$$\partial f = \sum_{j=1}^{n} \frac{\partial f}{\partial z_j}\, dz_j,$$

which is densely defined on $A^2(\mathbb{C}^n, e^{-|z|^2})$ and maps to $A^2_{1,0}(\mathbb{C}^n, e^{-|z|^2})$, the space of $(1,0)$-forms with coefficients in $A^2(\mathbb{C}^n, e^{-|z|^2})$. In general, we get the ∂-complex

$$A^2_{(p-1,0)}(\mathbb{C}^n, e^{-|z|^2}) \underset{\partial^*}{\overset{\partial}{\rightleftarrows}} A^2_{(p,0)}(\mathbb{C}^n, e^{-|z|^2}) \underset{\partial^*}{\overset{\partial}{\rightleftarrows}} A^2_{(p+1,0)}(\mathbb{C}^n, e^{-|z|^2}),$$

where $1 \le p \le n-1$, and ∂^* denotes the adjoint operator of ∂.

We will choose the domain $\operatorname{dom}(\partial)$ in such a way that ∂ becomes a closed operator on $A^2(\mathbb{C}^n, e^{-|z|^2})$. In addition, we get that the corresponding complex Laplacian

$$\tilde{\Box}_p = \partial^*\partial + \partial\partial^*$$

with $\operatorname{dom}(\tilde{\Box}_p) = \{f \in \operatorname{dom}(\partial) \cap \operatorname{dom}(\partial^*) : \partial f \in \operatorname{dom}(\partial^*) \text{ and } \partial f^* \in \operatorname{dom}(\partial)\}$ acts as an unbounded self-adjoint operator on $A^2_{(p,0)}(\mathbb{C}^n, e^{-|z|^2})$. We point out that in this case the complex Laplacian is a differential operator of order one. Nevertheless, we can use the general features of a Laplacian for these differential operators of order one.

Using an estimate analogous to the basic estimate for the $\overline{\partial}$-complex, we obtain that $\tilde{\Box}_p$ has a bounded inverse $\tilde{N}_p : A^2_{(p,0)}(\mathbb{C}^n, e^{-|z|^2}) \longrightarrow A^2_{(p,0)}(\mathbb{C}^n, e^{-|z|^2})$, and we show that $\tilde{N}_p$ is even compact. In addition, we compute the spectrum of $\tilde{\Box}_p$.

The inspiration for this comes from quantum mechanics, where the annihilation operator a_j can be represented by the differentiation with respect to z_j on $A^2(\mathbb{C}^n, e^{-|z|^2})$ and its adjoint, the creation operator a_j^*, the multiplication by z_j, both operators being unbounded densely defined, see Chapter 1. We can show that $A^2(\mathbb{C}^n, e^{-|z|^2})$ with this action of a_j and a_j^* is an irreducible representation M of the Heisenberg group; by the Stone–von Neumann theorem it is the only one up to unitary equivalence. Physically, M can be thought of as the Hilbert space of a harmonic oscillator with n degrees of freedom and Hamiltonian operator

https://doi.org/10.1515/9783111182926-016

$$H = \sum_{j=1}^{n} \frac{1}{2}(P_j^2 + Q_j^2) = \sum_{j=1}^{n} \frac{1}{2}(a_j^* a_j + a_j a_j^*).$$

Finally, we investigate operators of the form $Du = \sum_{j=1}^n p_j(u)\, dz_j$, where $u \in A^2(\mathbb{C}^n, e^{-|z|^2})$ and $p_j(\frac{\partial}{\partial z_1}, \dots, \frac{\partial}{\partial z_n})$ are polynomial differential operators with constant coefficients. Differential operators of polynomial type on the Segal–Bargmann space were also investigated by Newman and Shapiro [78, 79] and Render [81] in the real analytic setting. Replacing ∂ by D, we get the corresponding complex Laplacian $\tilde{\square}_D = DD^* + D^* D$, for which we can use duality and the machinery of the $\overline{\partial}$-Neumann operator [64, 65] to prove the existence and boundedness of the inverse to $\tilde{\square}_D$ and to find the canonical solutions to the inhomogeneous equations $Du = \alpha$ and $D^* v = \beta$.

16.1 Unbounded operators and the complex Laplacian

From Sections 1.4 and 1.5 we know that the operators $a(f) = \frac{\partial f}{\partial z}$ and $a^*(f) = zf$ are densely defined unbounded operators on $A^2(\mathbb{C}, e^{-|z|^2})$.

Lemma 16.1. *Let* $\mathrm{dom}(a) = \{f \in A^2(\mathbb{C}, e^{-|z|^2}) : a(f) \in A^2(\mathbb{C}, e^{-|z|^2})\}$. *The operators a and a^* are densely defined operators on $A^2(\mathbb{C}, e^{-|z|^2})$ with closed graphs.*

Proof. By general properties of unbounded operators it suffices to prove the statement for a (see Section 4.1). Let $(f_j)_j$ be a sequence in $\mathrm{dom}(a)$ such that $\lim_{j\to\infty} f_j = f$ and $\lim_{j\to\infty} a(f_j) = g$. We have to show that $f \in \mathrm{dom}(a)$ and $a(f) = g$. By (1.23) it follows that $(f_j)_j$ converges uniformly on each compact subset of $\mathbb{C}$ to f, and the same is true for the derivatives: $\lim_{j\to\infty} f_j' = f'$. This implies that $f' = a(f) = g$. We supposed that $g \in A^2(\mathbb{C}, e^{-|z|^2})$, and we know that taking primitives does not leave $A^2(\mathbb{C}, e^{-|z|^2})$. Therefore we get that $f \in \mathrm{dom}(a)$, which proves the statement. □

For the rest of this section, we consider the Segal–Bargmann space in several variables with weight $\varphi(z) = |z_1|^2 + \cdots + |z_n|^2$. We will denote the derivative with respect to z by ∂, and in the following, we will consider the ∂-complex for the Segal–Bargmann space in several variables

$$A^2_{(p-1,0)}(\mathbb{C}^n, e^{-|z|^2}) \underset{\partial^*}{\overset{\partial}{\rightleftarrows}} A^2_{(p,0)}(\mathbb{C}^n, e^{-|z|^2}) \underset{\partial^*}{\overset{\partial}{\rightleftarrows}} A^2_{(p+1,0)}(\mathbb{C}^n, e^{-|z|^2}),$$

where $A^2_{(p,0)}(\mathbb{C}^n, e^{-|z|^2})$ denotes the Hilbert space of $(p,0)$-forms with coefficients in $A^2(\mathbb{C}^n, e^{-|z|^2})$, and

$$\partial f = \sum_{|J|=p}{}' \sum_{j=1}^{n} \frac{\partial f_J}{\partial z_j}\, dz_j \wedge dz_J$$

for a $(p,0)$-form

$$f = \sum_{|J|=p}{}' f_J \, dz_J$$

with summation over increasing multiindices $J = (j_1, \ldots, j_p)$, $1 \le p \le n-1$; and we take

$$\mathrm{dom}(\partial) = \{f \in A^2_{(p,0)}(\mathbb{C}^n, e^{-|z|^2}) : \partial f \in A^2_{(p+1,0)}(\mathbb{C}^n, e^{-|z|^2})\}.$$

Now let

$$\tilde{\Box}_p = \partial^* \partial + \partial \partial^*$$

with $\mathrm{dom}(\tilde{\Box}_p) = \{f \in \mathrm{dom}(\partial) \cap \mathrm{dom}(\partial^*) : \partial f \in \mathrm{dom}(\partial^*) \text{ and } \partial^* f \in \mathrm{dom}(\partial)\}$.

Then $\tilde{\Box}_p$ acts as an unbounded self-adjoint operator on $A^2_{(p,0)}(\mathbb{C}^n, e^{-|z|^2})$ (see Proposition 4.54).

Remark 16.2. (a) We remark that a $(1,0)$-form $g = \sum_{j=1}^n g_j \, dz_j$ with holomorphic coefficients is invariant under the pull-back by a holomorphic map $F = (F_1, \ldots, F_n) : \mathbb{C}^n \longrightarrow \mathbb{C}^n$. We have

$$F^* g = \sum_{l=1}^n g_l \, dF_l = \sum_{j=1}^n \left(\sum_{l=1}^n g_l \frac{\partial F_l}{\partial z_j} \right) dz_j,$$

where we used the fact that

$$dF_l = \partial F_l + \overline{\partial} F_l = \sum_{j=1}^n \frac{\partial F_l}{\partial z_j} \, dz_j + \sum_{j=1}^n \frac{\partial F_l}{\partial \overline{z}_j} \, d\overline{z}_j = \sum_{j=1}^n \frac{\partial F_l}{\partial z_j} \, dz_j.$$

The expressions $\frac{\partial F_l}{\partial z_j}$ are holomorphic.

(b) For $p = 0$ and a function $f \in \mathrm{dom}(\tilde{\Box}_0)$, we have

$$\tilde{\Box}_0 f = \partial^* \partial f = \sum_{j=1}^n z_j \frac{\partial f}{\partial z_j}.$$

If $1 \le p \le n-1$ and $f = \sum_{|J|=p}{}' f_J \, dz_J \in \mathrm{dom}(\tilde{\Box}_p)$ is a $(p,0)$-form, then we have

$$\tilde{\Box}_p f = \sum_{|J|=p}{}' \left(\sum_{k=1}^n z_k \frac{\partial f_J}{\partial z_k} + p f_J \right) dz_J. \tag{16.1}$$

For $p = n$ and a $(n,0)$-form $F \in \mathrm{dom}(\tilde{\Box}_n)$ (here we identify the $(n,0)$-form with a function), we have

$$\tilde{\Box}_n F = \partial \partial^* F = \sum_{j=1}^n z_j \frac{\partial F}{\partial z_j} + nF.$$

16.2 Basic estimates and compactness

The $(p,0)$-forms with polynomial components are dense in $A^2_{(p,0)}(\mathbb{C}^n, e^{-|z|^2})$. In addition, we have

Lemma 16.3. *The $(p,0)$-forms with polynomial components are also dense in* $\operatorname{dom}(\partial) \cap \operatorname{dom}(\partial^*)$ *endowed with the graph norm*

$$u \mapsto (\|u\|^2 + \|\partial u\|^2 + \|\partial^* u\|^2)^{1/2}.$$

Proof. Let $u = \sum_{|J|=p}{}' u_J \, dz_J \in \operatorname{dom}(\partial) \cap \operatorname{dom}(\partial^*)$ and consider the partial sums of the Fourier series expansions of

$$u_J = \sum_\alpha u_{J,\alpha} \varphi_\alpha,$$

where

$$\varphi_\alpha(z) = \frac{z^\alpha}{\sqrt{\pi^n \alpha!}} \quad \text{and} \quad \sum_\alpha |u_{J,\alpha}|^2 < \infty \tag{16.2}$$

with $\alpha! = \alpha_1! \dots \alpha_n!$.

We have that $\partial u \in A^2_{(p+1,0)}(\mathbb{C}^n, e^{-|z|^2})$ and $\partial^* u \in A^2_{(p-1,0)}(\mathbb{C}^n, e^{-|z|^2})$. Hence the partial sums of the Fourier series of the components of ∂u converge to the components of ∂u in $A^2_{(p+1,0)}(\mathbb{C}^n, e^{-|z|^2})$, and the partial sums of the Fourier series of the components of $\partial^* u$ converge to the components of $\partial^* u$ in $A^2_{(p-1,0)}(\mathbb{C}^n, e^{-|z|^2})$. □

As an immediate consequence of (16.17), (6.11), and Lemma 16.3, we get the basic estimate.

Lemma 16.4. *Let $1 \le p \le n-1$, and let $u = \sum'_{|J|=p} u_J \, dz_J \in A^2_{(p,0)}(\mathbb{C}^n, e^{-|z|^2})$. Suppose that $u \in \operatorname{dom}(\partial) \cap \operatorname{dom}(\partial^*)$. Then*

$$\|u\|^2 \le \frac{1}{p}(\|\partial u\|^2 + \|\partial^* u\|^2). \tag{16.3}$$

Proof. Taking $(p,0)$-forms with polynomial components and replacing $\bar\partial$ by ∂ and taking into account that $\varphi(z) = |z|^2$, from (6.11) we derive that

$$\|\partial u\|^2 + \|\partial^* u\|^2 = \sum_{|J|=p}{}' \sum_{j=1}^n \left\| \frac{\partial u_J}{\partial z_j} \right\|^2 + p \sum_{|J|=p}{}' \int_{\mathbb{C}^n} |u_J|^2 e^{-|z|^2} \, d\lambda. \tag{16.4}$$

Since $(p,0)$-forms with polynomial components are dense in $\operatorname{dom}(\partial) \cap \operatorname{dom}(\partial^*)$ with the graph-norm (Lemma 16.3), we immediately get (16.3). □

At this point, we recall that (16.3) can be seen as a generalization of the uncertainty principle (see Section 1.5).

Now we can use the machinery of the classical $\bar{\partial}$-Neumann operator to show the following results.

Lemma 16.5. *Both operators ∂ and ∂^* have closed range.*

If we endow $\mathrm{dom}(\partial) \cap \mathrm{dom}(\partial^*)$ *with the graph-norm* $(\|\partial f\|^2 + \|\partial^* f\|^2)^{1/2}$, *then the dense subspace* $\mathrm{dom}(\partial) \cap \mathrm{dom}(\partial^*)$ *of* $A^2_{(p,0)}(\mathbb{C}^n, e^{-|z|^2})$ *becomes a Hilbert space.*

Proof. We notice that $\ker \partial = (\mathrm{im}\, \partial^*)^\perp$, which implies that

$$(\ker \partial)^\perp = \overline{\mathrm{im}\, \partial^*} \subseteq \ker \partial^*.$$

If $u \in \ker \partial \cap \ker \partial^*$, then by (16.3) we have that $u = 0$. Hence

$$(\ker \partial)^\perp = \ker \partial^*. \tag{16.5}$$

If $u \in \mathrm{dom}(\partial) \cap (\ker \partial)^\perp$, then $u \in \ker \partial^*$, and (16.3) implies

$$\|u\| \leq \frac{1}{p}\, \|\partial u\|.$$

Now we can use general results on unbounded operators on Hilbert spaces (see Chapter 4) to show that $\mathrm{im}\, \partial$ and $\mathrm{im}\, \partial^*$ are closed. The last statement follows again by (16.3); see also Chapter 4. □

The next result describes the implication of the basic estimates (16.3) for the $\tilde{\Box}$-operator.

Theorem 16.6. *The operator* $\tilde{\Box} : \mathrm{dom}(\tilde{\Box}) \longrightarrow A^2_{(p,0)}(\mathbb{C}^n, e^{-|z|^2})$ *is bijective and has a bounded inverse*

$$\tilde{N} : A^2_{(p,0)}(\mathbb{C}^n, e^{-|z|^2}) \longrightarrow \mathrm{dom}(\tilde{\Box}).$$

In addition,

$$\|\tilde{N} u\| \leq \frac{1}{p}\, \|u\| \tag{16.6}$$

for each $u \in A^2_{(p,0)}(\mathbb{C}^n, e^{-|z|^2})$.

Proof. The proof follows very closely the proof of Proposition 4.59 (see also [16, Section 4.4]). □

Using the classical $\bar{\partial}$-Neumann calculus, we obtain the following:

Theorem 16.7. *Let $\tilde{N}_p$ denote the inverse of $\tilde{\Box}$ on $A^2_{(p,0)}(\mathbb{C}^n, e^{-|z|^2})$. Then*

$$\tilde{N}_{p+1}\partial = \partial \tilde{N}_p \tag{16.7}$$

on $\operatorname{dom}(\partial)$, *and*

$$\tilde{N}_{p-1}\partial^* = \partial^* \tilde{N}_p \tag{16.8}$$

on $\operatorname{dom}(\partial^*)$.

In addition, we have that $\partial^* \tilde{N}_p$ is zero on $(\ker \partial)^\perp$.

Proof. See the proof of Proposition 4.62 (see also [16, Section 4.4]). □

Now we can also prove a solution formula for the equation $\partial u = \alpha$, where α is a given $(p,0)$-form in $A^2_{(p,0)}(\mathbb{C}^n, e^{-|z|^2})$ with $\partial \alpha = 0$.

Theorem 16.8. *Let $\alpha \in A^2_{(p,0)}(\mathbb{C}^n, e^{-|z|^2})$ with $\partial \alpha = 0$. Then $u_0 = \partial^* \tilde{N}_p \alpha$ is the canonical solution of $\partial u = \alpha$, which means that $\partial u_0 = \alpha$ and $u_0 \in (\ker \partial)^\perp = \operatorname{im} \partial^*$, and*

$$\|\partial^* \tilde{N}_p \alpha\| \le p^{-1/2} \, \|\alpha\|. \tag{16.9}$$

Proof. See the proof of Proposition 4.63. □

Now we discuss a different approach to the operator $\tilde{N}$, which is related to the quadratic form

$$Q(u,v) = (\partial u, \partial v) + (\partial^* u, \partial^* v).$$

For this purpose, we consider the embedding

$$\iota : \operatorname{dom}(\partial) \cap \operatorname{dom}(\partial^*) \longrightarrow A^2_{(p,0)}(\mathbb{C}^n, e^{-|z|^2}),$$

where $\operatorname{dom}(\partial) \cap \operatorname{dom}(\partial^*)$ is endowed with the graph norm

$$u \mapsto (\|\partial u\|^2 + \|\partial^* u\|^2)^{1/2}.$$

The graph norm stems from the inner product

$$Q(u,v) = (u,v)_Q = (\tilde{\Box} u, v) = (\partial u, \partial v) + (\partial^* u, \partial^* v).$$

The basic estimates (16.3) imply that ι is a bounded operator with operator norm

$$\|\iota\| \le \frac{1}{\sqrt{p}}.$$

By (16.3) it follows in addition that $\operatorname{dom}(\partial) \cap \operatorname{dom}(\partial^*)$ endowed with the graph norm $u \mapsto (\|\partial u\|^2 + \|\partial^* u\|^2)^{1/2}$ is a Hilbert space (see Lemma 16.5).

Since $(u,v) = (u, \iota v)$, we have that $(u,v) = (\iota^* u, v)_Q$.

For $u \in A^2_{(p,0)}(\mathbb{C}^n, e^{-|z|^2})$ and $v \in \operatorname{dom}(\partial) \cap \operatorname{dom}(\partial^*)$, we get

$$(u,v) = (\tilde{\Box}\tilde{N}u, v) = ((\partial\partial^* + \partial^*\partial)\tilde{N}u, v) = (\partial^*\tilde{N}u, \partial^* v) + (\partial\tilde{N}u, \partial v). \tag{16.10}$$

Equation (16.10) suggests that as an operator to $\operatorname{dom}(\partial) \cap \operatorname{dom}(\partial^*)$, $\tilde{N}$ coincides with ι^* and as an operator to $A^2_{(p,0)}(\mathbb{C}^n, e^{-|z|^2})$, $\tilde{N}$ is equal to $\iota \circ \iota^*$ (see (4.42) or [91, p. 30] for the details).

Hence $\tilde{N}$ is compact if and only if the embedding

$$\iota : \operatorname{dom}(\partial) \cap \operatorname{dom}(\partial^*) \longrightarrow A^2_{(p,0)}(\mathbb{C}^n, e^{-|z|^2})$$

is compact. We will use this to prove the following theorem.

Theorem 16.9. *The operator* $\tilde{N} : A^2_{(p,0)}(\mathbb{C}^n, e^{-|z|^2}) \longrightarrow A^2_{(p,0)}(\mathbb{C}^n, e^{-|z|^2})$, $1 \le p \le n$, *is compact.*

Proof. First, we consider the case $p = 1$. We use (16.17) for the graph norm on $\operatorname{dom}(\partial) \cap \operatorname{dom}(\partial^*)$ and indicate that it suffices to consider one component u_j of the $(1,0)$-form u. For this purpose, we will denote u_j by f. We have to handle

$$\sum_{k=1}^{n} \int_{\mathbb{C}^n} \left| \frac{\partial f}{\partial z_k} \right|^2 e^{-|z|^2}\, d\lambda + \int_{\mathbb{C}^n} |f|^2 e^{-|z|^2}\, d\lambda$$

for the graph norm. We use the complete orthonormal system (16.2) $(\varphi_\alpha)_\alpha$ of $A^2(\mathbb{C}^n, e^{-|z|^2})$. Let $f = \sum_\alpha f_\alpha \varphi_\alpha$ be an element of $\operatorname{dom}(\partial) \cap \operatorname{dom}(\partial^*)$. We have $\iota(f) = f$, and hence

$$\iota(f) = \sum_\alpha (f, \varphi_\alpha)\varphi_\alpha$$

in $A^2(\mathbb{C}^n, e^{-|z|^2})$. The basis elements φ_α are normalized in $A^2(\mathbb{C}^n, e^{-|z|^2})$. First, we have to compute the graph norm of the basis elements φ_α. Notice that

$$\frac{\partial \varphi_\alpha}{\partial z_k} = \frac{1}{\sqrt{\pi^n}} \frac{z_1^{\alpha_1}}{\sqrt{\alpha_1!}} \cdots \frac{z_{k-1}^{\alpha_{k-1}}}{\sqrt{\alpha_{k-1}!}} \frac{\alpha_k z_k^{\alpha_k - 1}}{\sqrt{\alpha_k!}} \frac{z_{k+1}^{\alpha_{k+1}}}{\sqrt{\alpha_{k+1}!}} \cdots \frac{z_n^{\alpha_n}}{\sqrt{\alpha_n!}} = \sqrt{\alpha_k}\, \varphi_{(\alpha k - 1)},$$

where $(\alpha k - 1) = (\alpha_1, \dots \alpha_{k-1}, \alpha_k - 1, \alpha_{k+1}, \dots, \alpha_n)$. Hence the graph norm of the basis elements φ_α equals

$$\|\varphi_\alpha\|_Q = \left(\|\varphi_\alpha\|^2 + \sum_{k=1}^{n} \left\| \frac{\partial \varphi_\alpha}{\partial z_k} \right\|^2 \right)^{1/2} = \sqrt{1 + |\alpha|},$$

where $|\alpha| = \alpha_1 + \cdots + \alpha_n$.

Now let

$$\psi_\alpha = \frac{\varphi_\alpha}{\sqrt{1+|\alpha|}}.$$

Then $(\psi_\alpha)_\alpha$ constitutes a complete orthonormal system in the Hilbert space $\operatorname{dom}(\partial) \cap \operatorname{dom}(\partial^*)$ endowed with the graph norm. Notice that

$$\begin{aligned}(f,\psi_\alpha)_Q = \sum_{k=1}^{n}\left(\frac{\partial f}{\partial z_k}, \frac{\partial \psi_\alpha}{\partial z_k}\right) + (f,\psi_\alpha) &= \sum_{k=1}^{n}\frac{\alpha_k}{\sqrt{1+|\alpha|}} f_\alpha + \frac{1}{\sqrt{1+|\alpha|}} f_\alpha \\ &= \sqrt{1+|\alpha|}\, f_\alpha,\end{aligned}$$

and we have

$$\iota(f) = f = \sum_\alpha (f,\psi_\alpha)_Q\, \psi_\alpha.$$

For the norm of $A^2(\mathbb{C}^n, e^{-|z|^2})$, we have

$$\begin{aligned}\Big\|\iota(f) - \sum_{|\alpha|\le N}(f,\psi_\alpha)_Q\,\psi_\alpha\Big\|^2 &= \Big\|\sum_{|\alpha|\ge N+1}(f,\psi_\alpha)_Q\,\psi_\alpha\Big\|^2 \\ &= \Big\|\sum_{|\alpha|\ge N+1}\frac{1}{\sqrt{1+|\alpha|}}(f,\psi_\alpha)_Q\,\varphi_\alpha\Big\|^2 \\ &= \sum_{|\alpha|\ge N+1}\left|\frac{1}{\sqrt{1+|\alpha|}}(f,\psi_\alpha)_Q\right|^2 \\ &\le \frac{\|f\|_Q^2}{N+2},\end{aligned}$$

where we finally used Bessel's inequality for the Hilbert space $\operatorname{dom}(\partial) \cap \operatorname{dom}(\partial^*)$ endowed with the graph norm. This proves that

$$\iota : \operatorname{dom}(\partial) \cap \operatorname{dom}(\partial^*) \longrightarrow A^2_{(1,0)}(\mathbb{C}^n, e^{-|z|^2})$$

is a compact operator and the same is true for

$$\tilde{N} : A^2_{(1,0)}(\mathbb{C}^n, e^{-|z|^2}) \longrightarrow A^2_{(1,0)}(\mathbb{C}^n, e^{-|z|^2}).$$

For arbitrary p between 1 and n, we can use (16.4) and the same reasoning as before to get the desired conclusion. □

Compare with the $\overline{\partial}$-Neumann operator N on $L^2(\mathbb{C}^n, e^{-|z|^2})$: in this case, N fails to be compact (see Chapter 14). This is related to the fact that the kernel of $\overline{\partial}$ is large (it is the

Segal–Bargmann space) in the case of the $\overline{\partial}$-complex, but the kernel of ∂ consists just of the constant functions in the case of the ∂-complex on the Segal–Bargmann space.

To compute the spectrum of the operator $\tilde{\Box}_p$, we will use Lemma 9.33.

Theorem 16.10. *The spectrum of $\tilde{\Box}_p$, where $0 \le p \le n$, consists of all numbers $m + p$ for $m = 0, 1, 2, \ldots$, where $m + p$ has the multiplicity $\binom{n+m-1}{n-1}\binom{n}{p}$.*

Proof. Recall that the monomials $(\varphi_\alpha)_\alpha$, where $\alpha = (\alpha_1, \ldots, \alpha_n) \in \mathbb{N}_0^n$ is a multi-index, constitute a complete orthonormal system in $A^2(\mathbb{C}^n, e^{-|z|^2})$. We use (16.1) and compute

$$\left(\sum_{k=1}^{n} z_k \frac{\partial \varphi_\alpha}{\partial z_k}\right) + p\varphi_\alpha = (|\alpha| + p)\varphi_\alpha.$$

We use Lemma 9.33 and indicate that there are $\binom{n+|\alpha|-1}{n-1}$ monomials of degree $|\alpha|$. Hence we get the statement about the multiplicity from the fact that $A^2_{(p,0)}(\mathbb{C}^n, e^{-|z|^2})$ is the direct sum of $\binom{n}{p}$ copies of $A^2(\mathbb{C}^n, e^{-|z|^2})$. □

The last result implies also that the inverse $\tilde{N}_p$ of $\tilde{\Box}_p$ is a compact operator with the eigenvalues $1/(m + p)$.

Note that the complex Laplacian $\Box_q$ of the $\overline{\partial}$-complex on $L^2(\mathbb{C}^n, e^{-|z|^2})$ has also the eigenvalues $q + m$ for $m = 0, 1, 2, \ldots$, but each of them has infinite multiplicity (see [72] and Chapter 14).

16.3 Properties of the graph norm and compactness

If we endow $\mathrm{dom}(\partial) \cap \mathrm{dom}(\partial^*)$ with the graph norm, then we obtain Hilbert spaces of entire functions with reproducing kernels. If $1 \le p < n$, then we write for the multi-index $J = (j_1(J), \ldots j_p(J))$ and denote the missing $n - p$ indices by $j_{p+1}(J), \ldots, j_n(J)$. Then we can write (16.4) in the form

$$\begin{aligned}\|\partial u\|^2 + \|\partial^* u\|^2 &= \sum_{|J|=p}{}' \left(\sum_{k=1}^{p}\left(\|u_J\|^2 + \left\|\frac{\partial u_J}{\partial z_{j_k(J)}}\right\|^2\right) + \sum_{k=p+1}^{n}\left\|\frac{\partial u_J}{\partial z_{j_k(J)}}\right\|^2\right)\\ &= \sum_{|J|=p}{}' \left(\sum_{k=1}^{p}\|z_{j_k(J)} u_J\|^2 + \sum_{k=p+1}^{n}\left\|\frac{\partial u_J}{\partial z_{j_k(J)}}\right\|^2\right),\end{aligned}$$

where we used (1.31). For $(p,0)$-forms $u = \sum'_{|J|=p} u_J\, dz_J \in A^2_{(p,0)}(\mathbb{C}^n, e^{-|z|^2})$ belonging to $\mathrm{dom}(\partial) \cap \mathrm{dom}(\partial^*)$, by (16.4), we have

$$\|u\|^2 = \sum_{|J|=p}{}' \|u_J\|^2 \le \frac{1}{p}(\|\partial u\|^2 + \|\partial^* u\|^2),$$

which implies

$$\|u_J\|^2 \leq \frac{1}{p}\left(\sum_{k=1}^{p} \|z_{j_k(J)} u_J\|^2 + \sum_{k=p+1}^{n} \left\|\frac{\partial u_J}{\partial z_{j_k(J)}}\right\|^2\right) \tag{16.11}$$

for each component u_J. For $p = n$, we get

$$\|u_J\|^2 \leq \frac{1}{p} \sum_{k=1}^{p} \|z_{j_k(J)} u_J\|^2. \tag{16.12}$$

The components $u_J dz_J \in \operatorname{dom}(\bar\partial) \cap \operatorname{dom}(\bar\partial^*)$ with the right-hand side of (16.11) or (16.12) as a norm form Hilbert spaces with reproducing kernel, since by (16.11) or (16.12) the point evaluations are continuous linear functionals. We denote these Hilbert spaces by $\mathcal{G}_p$.

The Bergman kernel of $A^2(\mathbb{C}^n, e^{-|z|^2})$ is of the form

$$K(z, w) = \frac{1}{\pi^2} \exp(\langle z, w\rangle),$$

where $\langle z, w\rangle = \sum_{j=1}^n z_j \bar{w}_j$.

Theorem 16.11. *Let $1 \leq p \leq n$. The Bergman kernel of $\mathcal{G}_p$ is of the form*

$$B_p(z, w) = \frac{p}{\pi^n} \frac{d^{p-1}}{d\zeta^{p-1}}\left(\frac{e^\zeta - \sum_{k=0}^{p-1} \frac{\zeta^k}{k!}}{\zeta}\right)\Bigg|_{\zeta = \langle z, w\rangle}. \tag{16.13}$$

Proof. We write $p_k(u)(w) = \frac{\partial u}{\partial w_k}$ and $p_k^*(u)(w) = w_k u(w)$ for $u \in A^2(\mathbb{C}^n, e^{-|z|^2})$ and $k = 1, \ldots, n$. Without loss of generality, we can assume that the graph norm is

$$\frac{1}{p}\left(\sum_{k=1}^{p} \|p_k^*(u)\|^2 + \sum_{k=p+1}^{n} \|p_k(u)\|^2\right).$$

Since $\mathcal{B}_p$ has the reproducing property, we have

$$\begin{aligned} u(z) &= \frac{1}{p} \int_{\mathbb{C}^n} \sum_{k=1}^{p} p_k^*(u)(w)\overline{[p_k^*(v_z)(w)]}\, e^{-|w|^2}\, d\lambda(w) \\ &\quad + \frac{1}{p} \int_{\mathbb{C}^n} \sum_{k=p+1}^{n} p_k(u)(w)\overline{[p_k(v_z)(w)]}\, e^{-|w|^2}\, d\lambda(w) \\ &= \frac{1}{p} \int_{\mathbb{C}^n} u(w) \overline{\left[\left(\sum_{k=1}^{p} p_k p_k^* + \sum_{k=p+1}^{n} p_k^* p_k\right)(v_z)(w)\right]}\, e^{-|w|^2}\, d\lambda(w) \end{aligned}$$

for a uniquely determined function $v_z \in \mathcal{G}_p$. Observe that

$$p\left[\left(\sum_{k=1}^{p} p_k p_k^* + \sum_{k=p+1}^{n} p_k^* p_k\right)(v_z)(w)\right]^- = K(z,w) = \frac{1}{\pi^n}\sum_{k=0}^{\infty}\frac{\langle z,w\rangle^k}{k!}.$$

Now we claim that

$$B_p(z,w) = \frac{p}{\pi^n}\sum_{k=0}^{\infty}\frac{\langle z,w\rangle^k}{(k+p)k!}. \tag{16.14}$$

For this aim, we compute

$$\begin{aligned}\left(\sum_{k=1}^{p} p_k p_k^* + \sum_{k=p+1}^{n} p_k^* p_k\right)((\langle w,z\rangle)^k &= p((\langle w,z\rangle)^k + k\sum_{j=1}^{p} w_j\bar z_j((\langle w,z\rangle)^{k-1} \\ &\quad + k\sum_{k=p+1}^{n} w_j\bar z_j((\langle w,z\rangle)^{k-1} \\ &= (k+p)((\langle w,z\rangle)^k,\end{aligned}$$

which proves (16.14). Comparison of coefficients in the Taylor series yields

$$B_p(z,w) = \frac{p}{\pi^n}\sum_{k=0}^{\infty}\frac{\langle z,w\rangle^k}{(k+p)k!} = \frac{p}{\pi^n}\frac{d^{p-1}}{d\zeta^{p-1}}\left(\frac{e^\zeta - \sum_{k=0}^{p-1}\frac{\zeta^k}{k!}}{\zeta}\right)\Bigg|_{\zeta=\langle z,w\rangle}. \qquad \square$$

For example,

$$B_1(z,w) = \frac{1}{\pi^n}\frac{e^{\langle z,w\rangle}-1}{\langle z,w\rangle} \quad\text{and}\quad B_2(z,w) = \frac{2}{\pi^n}\frac{\langle z,w\rangle e^{\langle z,w\rangle} - e^{\langle z,w\rangle}+1}{(\langle z,w\rangle)^2}.$$

Now we use the unbounded operators ∂^* and ∂ to exhibit compact subsets of $A^2(\mathbb{C}^n, e^{-|z|^2})$.

Theorem 16.12. *Let $n = 1$ and define*

$$\mathcal{U} = \{u \in \operatorname{dom}(\partial^*) : \|\partial^* u\|^2 \le 1\}.$$

Then $\mathcal{U}$ is a relatively compact set in $A^2(\mathbb{C}, e^{-|z|^2})$.

Proof. The basic estimate implies that $\mathcal{U}$ is a bounded subset of $A^2(\mathbb{C}^n, e^{-|z|^2})$. By Theorem 11.32 we have to show that for each $\epsilon > 0$, there exists a constant $R > 0$ such that

$$\int_{\mathbb{C}\setminus\mathbb{B}_R} |u(z)|^2 e^{-|z|^2}\, d\lambda(z) \le \epsilon\|\partial^*(u)\|^2, \tag{16.15}$$

for all $u \in \operatorname{dom}(\partial^*)$. For this purpose, we take the orthonormal basis $(\varphi_k)_k$ of $A^2(\mathbb{C}, e^{-|z|^2})$, where $\varphi_k(z) = \frac{z^k}{\sqrt{\pi k!}}$, $k = 0,1,2,\dots$. Then, for $u \in A^2(\mathbb{C}, e^{-|z|^2})$, we have $u(z) = \sum_{k=0}^{\infty} u_k\varphi_k(z)$, where $(u_k)_k \in l^2$.

For $u \in \operatorname{dom}(\partial^*)$, we get

$$\partial^* u(z) = zu(z) = \sum_{k=0}^{\infty} u_k \frac{z^{k+1}}{\sqrt{\pi k!}} = \sum_{k=0}^{\infty} u_k \sqrt{k+1} \frac{z^{k+1}}{\sqrt{\pi (k+1)!}}.$$

Then (16.15) reads as

$$\int_{\mathbb{C}\setminus D_R} \sum_{k=0}^{\infty} |u_k|^2 \frac{|z|^{2k}}{\pi k!} e^{-|z|^2} \, d\lambda(z) = 2 \sum_{k=0}^{\infty} \frac{|u_k|^2}{k!} \int_R^{\infty} r^{2k+1} e^{-r^2} \, dr \le \epsilon \sum_{k=0}^{\infty} |u_k|^2 (k+1),$$

where $D_R = \{z \in \mathbb{C} : |z| < R\}$. Comparing the terms in the sum from above, it remains to show that

$$\frac{2}{(k+1)!} \int_R^{\infty} r^{2k+1} e^{-r^2} \, dr \le \epsilon$$

for all $k = 0, 1, 2, \ldots$. It follows that

$$2 \int_R^{\infty} r^{2k+1} e^{-r^2} \, dr = \int_{R^2}^{\infty} s^k e^{-s} \, ds.$$

Now integrating k times by parts, we get

$$\int_{R^2}^{\infty} s^k e^{-s} \, ds = e^{-R^2} k! \sum_{j=0}^{k} \frac{R^{2j}}{j!}.$$

So we still have to prove that for each $\epsilon > 0$, there exists $t > 0$ such that

$$\frac{1}{k+1} e^{-t} \sum_{j=0}^{k} \frac{t^j}{j!} \le \epsilon \tag{16.16}$$

for all $k = 0, 1, 2, \ldots$. If $k \ge \frac{1}{\epsilon}$, then we obtain

$$\frac{1}{k+1} e^{-t} \sum_{j=0}^{k} \frac{t^j}{j!} \le \frac{1}{k+1} e^{-t} e^{t} = \frac{1}{k+1} \le \epsilon,$$

and if $k \le \frac{1}{\epsilon}$, then we can choose $t > 0$ large enough to get

$$\frac{1}{k+1} e^{-t} \sum_{j=0}^{k} \frac{t^j}{j!} \le \epsilon.$$

□

Remark 16.13.

(a) The operator $\tilde{\Box}_1 = \partial\partial^*$ has a bounded inverse

$$\tilde{N}_1 : A^2_{(1,0)}(\mathbb{C}, e^{-|z|^2}) \longrightarrow \operatorname{dom}(\tilde{\Box}_1).$$

Let $\tilde{j} : \operatorname{dom}(\partial^*) \longrightarrow A^2(\mathbb{C}, e^{-|z|^2})$ be the embedding, where $\operatorname{dom}(\partial^*)$ is endowed with the graph norm $\|\partial^* u\|$. It follows that $\tilde{N}_1 = \tilde{j} \circ \tilde{j}^*$. Since $\mathcal{U} = \{u \in \operatorname{dom}(\partial^*) : \|\partial^* u\|^2 \leq 1\}$ is a precompact set in $A^2(\mathbb{C}, e^{-|z|^2})$, the operator $\tilde{j}$ is compact. This implies $\tilde{N}_1$ is compact; see Theorem 16.9, where we showed directly that the operator $\tilde{j}$ can be approximated by finite-dimensional operators. It turns out that the method in the proof of Theorem 16.12 can be used in more general cases.

(b) In [58], we can find a different proof of compactness using coverings in the corresponding sequence spaces of the Taylor coefficients of the entire functions.

For several variables, for a (1, 0)-form $u = \sum_{j=1}^n u_j \, dz_j \in \operatorname{dom}(\partial) \cap \operatorname{dom}(\partial^*)$, we get

$$\|\partial u\|^2 + \|\partial^* u\|^2 = \sum_{j,k=1}^n \int_{\mathbb{C}^n} \left| \frac{\partial u_j}{\partial z_k} \right|^2 e^{-|z|^2} \, d\lambda + \sum_{j=1}^n \int_{\mathbb{C}^n} |u_j|^2 e^{-|z|^2} \, d\lambda. \tag{16.17}$$

For simplicity, we only handle the case $n = 2$; now we use (1.31), and (16.17) reads as

$$\begin{aligned}\|\partial u\|^2 + \|\partial^* u\|^2 &= \left\|\frac{\partial u_1}{\partial z_1}\right\|^2 + \left\|\frac{\partial u_1}{\partial z_2}\right\|^2 + \left\|\frac{\partial u_2}{\partial z_1}\right\|^2 + \left\|\frac{\partial u_2}{\partial z_2}\right\|^2 + \|u_1\|^2 + \|u_2\|^2 \\ &= \frac{1}{2}(\|z_1 u_1\|^2 + \|z_2 u_1\|^2 + \|z_1 u_2\|^2 + \|z_2 u_2\|^2) \\ &\quad + \frac{1}{2}\left(\left\|\frac{\partial u_1}{\partial z_1}\right\|^2 + \left\|\frac{\partial u_1}{\partial z_2}\right\|^2 + \left\|\frac{\partial u_2}{\partial z_1}\right\|^2 + \left\|\frac{\partial u_2}{\partial z_2}\right\|^2 \right).\end{aligned}$$

Theorem 16.14. *Let*

$$\mathcal{U} = \{u \in \operatorname{dom}(\partial) \cap \operatorname{dom}(\partial^*) : \|\partial u\|^2 + \|\partial^* u\|^2 \leq 1\}.$$

Then $\mathcal{U}$ is a relatively compact set in $A^2_{(1,0)}(\mathbb{C}^n, e^{-|z|^2})$.

Proof. We have to show that for each $\epsilon > 0$, there exists $R > 0$ such that

$$\int_{\mathbb{C}^2 \setminus D_R \times D_R} (|u_1(z)|^2 + |u_2(z)|^2) e^{-|z|^2} \, d\lambda(z) \leq \epsilon (\|\partial u\|^2 + \|\partial^* u\|^2)$$

for all $u \in \operatorname{dom}(\partial) \cap \operatorname{dom}(\partial^*)$, where $D_R = \{z \in \mathbb{C} : |z| < R\}$.

It suffices to show that for $u_j = v, j = 1, 2$, we have

$$\int\limits_{\mathbb{C}^2\setminus D_R\times D_R} |v(z)|^2 e^{-|z|^2}\, d\lambda(z) \le \epsilon(\|z_1 v\|^2 + \|z_2 v\|^2). \tag{16.18}$$

We use the orthonormal basis $\varphi_{\alpha,\beta}(z) = \frac{z_1^\alpha}{\sqrt{\pi\alpha!}}\frac{z_2^\beta}{\sqrt{\pi\beta!}}$ and write

$$v(z) = \sum_{\alpha,\beta} v_{\alpha,\beta}\varphi_{\alpha,\beta}(z),$$

where

$$\sum_{\alpha,\beta} |v_{\alpha,\beta}|^2 < \infty.$$

Then (16.18) reads as

$$\sum_{\alpha,\beta} |v_{\alpha,\beta}|^2 \Big(\int\limits_{\mathbb{C}\setminus D_R} \frac{|z_1|^{2\alpha}}{\pi\alpha!} e^{-|z_1|^2}\, d\lambda(z_1) \int\limits_{\mathbb{C}\setminus D_R} \frac{|z_2|^{2\beta}}{\pi\beta!} e^{-|z_2|^2}\, d\lambda(z_2)\Big)$$
$$= 4\sum_{\alpha,\beta} \frac{|v_{\alpha,\beta}|^2}{\alpha!\beta!}\Big(\int_R^\infty r^{2\alpha+1} e^{-r^2}\, dr \int_R^\infty r^{2\beta+1} e^{-r^2}\, dr\Big) \le \epsilon \sum_{\alpha,\beta} |v_{\alpha,\beta}|^2(\alpha+\beta+2).$$

Since

$$\frac{1}{\alpha!}\int_R^\infty r^{2\alpha+1} e^{-r^2}\, dr \le 1$$

for all $\alpha = 0, 1, 2, \dots$, we can proceed as in the proof of Theorem 16.12 to show (16.18). □

In general, for $1 \le p \le n-1$ and for a $(p,0)$-form $u = \sum_{|J|=p}{}' u_J\, dz_J \in \operatorname{dom}(\partial) \cap \operatorname{dom}(\partial^*)$, we have that

$$\|\partial u\|^2 + \|\partial^* u\|^2 = \sum_{|J|=p}{}' \sum_{j=1}^n \Big\|\frac{\partial u_J}{\partial z_j}\Big\|^2 + p\sum_{|J|=p}{}' \int\limits_{\mathbb{C}^n} |u_J|^2 e^{-|z|^2}\, d\lambda. \tag{16.19}$$

As an immediate consequence of (16.19), we get the basic estimate

$$\|u\|^2 \le \frac{1}{p}(\|\partial u\|^2 + \|\partial^* u\|^2). \tag{16.20}$$

It is now clear how to adopt the methods we developed in the case $n = 2$ for general n and $(p,0)$-forms in $\operatorname{dom}(\partial) \cap \operatorname{dom}(\partial^*)$, and we get that the set

$$\mathcal{U} = \{u \in \operatorname{dom}(\partial) \cap \operatorname{dom}(\partial^*) : \|\partial u\|^2 + \|\partial^* u\|^2 \le 1\}$$

is precompact in $A^2_{(p,0)}(\mathbb{C}^n, e^{-|z|^2})$ and that the corresponding complex Laplacian

$$\tilde{\Box}_p = \partial^* \partial + \partial \partial^*$$

has a compact inverse

$$\tilde{N}_p : A^2_{(p,0)}(\mathbb{C}^n, e^{-|z|^2}) \longrightarrow A^2_{(p,0)}(\mathbb{C}^n, e^{-|z|^2}).$$

16.4 The generalized ∂-complex

We consider the classical Segal–Bargmann space $A^2(\mathbb{C}^n, e^{-|z|^2})$ and replace a single derivative with respect to z_j by a differential operator of the form $p_j(\frac{\partial}{\partial z_1}, \dots, \frac{\partial}{\partial z_n})$, where p_j is a complex polynomial on $\mathbb{C}^n$ (see [78, 79]).

We write $p_j(u)$ for $p_j(\frac{\partial}{\partial z_1}, \dots, \frac{\partial}{\partial z_n})u$, where $u \in A^2(\mathbb{C}^n, e^{-|z|^2})$, and consider the densely defined operators

$$Du = \sum_{j=1}^{n} p_j(u)\, dz_j, \tag{16.21}$$

where $u \in A^2(\mathbb{C}^n, e^{-|z|^2})$ and $p_j(\frac{\partial}{\partial z_1}, \dots, \frac{\partial}{\partial z_n})$ are polynomial differential operators with constant coefficients [44].

More generally, we define

$$Du = \sum_{|J|=p}{}' \sum_{k=1}^{n} p_k(u_J)\, dz_k \wedge dz_J, \tag{16.22}$$

where $u = \sum_{|J|=p}{}' u_J\, dz_J$ is a $(p,0)$-form with coefficients in $A^2(\mathbb{C}^n, e^{-|z|^2})$, $J = (j_1, \dots, j_p)$ is a multi-index, $dz_J = dz_{j_1} \wedge \dots \wedge dz_{j_p}$, and the summation is taken only over increasing multiindices.

It is clear that $D^2 = 0$ and

$$(Du, v) = (u, D^* v), \tag{16.23}$$

where $u \in \operatorname{dom}(D) = \{u \in A^2_{(p,0)}(\mathbb{C}^n, e^{-|z|^2}) : Du \in A^2_{(p+1,0)}(\mathbb{C}^n, e^{-|z|^2})\}$,

$$D^* v = \sum_{|K|=p-1}{}' \sum_{j=1}^{n} p_j^*\, v_{jK}\, dz_K$$

for $v = \sum_{|J|=p}{}' v_J\, dz_J$, and $p_j^*(z_1, \dots, z_n)$ is the polynomial p_j with complex conjugate coefficients, taken as the multiplication operator.

Now the corresponding D-complex has the form

$$A^2_{(p-1,0)}(\mathbb{C}^n, e^{-|z|^2}) \underset{D^*}{\overset{D}{\rightleftarrows}} A^2_{(p,0)}(\mathbb{C}^n, e^{-|z|^2}) \underset{D^*}{\overset{D}{\rightleftarrows}} A^2_{(p+1,0)}(\mathbb{C}^n, e^{-|z|^2}).$$

Similarly to the classical $\overline{\partial}$-complex (see [44]), we consider the generalized box operator $\tilde{\square}_{D,p} := D^*D + DD^*$ as a densely defined self-adjoint operator on $A^2_{(p,0)}(\mathbb{C}^n, e^{-|z|^2})$ with

$$\operatorname{dom}(\tilde{\square}_{D,p}) = \{f \in \operatorname{dom}(D) \cap \operatorname{dom}(D^*) : Df \in \operatorname{dom}(D^*) \text{ and } D^*f \in \operatorname{dom}(D)\};$$

see [45] for more detail.

The $(p, 0)$-forms with polynomial components are dense in $A^2_{(p,0)}(\mathbb{C}^n, e^{-|z|^2})$. In addition, we have the following:

Lemma 16.15. *The $(p, 0)$-forms with polynomial components are also dense in* $\operatorname{dom}(D) \cap \operatorname{dom}(D^*)$ *endowed with the graph norm*

$$u \mapsto (\|u\|^2 + \|Du\|^2 + \|D^*u\|^2)^{1/2}.$$

Proof. Let $u = \sum'_{|J|=p} u_J \, dz_J \in \operatorname{dom}(D) \cap \operatorname{dom}(D^*)$ and consider the partial sums of the Fourier series expansions of

$$u_J = \sum_\alpha u_{J,\alpha} \varphi_\alpha,$$

where

$$\varphi_\alpha(z) = \frac{z^\alpha}{\sqrt{\pi^n \alpha!}} \quad \text{and} \quad \sum_\alpha |u_{J,\alpha}|^2 < \infty \tag{16.24}$$

with $\alpha! = \alpha_1! \dots \alpha_n!$.

We have that $Du \in A^2_{(p+1,0)}(\mathbb{C}^n, e^{-|z|^2})$ and $D^*u \in A^2_{(p-1,0)}(\mathbb{C}^n, e^{-|z|^2})$. Hence the partial sums of the Fourier series of the components of Du converge to the components of Du in $A^2_{(p+1,0)}(\mathbb{C}^n, e^{-|z|^2})$, and the partial sums of the Fourier series of the components of D^*u converge to the components of D^*u in $A^2_{(p-1,0)}(\mathbb{C}^n, e^{-|z|^2})$. □

We want to find conditions under which $\tilde{\square}_{D,1}$ has a bounded inverse. For this purpose, we have to consider the graph norm $(\|u\|^2 + \|Du\|^2 + \|D^*u\|^2)^{1/2}$ on $\operatorname{dom}(D) \cap \operatorname{dom}(D^*)$.

Theorem 16.16. *Suppose that there exists a constant $C > 0$ such that*

$$\|u\|^2 \le C \sum_{j,k=1}^n ([p_k, p_j^*]u_j, u_k) \tag{16.25}$$

for all $(1, 0)$-forms $u = \sum_{j=1}^n u_j dz_j$ with polynomial components. Then

$$\|u\|^2 \le C(\|Du\|^2 + \|D^*u\|^2) \tag{16.26}$$

for all $u \in \mathrm{dom}(D) \cap \mathrm{dom}(D^*)$.

Proof. First, we have

$$Du = \sum_{j<k} (p_j(u_k) - p_k(u_j))\, dz_j \wedge dz_k \quad \text{and} \quad D^*u = \sum_{j=1}^n p_j^* u_j,$$

and hence

$$\begin{aligned}
\|Du\|^2 + \|D^*u\|^2 &= \int_{\mathbb{C}^n} \sum_{j<k} |p_k(u_j) - p_j(u_k)|^2 e^{-|z|^2}\, d\lambda \\
&\quad + \int_{\mathbb{C}^n} \sum_{j,k=1}^n p_j^* u_j \overline{p_k^* u_k}\, e^{-|z|^2}\, d\lambda \\
&= \sum_{j,k=1}^n \int_{\mathbb{C}^n} |p_k(u_j)|^2 e^{-|z|^2}\, d\lambda + \sum_{j,k=1}^n \int_{\mathbb{C}^n} (p_j^* u_j \overline{p_k^* u_k} - p_k(u_j)\overline{p_j(u_k)})\, e^{-|z|^2}\, d\lambda \\
&= \sum_{j,k=1}^n \int_{\mathbb{C}^n} |p_k(u_j)|^2 e^{-|z|^2}\, d\lambda + \sum_{j,k=1}^n \int_{\mathbb{C}^n} [p_k, p_j^*] u_j \overline{u_k}\, e^{-|z|^2}\, d\lambda,
\end{aligned}$$

where we used (16.23). Note that the expression

$$\sum_{j,k=1}^n \int_{\mathbb{C}^n} |p_k(u_j)|^2 e^{-|z|^2}\, d\lambda$$

is finite, since the components u_j are polynomials, and it follows that the expression $\sum_{j,k=1}^n ([p_k, p_j^*]u_j, u_k)$ is a real number.

Now assumption (16.25) implies that (16.26) holds for $(1,0)$-forms with polynomial components, and by Lemma 16.15 we obtain (16.26) for all $u \in \mathrm{dom}(D) \cap \mathrm{dom}(D^*)$. □

Remark 16.17. In Theorem 16.16, we implicitly suppose that the expression

$$\sum_{j,k=1}^n ([p_k, p_j^*]u_j, u_k)$$

is nonnegative.

Let $1 \le p \le n-1$, let $u = \sum'_{|J|=p} u_J\, dz_J \in A^2_{(p,0)}(\mathbb{C}^n, e^{-|z|^2})$, and suppose that $u \in \mathrm{dom}(\tilde{\Box}_{D,p})$. In a similar way as in (16.4), we get

$$\|Du\|^2 + \|D^*u\|^2 = \sum_{|J|=p}{}' \sum_{k=1}^n \|p_k(u_J)\|^2 \tag{16.27}$$

$$+\sum_{|K|=p-1}{}' \sum_{j,k=1}^{n} \int_{\mathbb{C}^n} [p_k, p_j^*] u_{jK} \overline{u_{kK}}\, e^{-|z|^2}\, d\lambda,$$

and if we suppose that

$$\|u\|^2 \leq C \sum_{|K|=p-1}{}' \sum_{j,k=1}^{n} ([p_k, p_j^*] u_{jK}, u_{kK}), \tag{16.28}$$

then we get the basic estimate (16.26), which also implies that both $\operatorname{im} D$ and $\operatorname{im} D^*$ are closed (see, for instance, [44, Chapter 5]). With the basic estimate (16.26), we are now able to use the machinery of the corresponding Neumann operator, the bounded inverse of $\tilde{\Box}_{D,p}$ (see Theorems 16.6 and 16.8), and get the following results.

Theorem 16.18. *Let D be as in* (16.22), *and let $u \in \operatorname{dom}(\tilde{\Box}_{D,p})$. Suppose that there exists a constant $C > 0$ such that*

$$\|u\|^2 \leq C \sum_{|K|=p-1}{}' \sum_{j,k=1}^{n} ([p_k, p_j^*] u_{jK}, u_{kK}).$$

Then $\tilde{\Box}_{D,p}$ has a bounded inverse

$$\tilde{N}_{D,p} : A^2_{(p,0)}(\mathbb{C}^n, e^{-|z|^2}) \longrightarrow \operatorname{dom}(\tilde{\Box}_{D,p}).$$

If $\alpha \in A^2_{(p,0)}(\mathbb{C}^n, e^{-|z|^2})$ satisfies $D\alpha = 0$, then $u_0 = D^* \tilde{N}_{D,p} \alpha$ is the canonical solution of $Du = \alpha$, which means $Du_0 = \alpha$ and $u_0 \in (\ker D)^\perp = \operatorname{im} D^*$, and $\|D^* \tilde{N}_{D,p} \alpha\| \leq C \|\alpha\|$ for some constant $C > 0$ independent of α.

In the following, we discuss some examples. First, we consider the one-dimensional case. Let p denote the polynomial differential operator

$$p = a_0 + a_1 \frac{\partial}{\partial z} + \cdots + a_m \frac{\partial^m}{\partial z^m}$$

with constant coefficients $a_0, a_1, \ldots, a_m \in \mathbb{C}$, and let p^* denote the polynomial

$$p^*(z) = \overline{a}_0 + \overline{a}_1 z + \cdots + \overline{a}_m z^m$$

with the complex conjugate coefficients $\overline{a}_0, \overline{a}_1, \ldots, \overline{a}_m \in \mathbb{C}$.

We consider the densely defined operator

$$Du = p(u)\, dz, \tag{16.29}$$

where $u \in A^2(\mathbb{C}, e^{-|z|^2})$, and $p(u)\, dz$ is considered as a $(1,0)$-form.

It is clear that $D^2 = 0$, as all $(2,0)$-forms are zero if $n = 1$, and that

$$(Du, v) = (u, D^* v), \tag{16.30}$$

where $u \in \operatorname{dom}(D) = \{u \in A^2(\mathbb{C}, e^{-|z|^2}) : Du \in A^2_{(1,0)}(\mathbb{C}, e^{-|z|^2})\}$ and

$$D^* v \, dz = p^* v.$$

In the sequel, we consider the generalized box operator

$$\tilde{\Box}_{D,1} := DD^*$$

as a densely defined self-adjoint operator on $A^2_{(1,0)}(\mathbb{C}, e^{-|z|^2})$ with $\operatorname{dom}(\tilde{\Box}_{D,1}) = \{f \in \operatorname{dom}(D^*) : D^* f \in \operatorname{dom}(D)\}$.

Lemma 16.19. *Let u be an arbitrary polynomial. Then*

$$\int_{\mathbb{C}} [p, p^*] u(z) \overline{u(z)} \, e^{-|z|^2} \, d\lambda(z)$$

$$= \sum_{\ell=1}^{m} \ell! \int_{\mathbb{C}} \left| \sum_{k=\ell}^{m} \binom{k}{\ell} a_k u^{(k-\ell)}(z) \right|^2 e^{-|z|^2} \, d\lambda(z). \tag{16.31}$$

Proof. On the right-hand side of (16.31), we have the integrand

$$\sum_{\ell=1}^{m} \ell! \sum_{k,j=\ell}^{m} \binom{k}{\ell} \binom{j}{\ell} a_k \overline{a}_j u^{(k-\ell)} \overline{u}^{(j-\ell)}. \tag{16.32}$$

For the left-hand side of (16.31), we first compute

$$[p, p^*] u = \sum_{k=0}^{m} a_k \frac{\partial^k}{\partial z^k} \left(\sum_{j=0}^{m} \overline{a}_j z^j u \right) - \left(\sum_{j=0}^{m} \overline{a}_j z^j \right) \left(\sum_{j=0}^{m} a_j u^{(j)} \right). \tag{16.33}$$

We use the Leibniz rule to get

$$\frac{\partial^k}{\partial z^k} \left(\sum_{j=0}^{m} \overline{a}_j z^j u \right) = \sum_{j=0}^{m} \overline{a}_j \frac{\partial^k}{\partial z^k} (z^j u) = \sum_{j=0}^{m} \overline{a}_j \sum_{\ell=0}^{j} \ell! \binom{k}{\ell} \binom{j}{\ell} u^{(k-\ell)} z^{j-\ell}, \tag{16.34}$$

noticing that $\binom{k}{\ell} = 0$ in the case $k < \ell$. Hence we obtain

$$[p, p^*] u = \sum_{j,k=1}^{m} a_k \overline{a}_j \sum_{\ell=1}^{j} \ell! \binom{k}{\ell} \binom{j}{\ell} u^{(k-\ell)} z^{j-\ell}. \tag{16.35}$$

After integration, we obtain

$$\int_{\mathbb{C}} [p, p^*] u(z) \overline{u(z)} \, e^{-|z|^2} \, d\lambda(z)$$

$$= \sum_{j,k=1}^{m} a_k \bar{a}_j \sum_{\ell=1}^{j} \ell! \binom{k}{\ell} \binom{j}{\ell} \int_{\mathbb{C}} u^{(k-\ell)} \overline{u}^{(j-\ell)} e^{-|z|^2} \, d\lambda(z). \tag{16.36}$$

Now it is easy to show that integration of (16.32) coincides with (16.36), and we are done. □

Lemma 16.20. *There exists a constant $C > 0$ such that*

$$\|u\| \leq C\|D^* u\| \tag{16.37}$$

for each $u \in \mathrm{dom}(D^*)$.

Proof. First, let u be a polynomial and note that the last term in (16.31) equals

$$m! \int_{\mathbb{C}} |a_m u(z)|^2 e^{-|z|^2} \, d\lambda(z) = m! \, |a_m|^2 \|u\|^2$$

and all the other terms are nonnegative; see Lemma 16.19. Now we get that

$$\|D^* u\|^2 = (p^* u, p^* u) = ([p, p^*]u, u) + (p(u), p(u)) \geq \frac{1}{C} \|u\|^2,$$

where

$$C = \frac{1}{m! \, |a_m|^2}$$

if we suppose that $a_m \neq 0$, and we are done. Finally, apply Lemma 16.15 to obtain the desired result. □

Theorem 16.21. *Let D be as in (16.29). Then $\tilde{\Box}_{D,1} = DD^*$ has a bounded inverse*

$$\tilde{N}_{D,1} : A^2_{(1,0)}(\mathbb{C}, e^{-|z|^2}) \longrightarrow \mathrm{dom}(\tilde{\Box}_{D,1}).$$

If $\alpha \in A^2_{(1,0)}(\mathbb{C}, e^{-|z|^2})$, then $u_0 = D^* \tilde{N}_{D,1} \alpha$ is the canonical solution of $Du = \alpha$, which means $Du_0 = \alpha$ and $u_0 \in (\ker D)^\perp = \mathrm{im}\, D^*$, and $\|D^* \tilde{N}_{D,1} \alpha\| \leq C\|\alpha\|$ for some constant $C > 0$ independent of α.

Proof. Using Lemma 16.20, we obtain that

$$\tilde{\Box}_{D,1} : \mathrm{dom}(\tilde{\Box}_{D,1}) \longrightarrow A^2_{(1,0)}(\mathbb{C}, e^{-|z|^2})$$

is bijective and has the bounded inverse $\tilde{N}_{D,1}$, and the rest follows from Theorem 16.18. □

To show that $\tilde{N}_{D,1}$ is even compact, we first consider the case $m = 1$, taking $p^*(z) = \bar{a}_0 + \bar{a}_1 z$. We suppose that $a_0, a_1 \neq 0$ and choose a complex number c such that

$$\frac{|a_1|^2}{1+\frac{|a_1|^2}{|a_0|^2}} < |c|^2 < |a_1|^2. \tag{16.38}$$

Now let

$$d := \frac{a_0\bar{a}_1}{\bar{c}}. \tag{16.39}$$

Then using (1.31), for an arbitrary polynomial u, we get that

$$\begin{aligned}
\|D^*u\|^2 &= \|\bar{a}_0 u + \bar{a}_1 zu\|^2 \\
&= |a_0|^2\|u\|^2 + \bar{a}_0 a_1(u, zu) + \bar{a}_1 a_0(zu, u) + |a_1|^2\|zu\|^2 \\
&= |a_0|^2\|u\|^2 + \bar{a}_0 a_1(u', u) + \bar{a}_1 a_0(u, u') + |a_1|^2(\|u\|^2 + \|u'\|^2) \\
&\geq (|a_0|^2 + |a_1|^2)\|u\|^2 + \bar{a}_0 a_1(u', u) + \bar{a}_1 a_0(u, u') + (|a_1|^2)\|u'\|^2 - \|du + cu'\|^2 \\
&= (|a_0|^2 + |a_1|^2 - |d|^2)\|u\|^2 + (|a_1|^2 - |c|^2)\|u'\|^2 \\
&\geq \alpha(\|u\|^2 + \|u'\|^2) \\
&= \alpha\|zu\|^2,
\end{aligned}$$

where

$$\alpha = \min(|a_0|^2 + |a_1|^2 - |d|^2, |a_1|^2 - |c|^2). \tag{16.40}$$

Note that (16.38) and (16.39) imply $\alpha > 0$.

For a general polynomial p, from [45] we obtain that there exists a constant $C > 0$ such that $\|u\| \leq C\|D^*u\|$ for each $u \in \operatorname{dom}(D^*)$ and

$$\begin{aligned}
&\int_{\mathbb{C}} [p, p^*]u(z)\,\overline{u(z)}\,e^{-|z|^2}\,d\lambda(z) \\
&\quad = \sum_{\ell=1}^{m} \ell! \int_{\mathbb{C}} \left|\sum_{k=\ell}^{m} \binom{k}{\ell} a_k u^{(k-\ell)}(z)\right|^2 e^{-|z|^2}\,d\lambda(z).
\end{aligned} \tag{16.41}$$

Now from (16.41), taking $\ell = m$ and $\ell = m - 1$, we get

$$\begin{aligned}
\|D^*u\|^2 &= (p^*u, p^*u) = ([p, p^*]u, u) + (p(u), p(u)) \\
&\geq m!|a_m|^2\|u\|^2 + (m-1)!\|a_{m-1}u + ma_m u'\|^2.
\end{aligned}$$

If $a_{m-1} = 0$, then we get

$$\|D^*u\|^2 \geq m!|a_m|^2(\|u\|^2 + \|u'\|^2) = m!|a_m|^2\|zu\|^2. \tag{16.42}$$

If $a_{m-1} \neq 0$, then we set $a = \sqrt{m!}\, a_m$ and $b = \sqrt{(m-1)!}\, a_{m-1}$ and get

$$\begin{aligned}\|D^* u\|^2 &\geq m!|a_m|^2 \|u\|^2 + (m-1)!\|a_{m-1}u + m a_m u'\|^2 \\ &= (|a|^2 + |b|^2)\|u\|^2 + b\bar{a}(u, u') + a\bar{b}(u', u) + |a|^2 \|u'\|^2,\end{aligned}$$

and we can proceed as before to get

$$\|D^* u\|^2 \geq \alpha \|zu\|^2 \tag{16.43}$$

for $\alpha > 0$.

Using the orthonormal basis $(\varphi_k)_k$, we write

$$u = \sum_{k=0}^{\infty} u_k \varphi_k$$

and get

$$\|zu\|^2 = \sum_{k=0}^{\infty} |u_k|^2 (k+1).$$

Now we want to show that for each $\epsilon > 0$, there exists a constant $R > 0$ such that

$$\int_{\mathbb{C}\setminus\mathbb{B}_R} |u(z)|^2 e^{-|z|^2}\, d\lambda(z) \leq \epsilon \|D^* u\|^2 \tag{16.44}$$

for all $u \in \operatorname{dom}(D^*)$. As for the operator ∂^*, we get

$$\int_{\mathbb{C}\setminus\mathbb{B}_R} |u(z)|^2 e^{-|z|^2}\, d\lambda(z) \leq \alpha\epsilon \|\partial^* u\|^2;$$

by (16.42) and (16.43)

$$\int_{\mathbb{C}\setminus\mathbb{B}_R} |u(z)|^2 e^{-|z|^2}\, d\lambda(z) \leq \alpha\epsilon \|D^* u\|^2,$$

and so $\{u \in \operatorname{dom}(D^*) : \|D^* u\|^2 \leq 1\}$ is a relatively compact set in $A^2(\mathbb{C}, e^{-|z|^2})$.

We can now summarize in the following:

Theorem 16.22. *Let $Du = p_m(u)\, dz$, where*

$$p = a_0 + a_1 \frac{\partial}{\partial z} + \cdots + a_m \frac{\partial^m}{\partial z^m}, \quad a_m \neq 0,\ m \geq 1.$$

Then the operator $\tilde{\Box}_{D,1} = DD^*$ has a compact inverse

$$\tilde{N}_{D,1} : A^2_{(1,0)}(\mathbb{C}, e^{-|z|^2}) \longrightarrow \operatorname{dom}(\tilde{\Box}_{D,1}).$$

Remark 16.23. We give a simple example in $\mathbb{C}^2$, where the basic estimate does not hold. Let p be a polynomial differential operator without constant summand and set $D(u) = p(u)dz_1 - p(u)dz_2$ for a polynomial u. If $u = u_1 dz_1 + u_2 dz_2$ is a $(1,0)$-form with polynomial components, then we have $D(u) = (p(u_2) - p(u_1))dz_1 \wedge dz_2$ and $D^* u = p^* u_1 + p^* u_2$. Now take $u = dz_1 + dz_2$. Then $D(u) = 0$ and $D^* u = p^* 1 - p^* 1 = 0$. As $\|u\| \neq 0$, the basic estimate does not hold.

Remark 16.24. The basic estimates can be viewed as generalizations of the uncertainty principle; see Section 1.5. Let $n = 1$ and $p = \frac{\partial^m}{\partial z^m}$. Then we have

$$\|z^m u\|^2 = \|u\|^2 + \sum_{j=0}^{m-1} \left\| \frac{\partial}{\partial z}(z^j u) \right\|^2 .$$

So we get

$$\|u\|^2 \le \|z^m u\|^2 + \left\| \frac{\partial^m u}{\partial z^m} \right\|^2 .$$

For a general polynomial differential operator p of degree m with leading coefficient a_m, we obtain by Lemma 16.20 that there exists a constant $C > 0$ such that $\|u\| \le C\|D^* u\| = C\|p^* u\|$ for each $u \in \operatorname{dom}(D^*)$, where $C = \sqrt{\frac{1}{m!|a_m|^2}}$. Hence we have again

$$\|u\|^2 \le C(\|p^* u\|^2 + \|p(u)\|^2).$$

We remark that we can interchange the roles of D and D^* and obtain the following:

Theorem 16.25. *Suppose that $n > 1$ and $1 \le p \le n-1$. Let D be as in (16.22) and suppose that*

$$\|u\|^2 \le C \sum_{|K|=p-1}{}' \sum_{j,k=1}^{n} ([p_k, p_j^*] u_{jK}, u_{kK})$$

for all $u \in \operatorname{dom}(\tilde{\Box}_{D,p})$. If $\beta \in A^2_{(p,0)}(\mathbb{C}^n, e^{-|z|^2})$ satisfies $D^*\beta = 0$, then $v_0 = D\tilde{N}_{D,p}\beta \in A^2_{(p+1,0)}(\mathbb{C}^n, e^{-|z|^2})$ is the canonical solution of $D^* v = \beta$, which means that $D^* v_0 = \beta$ and $v_0 \in (\ker D^*)^\perp = \operatorname{im} D$, and $\|D\tilde{N}_{D,p}\beta\| \le C\|\beta\|$ for some constant $C > 0$ independent of β.

Proof. As in Theorem 16.18, we get that $\tilde{\Box}_{D,p}$ has a bounded inverse

$$\tilde{N}_{D,p} : A^2_{(p,0)}(\mathbb{C}^n, e^{-|z|^2}) \longrightarrow \operatorname{dom}(\tilde{\Box}_{D,p}).$$

Now using the $\overline{\partial}$-Neumann calculus, we obtain that

$$0 = D^*\beta = D^*(D^*D + DD^*)\tilde{N}_{D,p}\beta = D^*DD^*\tilde{N}_{D,p}\beta,$$

and hence

$$0 = (D^*DD^*\tilde{N}_{D,p}\beta, D^*\tilde{N}_{D,p}\beta) = (DD^*\tilde{N}_{D,p}\beta, DD^*\tilde{N}_{D,p}\beta).$$

This implies that $DD^*\tilde{N}_{D,p}\beta = 0$, and we get

$$D^*v_0 = D^*D\tilde{N}_{D,p}\beta = (DD^* + D^*D)\tilde{N}_{D,p}\beta = \beta$$

and $(v_0, f) = (D\tilde{N}_{D,p}\beta, f) = (\tilde{N}_{D,p}\beta, D^*f) = 0$ for all $f \in \ker D^*$. □

Example 16.26. We take Example 16.28(b) and consider $f = f_1\,dz_1 + f_2\,dz_2 \in A^2_{(1,0)}(\mathbb{C}^2, e^{-|z|^2})$ such that $D^*f = p_1^*f_1 + p_2^*f_2 = 0$. By Theorem 16.25 we get

$$g = g\,dz_1 \wedge dz_2 = D\tilde{N}_{D,1}f \in A^2_{(2,0)}(\mathbb{C}^2, e^{-|z|^2})$$

such that $D^*g = -p_2^*g\,dz_1 + p_1^*g\,dz_2 = f$ and $\|D\tilde{N}_{D,1}f\| \le C\|f\|$ for some constant $C > 0$.

Remark 16.27. Finally, we point out that the ∂-Neumann operator $\tilde{N}_{D,p}$ exists and is bounded on $A^2_{(p,0)}(\mathbb{C}^n, e^{-|z|^2})$ if and only if the basic estimate (16.26) holds (see Remark 9.12 for the details).

In the following, we exhibit some examples in $\mathbb{C}^2$, where the basic estimates hold.

Example 16.28.

(a) Let $p_k = \frac{\partial^2}{\partial z_k^2}$. Then $p_j^*(z) = z_j^2$, and we have

$$\begin{aligned}\sum_{j,k=1}^n ([p_k, p_j^*]u_j, u_k) &= \sum_{j,k=1}^n (2\delta_{j,k}u_j, u_k) + \sum_{j,k=1}^n \left(4\delta_{jk}z_j\frac{\partial u_j}{\partial z_k}, u_k\right)\\ &= 2\|u\|^2 + 4\sum_{j=1}^n \left\|\frac{\partial u_j}{\partial z_j}\right\|^2\end{aligned}$$

for $u = \sum_{j=1}^n u_j dz_j \in \mathrm{dom}(\tilde{\square}_{D,1})$. Hence (16.7) is satisfied.

(b) Let $n = 2$ and take $p_1 = \frac{\partial^2}{\partial z_1 \partial z_2}$ and $p_2 = \frac{\partial^2}{\partial z_1^2} + \frac{\partial^2}{\partial z_2^2}$. Then $p_1^*(z) = z_1z_2$ and $p_2^*(z) = z_1^2 + z_2^2$, and we have

$$\begin{aligned}([p_1, p_1^*]u_1, u_1) &= (u_1, u_1) + \left(\frac{\partial u_1}{\partial z_1}, \frac{\partial u_1}{\partial z_1}\right) + \left(\frac{\partial u_1}{\partial z_2}, \frac{\partial u_1}{\partial z_2}\right),\\ ([p_1, p_2^*]u_2, u_1) &= 2\left(\frac{\partial u_2}{\partial z_1}, \frac{\partial u_1}{\partial z_2}\right) + 2\left(\frac{\partial u_2}{\partial z_2}, \frac{\partial u_1}{\partial z_1}\right),\\ ([p_2, p_1^*]u_1, u_2) &= 2\left(\frac{\partial u_1}{\partial z_1}, \frac{\partial u_2}{\partial z_2}\right) + 2\left(\frac{\partial u_1}{\partial z_2}, \frac{\partial u_2}{\partial z_1}\right),\end{aligned}$$

$$([p_2, p_2^*]u_2, u_2) = 4(u_2, u_2) + 4\left(\frac{\partial u_2}{\partial z_1}, \frac{\partial u_2}{\partial z_1}\right) + 4\left(\frac{\partial u_2}{\partial z_2}, \frac{\partial u_2}{\partial z_2}\right).$$

So we obtain

$$\sum_{j,k=1}^{2} ([p_k, p_j^*]u_j, u_k)$$
$$= \int_{\mathbb{C}^2} \left(|u_1|^2 + 4|u_2|^2 + \left|\frac{\partial u_1}{\partial z_1} + 2\frac{\partial u_2}{\partial z_2}\right|^2 + \left|\frac{\partial u_1}{\partial z_2} + 2\frac{\partial u_2}{\partial z_1}\right|^2\right) e^{-|z|^2}\, d\lambda$$

for $u = \sum_{j=1}^{2} u_j dz_j \in \operatorname{dom}(\tilde{\Box}_{D,1})$. Again, (16.7) is satisfied.

(c) We easily see that for $p_1^*(z_1, z_2) = z_1^k$ and $p_2^*(z_1, z_2) = z_2^m$, where $k, m \in \mathbb{N}$, and we get $[p_2, p_1^*] = 0$ and

$$\|u\|^2 \leq C \sum_{j,k=1}^{2} ([p_k, p_j^*]u_j, u_k),$$

where $C > 0$.

(d) Let

$$p_1^*(z_1, z_2) = az_1 + bz_2 \quad \text{and} \quad p_2^*(z_1, z_2) = cz_1 + dz_2$$

for $a, b, c, d \in \mathbb{R}$. A computation shows that

$$([p_1, p_1^*]u_1, u_1) = (a^2 + b^2)\|u_1\|^2 \quad \text{and} \quad ([p_2, p_2^*]u_2, u_2) = (c^2 + d^2)\|u_2\|^2,$$
$$([p_1, p_2^*]u_2, u_1) = (ac + bd)(u_2, u_1) \quad \text{and} \quad ([p_2, p_1^*]u_1, u_2) = (ac + bd)(u_1, u_2).$$

Given $a, b \in \mathbb{R}$, all nonzero, we suppose that $c = \mu b$ and $d = -\mu a$ for $\mu \in \mathbb{R}$. Then we have $ac + bd = 0$, and we get

$$\|u\|^2 \leq C \sum_{j,k=1}^{2} ([p_k, p_j^*]u_j, u_k),$$

where $C = \min\{(a^2 + b^2), (c^2 + d^2)\}^{-1}$.

(e) If we take

$$p_1^*(z_1, z_2) = az_1^2 + bz_2^2 \quad \text{and} \quad p_2^*(z_1, z_2) = cz_1^2 + dz_2^2,$$

for $a, b, c, d \in \mathbb{R}$, then we get

$$\sum_{j,k=1}^{2} ([p_k, p_j^*]u_j, u_k) = 2\|au_1 + cu_2\|^2 + 2\|bu_1 + du_2\|^2$$

$$+\left\|2a\frac{\partial u_1}{\partial z_1}+2c\frac{\partial u_2}{\partial z_1}\right\|^2+\left\|2b\frac{\partial u_1}{\partial z_2}+2d\frac{\partial u_2}{\partial z_2}\right\|^2.$$

Given a, b, c, all nonzero, we choose d such that $ac = -bd$. Then

$$2\|au_1 + cu_2\|^2 + 2\|bu_1 + du_2\|^2 = 2(a^2 + b^2)\|u_1\|^2 + 2(c^2 + d^2)\|u_2\|^2,$$

which implies that

$$\|u\|^2 \leq C \sum_{j,k=1}^{2} ([p_k, p_j^*]u_j, u_k),$$

where $C = \min\{2(a^2 + b^2), 2(c^2 + d^2)\}^{-1}$.

16.5 Notes

Most of the material of this chapter was taken from [44] and [45]. Weighted Bergman spaces on Hermitian manifolds, which carry a similar duality of differentiation and multiplication as the Segal–Bargmann space, are studied in [48] and [49].

Bibliography

[1] R. A. Adams and J. J. F. Fournier, *Sobolev spaces*, Pure and Applied Math. 140, Academic Press, 2006.

[2] P. Auscher and B. B. Ali, Maximal inequalities and Riesz transform estimates on L^p spaces for Schrödinger operators with nonnegative potentials, *Ann. Inst. Fourier (Grenoble)* 57 (2007), 1975–2013.

[3] J. Avron, I. Herbst and B. Simon, Schrödinger operators with magnetic fields I, *Duke Math. J.* 45 (1978), 847–883.

[4] S. Bell, *The Cauchy Transform, Potential Theory, and Conformal Mapping*, Studies in Advanced Mathematics, CRC Press, 1992.

[5] F. Berger, *The inhomogeneous Cauchy-Riemann equation*, Bachelorarbeit, Universitaet Wien, 2012.

[6] F. Berger, Essential spectra of tensor product Hilbert complexes, and the $\overline{\partial}$-Neumann problem on product manifolds, *J. Funct. Anal.* 271 (2016), 1434–1461.

[7] F. Berger and F. Haslinger, On some spectral properties of the weighted $\overline{\partial}$-Neumann operator, *J. Math. Kyoto Univ.* 59 (2019), 441–453.

[8] B. Berndtsson, $\overline{\partial}$ and Schrödinger operators, *Math. Z.* 221 (1996), 401–413.

[9] H. P. Boas and E. J. Straube, Global regularity of the $\overline{\partial}$-Neumann problem: a survey of the L^2-Sobolev theory, *Several Complex Variables* (M. Schneider and Y.-T. Siu, eds.), MSRI Publications **37**, Cambridge University Press (1999), 79–111.

[10] P. Bolley, M. Dauge and B. Helffer, Conditions suffisantes pour l'injection compacte d'espaces de Sobolev à poids, Séminaire équation aux dérivées partielles (France), *Univ. Nantes* 1 (1989), 1–14.

[11] A. Borichev, L. Van An, and E.-H. Youssfi, On the dimension of the Fock type spaces, *J. Math. Anal. Appl.* 503 (2021), 125288.

[12] H. Brezis, *Analyse fonctionnelle, Théorie et applications*, Masson, Paris, 1983.

[13] D. W. Catlin, Global regularity of the $\overline{\partial}$-Neumann operator, *Proc. Symp. Pure Math.* 41 (1984), 39–49.

[14] D. W. Catlin and J. P. D'Angelo, Positivity conditions for bihomogeneous polynomials, *Math. Res. Lett.* 4 (1997), 555–567.

[15] M. Celik and S. Sahutoglu, Compactness of the $\overline{\partial}$-Neumann operator and commutators of the Bergman projection with continuous functions, *J. Math. Anal. Appl.* 409 (2014), 393–398.

[16] S.-C. Chen and M.-C. Shaw, *Partial differential equations in several complex variables*, Studies in Advanced Mathematics 19, Amer. Math. Soc., 2001.

[17] M. Christ, On the $\overline{\partial}$ equation in weighted L^2 norms in $\mathbb{C}^1$, *J. Geom. Anal.* 1 (1991), 193–230.

[18] M. Christ and S. Fu, Compactness in the $\overline{\partial}$-Neumann problem, magnetic Schrödinger operators, and the Aharonov-Bohm effect, *Adv. Math.* 197 (2005), 1–40.

[19] O. Constantin, A Volterra-type integration operator on Fock spaces, *Proc. Am. Math. Soc.* 140 (2012), 4247–4257.

[20] O. Constantin and J. Ortega-Cerdà, Some spectral properties of the canonical solution operator to $\overline{\partial}$ on weighted Fock spaces, *J. Math. Anal. Appl.* 377 (2011), 353–361.

[21] H. L. Cycon, R. G. Froese, W. Kirsch and B. Simon, *Schrödinger Operators with Applications to Quantum Mechanics and Global Geometry*, Text and Monographs in Physics, Springer, 1987.

[22] G.-M. Dall'Ara, Discreteness of the spectrum of Schrödinger operators with nonnegative matrix-valued potentials, *J. Funct. Anal.* 268 (2015), 3649–3679.

[23] J. P. D'Angelo, *Inequalities from complex analysis*, Carus Mathematical Monographs 28, Mathematical Association of America, Washington, DC, 2002.

[24] E. B. Davies, *Spectral Theory and Differential Operators*, Cambridge studies in advanced mathematics 42, Cambridge University Press, Cambridge, 1995.

[25] M. Dörfler, H. Feichtinger and K. Gröchenig, Compactness criteria in function spaces, *Colloq. Math.* 94 (2002), 37–50.

[26] J. Elstrodt, *Maß- und Integrationstheorie*, Springer, Berlin, 1996.

[27] L. C. Evans, *Partial Differential Equations*, Graduate Studies in Mathematics 19, Amer. Math. Soc., 1998.

https://doi.org/10.1515/9783111182926-017

[28] G. B. Folland, The tangential Cauchy-Riemann complex on spheres, *Trans. Am. Math. Soc.* 171 (1972), 83–133.

[29] G. B. Folland, *Real analysis, modern techniques and their applications*, John Wiley & Sons, 1984.

[30] G. B. Folland, *Harmonic Analysis in Phase Space*, Ann. of Math. Stud. 122, Princeton University Press, Princeton, NJ, 1989.

[31] G. B. Folland, *Introduction to partial differential equations*, Princeton University Press, Princeton, 1995.

[32] S. Fu, Spectrum of the $\overline{\partial}$-Neumann Laplacian on polydiscs, *Proc. Am. Math. Soc.* 135 (2007), 725–730.

[33] S. Fu and E. J. Straube, Compactness of the $\overline{\partial}$-Neumann problem on convex domains, *J. Funct. Anal.* 159 (1998), 629–641.

[34] S. Fu and E. J. Straube, Compactness in the $\overline{\partial}$-Neumann problem, *Complex Analysis and Geometry* (J. McNeal, ed.), Ohio State Math. Res. Inst. Publ. **9**, (2001), 141–160.

[35] K. Gansberger, *Compactness of the $\overline{\partial}$-Neumann operator*, Dissertation, University of Vienna, 2009.

[36] K. Gansberger and F. Haslinger, Compactness estimates for the $\overline{\partial}$- Neumann problem in weighted L^2 - spaces, *Complex Analysis* (P. Ebenfelt, N. Hungerbühler, J. J. Kohn, N. Mok and E. J. Straube, eds.), Trends in Mathematics, Birkhäuser (2010), 159–174.

[37] L. Grafakos, *Classical Fourier Analysis*, Graduate Texts in Mathematics 249, Springer, 2014.

[38] F. Haslinger, The canonical solution operator to $\overline{\partial}$ restricted to Bergman spaces, *Proc. Am. Math. Soc.* 129 (2001), 3321–3329.

[39] F. Haslinger, The canonical solution operator to $\overline{\partial}$ restricted to spaces of entire functions, *Ann. Fac. Sci. Toulouse Math.* 11 (2002), 57–70.

[40] F. Haslinger, Schrödinger operators with magnetic fields and the $\overline{\partial}$-equation, *J. Math. Kyoto Univ.* 46 (2006), 249–257.

[41] F. Haslinger, The $\overline{\partial}$-Neunmann operator and commutators between multiplication operators and the Bergman projection, *Czechoslov. Math. J.* 58 (2008), 1247–1256.

[42] F. Haslinger, Compactness for the $\overline{\partial}$- Neumann problem- a functional analysis approach, *Collect. Math.* 62 (2011), 121–129.

[43] F. Haslinger, *Complex analysis - a functional analytic approach*, de Gruyter Graduate, Walter de Gruyter, Berlin/Boston, 2018.

[44] F. Haslinger, The ∂-complex on the Segal-Bargmann space, *Ann. Pol. Math.* 123 (2019), 295–317.

[45] F. Haslinger, Basic estimates for the generalized ∂-complex, *Pure Appl. Math. Q.* 18 (2022), 583–597.

[46] F. Haslinger and B. Helffer, Compactness of the solution operator to $\overline{\partial}$ in weighted L^2 - spaces, *J. Funct. Anal.* 243 (2007), 679–697.

[47] F. Haslinger and B. Lamel, Spectral properties of the canonical solution operator to $\overline{\partial}$, *J. Funct. Anal.* 255 (2008), 13–24.

[48] F. Haslinger and D. Son, The ∂-complex on weighted Bergman spaces on Hermitian manifolds, *J. Math. Anal. Appl.* 487, 123994 (2020).

[49] F. Haslinger and D. Son, The ∂-operator and real holomorphic vector fields, *Pure Appl. Math. Q.* 18 (2022), 793–833.

[50] B. Helffer, *Semiclassical Analysis, Witten Laplacians, and Statistical Mechanics*, World Scientific, 2002.

[51] B. Helffer, *Spectral Theory and its Applications*, Cambridge studies in advanced mathematics 139, Cambridge University Press, 2013.

[52] B. Helffer and A. Mohamed, Caractérisation du spectre essentiel de l'opérateur de Schrödinger avec un champ magnétique, *Ann. Inst. Fourier (Grenoble)* 38 (1988), 95–112.

[53] B. Helffer and F. Nier, *Hypoelliptic Estimates and Spectral Theory for Fokker-Planck Operators and Witten Laplacians*, Lecture Notes in Mathematics 1862, Springer, 2005.

[54] B. Helffer, J. Nourrigat and X. P. Wang, Sur le spectre de l'équation de Dirac (dans $\mathbb{R}^2$ ou $\mathbb{R}^3$) avec champ magnétique, *Ann. Sci. Éc. Norm. Supér.* 22 (1989), 515–533.

[55] B. Helffer and J. Sjöstrand, Puits multiples en limite semi-classique IV- Etude du complex de Witten, *Commun. Partial Differ. Equ.* 10 (1985), 245–340.

[56] B. Helffer and J. Sjöstrand, in *Algebraic Analysis*, Volume in honor of M. Sato, Algebraic Analysis, Vol. 1, Academic Press (1988), 171–183.

[57] L. Hörmander, *An introduction to complex analysis in several variables*, North Holland, 1990.

[58] A. Intissar, Spectral Analysis of Non-self-adjoint Jacobi-Gribov Operators and Asymptotic Analysis of its Generalized Eigenvectors, *Adv. Math.(China)* 44 (2015), 335–353.

[59] A. Iwatsuka, Magnetic Schrödinger operators with compact resolvent, *J. Math. Kyoto Univ.* 26 (1986), 357–374.

[60] M. Jarnicki and P. Pflug, *Extensions of Holomorphic Functions*, Expositions in Mathematics 34, Walter De Gruyter, 2000.

[61] J. Johnsen, On the spectral properties of Witten Laplacians, their range projections and Brascamp-Lieb's inequality, *Integral Equ. Oper. Theory* 36 (2000), 288–324.

[62] J.-M. Kneib and F. Mignot, Equation de Schmoluchowski généralisée, *Ann. Math. Pura Appl. (IV)* 167 (1994), 257–298.

[63] W. Knirsch and G. Schneider, Continuity and Schatten-von Neumann p-class membership of Hankel operators with anti-holomorphic symbols on (generalized) Fock spaces, *J. Math. Anal. Appl.* 320 (2006), 403–414.

[64] J. Kohn, Harmonic integrals on strongly pseudoconvex manifolds, I, *Ann. Math.* 78 (1963), 112–148.

[65] J. Kohn, Harmonic integrals on strongly pseudoconvex manifolds, II, *Ann. Math.* 79 (1964), 450–472.

[66] J. Kohn and L. Nirenberg, Non-coercive boundary value problems, *Commun. Pure Appl. Math.* 18 (1965), 443–492.

[67] V. Kondratiev and M. Shubin, Discreteness of spectrum for the magnetic Schrödinger operators, *Commun. Partial Differ. Equ.* 27 (2002), 477–525.

[68] S. Krantz, Compactness of the $\overline{\partial}$-Neumann operator, *Proc. Am. Math. Soc.* 103 (1988), 1136–1138.

[69] S. Krantz, *Function Theory of Several Complex Variables*, AMS Chelsea Publishing Series, Amer. Math. Soc., 2001.

[70] E. Ligocka, in *The regularity of the weighted Bergman projections*, Lecture Notes in Math. **1165**, Seminar on deformations, Proceedings, Lodz-Warsaw, 1982/84, Springer, Berlin (1985), 197–203.

[71] S. Lovera and E. H. Youssfi, Spectral properties of the $\overline{\partial}$-canonical solution operator, *J. Funct. Anal.* 208 (2004), 360–376.

[72] X. Ma and G. Marinescu, *Holomorphic Morse Inequalities and Bergman Kernels*, Progress in Mathematics, Birkhäuser Verlag, 2007.

[73] X. Ma and G. Marinescu, Generalized Bergman kernels on symplectic manifolds, *Adv. Math.* 217 (2008), 1756–1815.

[74] J. Marzo and J. Ortega-Cerdá, Pointwise estimates for the Bergman kernel of the weighted Fock space, *J. Geom. Anal.* 19 (2009), 890–910.

[75] J. D. McNeal, A sufficient condition for compactness of the $\overline{\partial}$-Neumann operator, *J. Funct. Anal.* 195 (2002), 190–205.

[76] J. D. McNeal, L^2 estimates on twisted Cauchy-Riemann complexes, 150 years of mathematics at Washington University in St. Louis. Sesquicentennial of mathematics at Washington University, St. Louis, MO, USA, October 3–5, 2003. Providence, RI: American Mathematical Society (AMS), *Contemp. Math.* 395 (2006), 83–103.

[77] R. Meise and D. Vogt, *Einführung in die Funktionalanalysis*, Vieweg Studium, 1992.

[78] D. Newman and H. Shapiro, Certain Hilbert spaces of entire functions, *Bull. Am. Math. Soc.* 72 (1966), 971–977.

[79] D. Newman and H. Shapiro, Fischer spaces of entire functions, *Proc. Symp. Pure Math.* 11 (1968), 360–369.

[80] J. Peetre , S. Thangavelu and N.-O. Wallin , Generalized Fock spaces, interpolation, multipliers, circle geometry, *Rev. Mat. Iberoamericana* 12 (1996), 63–110.

[81] H. Render, Real Bargmann spaces, Fischer decompositions and sets of uniqueness for polyharmonic functions, *Duke Math. J.* 142 (2008), 313–352.

[82] G. Rozenblum and N. Shirokov, Infiniteness of zero modes for the Pauli operator with singular magnetic field, *J. Funct. Anal.* 233 (2006), 135–172.

[83] W. Rudin, *Real and complex analysis*, Springer, 1974.

[84] W. Rudin, *Function Theory in the Unit Ball in* $\mathbb{C}^n$, McGraw-Hill, 1980.

[85] W. Rudin, *Functional Analysis*, McGraw-Hill, 1991.

[86] N. Salinas, Noncompactness of the $\overline{\partial}$-Neumann problem and Toeplitz C^*- algebras, in *Several complex variables and complex geometry, Part 3 (Santa Cruz, CA, 1989)*, Proc. Sympos. Pure Math. **52**, Amer. Math. Soc., Providence, R. I. (1991), 329–334.

[87] N. Salinas, A. Sheu and H. Upmeier, Toeplitz operators on pseudoconvex domains and foliation C^*- algebras, *Ann. Math.* 130 (1989), 531–565.

[88] G. Schneider, Non-compactness of the solution operator to $\overline{\partial}$ on the Fock-space in several dimensions, *Math. Nachr.* 278 (2005), 312–317.

[89] I. Shigekawa, Spectral properties of Schrödinger operators with magnetic fields for a spin 1/2 particle, *J. Funct. Anal.* 101 (1991), 255–285.

[90] E. Stein, *Harmonic Analysis: Real-Variable Methods, Orthogonality, and Oscillatory Integrals*, Princeton University Press, Princeton, New Jersey, 1993.

[91] E. J. Straube, *The* L^2*-Sobolev theory of the* $\overline{\partial}$*-Neumann problem*, ESI Lectures in Mathematics and Physics, EMS, 2010.

[92] B. Thaller, *The Dirac Equation*, Texts Monogr. Phys., Springer, 1991.

[93] J. Weidmann, *Lineare Operatoren in Hilberträumen, Teil I Grundlagen*, B. G. Teubner, 2000.

[94] R. O. Wells, *Differential Analysis on Complex Manifolds*, Prentice-Hall, 1973.

[95] E. Witten, Supersymmetry and Morse inequalities, *J. Differ. Geom.* 17 (1982), 661–692.

[96] E. Witten, Holomorphic Morse inequalities, *Teubner-Texte Math.* 70 (1984), 318–333.

[97] K. Zhu, *Analysis on Fock spaces*, Graduate Texts in Mathematics 263, Springer, 2012.

Index

https://doi.org/10.1515/9783111182926-018

De Gruyter Expositions in Mathematics

Volume 69/3
Mariusz Urbański, Mario Roy, Sara Munday
Non-Invertible Dynamical Systems. Volume 3: Analytic Endomorphisms of the Riemann Sphere, 2023
ISBN 978-3-11-076984-5, e-ISBN 978-3-11-076987-6

Volume 73
Sergei S. Akbarov
Stereotype Spaces and Algebras, 2022
ISBN 978-3-11-078086-4, e-ISBN 978-3-11-078091-8

Volume 72
Ali Baklouti
Deformation Theory of Discontinuous Groups, 2022
ISBN 978-3-11-076529-8, e-ISBN 978-3-11-076530-4

Volume 71
Victor P. Maslov, Yurievich Oleg Shvedov†
The Canonical Operator in Multiparticle Problems and Quantum Field Theory, 2022
ISBN 978-3-11-076238-9, e-ISBN 978-3-11-076270-9

Volume 70
Aron Simis, Zaqueu Ramos
Graded Algebras in Algebraic Geometry, 2022
ISBN 978-3-11-063754-0, e-ISBN 978-3-11-064069-4

Volume 69/2
Mariusz Urbański, Mario Roy, Sara Munday
Non-Invertible Dynamical Systems. Volume 1: Finer Thermodynamic Formalism – Distance Expanding Maps and Countable State Subshifts of Finite Type, Conformal GDMSs, Lasota-Yorke Maps and Fractal Geometry, 2022
ISBN 978-3-11-070061-9, e-ISBN 978-3-11-070269-9

Volume 69/1
Mariusz Urbański, Mario Roy, Sara Munday
Non-Invertible Dynamical Systems. Volume 1: Ergodic Theory – Finite and Infinite, Thermodynamic Formalism, Symbolic Dynamics and Distance Expanding Maps, 2021
ISBN 978-3-11-070264-4, e-ISBN 978-3-11-070268-2

Printed and bound by CPI Group (UK) Ltd, Croydon, CR0 4YY
13/09/2026
14961093-0001